FROM THE LIBRARY OF
SY/Partners
SAN FRANCISCO

Data Points

D1003219

Data Points

Visualization That Means Something

Nathan Yau

WILEY

Data Points: Visualization That Means Something

Executive Editor: Carol Long
Senior Project Editor: Adaobi Obi Tulton
Technical Editor: Jen Lowe
Production Editor: Kathleen Wisor
Copy Editor: Apostrophe Editing Services
Editorial Manager: Mary Beth Wakefield
Freelancer Editorial Manager: Rosemarie Graham
Associate Director of Marketing: David Mayhew
Marketing Manager: Ashley Zurcher
Business Manager: Amy Knies
Production Manager: Tim Tate
Vice President and Executive Group Publisher: Richard Swadley
Vice President and Executive Publisher: Neil Edde
Associate Publisher: Jim Minatel
Project Coordinator, Cover: Katie Crocker
Compositor: James D. Kramer, Happenstance Type-O-Rama
Proofreader: James Saturino, Word One
Indexer: Johnna VanHoose Dinse
Cover Designer: Ryan Sneed
Cover Image: Nathan Yau
Chapter Photographs: Serena Yau

Published by
John Wiley & Sons, Inc.
10475 Crosspoint Boulevard
Indianapolis, IN 46256
www.wiley.com

Copyright © 2013 by John Wiley & Sons, Inc., Indianapolis, Indiana

Published simultaneously in Canada

ISBN: 978-1-118-46219-5
ISBN: 978-1-118-46222-5 (ebk)
ISBN: 978-1-118-65493-4 (ebk)
ISBN: 978-1-118-65493-4 (ebk)

Manufactured in the United States of America

10 9 8 7 6 5 4 3 2 1

No part of this publication may be reproduced, stored in a retrieval system or transmitted in any form or by any means, electronic, mechanical, photocopying, recording, scanning or otherwise, except as permitted under Sections 107 or 108 of the 1976 United States Copyright Act, without either the prior written permission of the Publisher, or authorization through payment of the appropriate per-copy fee to the Copyright Clearance Center, 222 Rosewood Drive, Danvers, MA 01923, (978) 750-8400, fax (978) 646-8600. Requests to the Publisher for permission should be addressed to the Permissions Department, John Wiley & Sons, Inc., 111 River Street, Hoboken, NJ 07030, (201) 748-6011, fax (201) 748-6008, or online at http://www.wiley.com/go/permissions.

Limit of Liability/Disclaimer of Warranty: The publisher and the author make no representations or warranties with respect to the accuracy or completeness of the contents of this work and specifically disclaim all warranties, including without limitation warranties of fitness for a particular purpose. No warranty may be created or extended by sales or promotional materials. The advice and strategies contained herein may not be suitable for every situation. This work is sold with the understanding that the publisher is not engaged in rendering legal, accounting, or other professional services. If professional assistance is required, the services of a competent professional person should be sought. Neither the publisher nor the author shall be liable for damages arising herefrom. The fact that an organization or Web site is referred to in this work as a citation and/or a potential source of further information does not mean that the author or the publisher endorses the information the organization or website may provide or recommendations it may make. Further, readers should be aware that Internet websites listed in this work may have changed or disappeared between when this work was written and when it is read.

For general information on our other products and services please contact our Customer Care Department within the United States at (877) 762-2974, outside the United States at (317) 572-3993 or fax (317) 572-4002.

Wiley publishes in a variety of print and electronic formats and by print-on-demand. Some material included with standard print versions of this book may not be included in e-books or in print-on-demand. If this book refers to media such as a CD or DVD that is not included in the version you purchased, you may download this material at http://booksupport.wiley.com. For more information about Wiley products, visit www.wiley.com.

Library of Congress Control Number: 2012956416

Trademarks: Wiley and the Wiley logo are trademarks or registered trademarks of John Wiley & Sons, Inc. and/or its affiliates, in the United States and other countries, and may not be used without written permission. All other trademarks are the property of their respective owners. John Wiley & Sons, Inc. is not associated with any product or vendor mentioned in this book.

To Bea

About the Author

Nathan Yau has written and created graphics for FlowingData since 2007, a site on visualization, statistics, and design and believes that visualization is a medium that can be used as both a tool and a way to express data. He is the author of *Visualize This: The FlowingData Guide to Design, Visualization, and Statistics*, also published by Wiley.

Yau has a master's degree in statistics from the University of California, Los Angeles, and has a Ph.D. in statistics with a focus on visualization and personal data. He is a statistician.

About the Technical Editor

Jen Lowe is an Associate Research Scholar at the Spatial Information Design Lab at Columbia University where she experiments with new forms of data visualization and communication. She has a bachelor's degree in applied mathematics and a master's degree in information science. She has publications in applied meteorology and astronomy in journals such as the *Journal of Atmospheric and Oceanic Technology*, and has spoken internationally on the relationship between data, visualization, and identity. A sucker for a good argument, she'll nearly always side with complexity.

Acknowledgments

Thank you to everyone I spoke to and received feedback from in the process of writing this book. You provided valuable insight and gave me fresh perspectives. Thank you to those who graciously let me feature their beautiful work.

Thank you to FlowingData readers who continue to support and always give me something to be excited about. This book would not exist without you.

Thank you to Wiley Publishing for helping me write a book worth reading and to Jen Lowe for making sure my words meant something.

Thank you to Serena Yau for taking pictures.

Thank you to the statisticians, computer scientists, data scientists, designers, and everyone else involved in this hodge-podge data world for making it fun and always keeping it interesting.

Thank you to all those who have taught me.

Thank you to my wife, my parents, and all other family and friends for their support and encouragement.

Contents

Introduction

What is good visualization? It is a representation of data that helps you see what you otherwise would have been blind to if you looked only at the naked source. It enables you to see trends, patterns, and outliers that tell you about yourself and what surrounds you. The best visualization evokes that moment of bliss when seeing something for the first time, knowing that what you see has been right in front of you, just slightly hidden. Sometimes it is a simple bar graph, and other times the visualization is complex because the data requires it.

THE PROCESS

A data set is a snapshot in time that captures something that moves and changes. Collectively, data points form aggregates and statistical summaries that can tell you what to expect. These are your means, medians, and standard deviations. They describe the world, countries, and populations and enable you to compare and contrast things. When you push down on the data, you get details about the individuals and objects within the population. These are the stories that make a data set human and relatable.

Data in an abstract sense that includes information and facts is the foundation of every visualization. The more you understand about the source and the stronger base that you build, the greater the potential for a compelling data graphic. This is the part that a lot of people miss: Good visualization is a winding process that requires statistics and design knowledge. Without the former, the visualization becomes an exercise only in illustration and aesthetics, and without the latter, one of only analyses. On their own, these are fine skills, but they make for incomplete data graphics. Having skills in both provides you with the luxury—which is growing into a necessity—to jump back and forth between data exploration and storytelling.

This book is for those interested in the process of design and analysis, where each chapter represents a step toward visualization that means something. It is about visualization that is more than large printed numbers with clipart. It is about making sense of data. Visualization creation is iterative, and the cycle is always a little different for each new dataset.

The first part of *Data Points* helps you know your data and what it means to visualize it. Because of what data represents—people, places, and things—there is always important context attached to the factual numbers. Who is the data about? Where is the data from? When was it collected? We are responsible for this human part of the computer-generated output, too. On top of that, most datasets are estimated, so they are not the absolute truth; there is uncertainty and variability attached, just like in real life.

In the middle of the book, you go into exploration mode. You have the freedom to ask questions and try to answer them by digging through the data. Look for patterns, relationships, and anything that does not look right. Missing values are common, as are typos. This is a great time to play and experiment, to look around from different angles, and maybe you'll find something unexpected. Maybe that bit ends up being the most interesting part of the story. For whatever reason, the exploration stage is skipped too often, and lack of understanding shows in the final product. Take the time to get to know your data and what it represents, and the visualization improves exponentially.

When you find the underlying narratives, the next step is typically to communicate your results to a wider audience. This is the last section, when you put your design hat on. A graphic for a small audience of four people who are familiar with the subject matter and have read every significant paper on the topic will be different than a graphic for a large, general audience unfamiliar with the complex background implied by the numbers.

Again, these stages are not meant as a step-by-step guide. If you work with data already, you know that it is common to discover a need for new data as you explore what you already have. Similarly, the design process can force you to see details that you didn't notice before, which takes you back to exploration or to the beginning. If you are new to data, you must learn this process while reading this book to feel confident enough to use it in your own projects. The back and forth between data and story is the fun stuff.

Data Points is a complement to my previous book *Visualize This*. The first book serves as an introduction to the tools available and offers concrete programming examples, whereas this book describes the full process and thinking that goes into larger data projects and is software-independent. In other words, the two books feed off each other. *Visualize This* provides technical guidance for those ready to make their own graphics, and *Data Points* describes a process for data and visualization so that you can create better and more thoughtful things.

MORE THAN A TOOL

Throughout this book, visualization is referred to as a medium rather than a specific tool. When you approach visualization as an unyielding tool, it is easy to get caught thinking that almost every graphic would be better as a bar graph. This is true for a lot of charts, but it must be in the right context. In an analysis setting, yes, you often want graphs that read the quickest and most accurately; however, from another point of view, such a comment might be premature. What if emotion and curiosity are the goals? Visualization is a way to represent data, an abstraction of the real world, in the same way that the written word can be used to tell different kinds of stories. Newspaper articles aren't judged on the same criteria as novels, and data art should be critiqued differently than a business dashboard.

That said, there are rules to follow regardless of the visualization type. These aren't dictated by design or statistics. Rather they are governed by human perception, and they ensure accuracy when readers interpret encoded data. There are only a handful of these, such as to properly size by area when that aspect is the actual visual cue, and all the rest are suggestions.

You must distinguish between rules and suggestions. You should follow the former almost always, whereas suggestions are rooted in opinion and vary among individuals and situations. Many beginners make the mistake that advice is concrete, and they lose the context in which the data is presented in. For example, Edward Tufte suggests stripping charts of all junk, but the definition of junk can change. What needs to be stripped from one chart might need to stay in another. In the words of Tufte, "Most principles of design should be greeted with some skepticism."

Similarly, people often cite the work of statisticians William Cleveland and Robert McGill on perception and accuracy. They found that position along a common scale, such as with a scatterplot, was decoded most accurately, followed by length, angle, and then slope. These results are based on research trials and were reproduced in other studies, so it is easy to mistake Cleveland and McGill's findings as rules. However, Cleveland also notes that the mark of a good graph is not only how fast you can read it, but also what it shows. Does it enable you to see what you could not see before?

You must come back to the data for worthwhile visualization. Fortunately, you have plenty of data to play with, and the source keeps growing. Every week for the past few years, there is an article that describes the flood of data and the risk of drowning in it, but you see, the amount is controlled, and you can easily filter and aggregate it. Storage is cheap and practically infinite, which means more potential happy feelings for those who know how to swim. The challenge is learning to dive deeper.

Okay, I'm psyched. Let's have some fun.

Data Points

Understanding Data

When you ask people what data is, most reply with a vague description of something that resembles a spreadsheet or a bucket of numbers. The more technically savvy might mention databases or warehouses. However, this is just the format that the data comes in and how it is stored, and it doesn't say anything about what data is or what any particular dataset represents. It's an easy trap to fall in because when you ask for data, you usually get a computer file, and it's hard to think of computer output as anything but just that. Look beyond the file though, and you get something more meaningful.

WHAT DATA REPRESENTS

Data is more than numbers, and to visualize it, you must know what it represents. Data represents real life. It's a snapshot of the world in the same way that a photograph captures a small moment in time.

Look at Figure 1-1. If you were to come across this photo, isolated from everything else, and I told you nothing about it, you wouldn't get much out of it. It's just another wedding photo. For me though, it's a happy moment during one of the best days of my life. That's my wife on the left, all dolled up, and me on the right, wearing something other than jeans and a T-shirt for a change. The

FIGURE 1-1 *A single photo, a single data point*

pastor who is marrying us is my wife's uncle, who added a personal touch to the ceremony, and the guy in the back is a family friend who took it upon himself to record as much as possible, even though we hired a photographer. The flowers and archway came from a local florist about an hour away from the venue, and the wedding took place during early summer in Los Angeles, California.

That's a lot of information from just one picture, and it works the same with data. (For some, me included, pictures are data, too.) A single data point can have a who, what, when, where, and why attached to it, so it's easy for a digit to become more than a toss in a bucket. Extracting information from a data point isn't as easy as looking at a photo, though. You can guess what's going on in the photo, but when you make assumptions about data, such as how accurate it is or how it relates to its surroundings, you can end up with a skewed view of what your data actually represents. You need to look at everything around, find context, and see what your dataset looks like as a whole. When you see the full picture, it's much easier to make better judgments about individual points.

Imagine that I didn't tell you those things about my wedding photo. How could you find out more? What if you could see pictures that were taken before and after?

FIGURE 1-2 *Grid of photos*

Now you have more than just a moment in time. You have several moments, and together they represent the part of the wedding when my wife first walked out, the vows, and the tea drinking ceremony with the parents and my grandma, which is customary for Chinese weddings. Like the first photo, each of these has its own story, such as my father-in-law welling up as he gave away his daughter or how happy I felt when I walked down the aisle with my bride. Many of the photos captured moments that I didn't see from my point of view during the wedding, so I almost feel like an outsider looking in, which is probably how you feel. But the more I tell you about that day, the less obscure each point becomes.

Still though, these are snapshots, and you don't know what happened in between each photo. (Although you could guess.) For the complete story, you'd either need to be there or watch a video. Even with that, you'd still see only the ceremony from a certain number of angles because it's often not feasible to record every single thing. For example, there was about five minutes of confusion during the ceremony when we tried to light a candle but the wind kept blowing it out. We eventually ran out of matches, and the wedding planner went on a scramble to find something, but luckily one of our guests was a smoker, so he busted out his lighter. This set of photos doesn't capture that, though, because again, it's an abstraction of the real thing.

This is where sampling comes in. It's often not possible to count or record everything because of cost or lack of manpower (or both), so you take bits and pieces, and then you look for patterns and connections to make an educated guess about what your data represents. The data is a simplification—an abstraction—of the real world. So when you visualize data, you visualize an abstraction of the world, or at least some tiny facet of it. Visualization is an abstraction of data, so in the end, you end up with an abstraction of an abstraction, which creates an interesting challenge.

However, this is not to say that visualization obscures your view—far from it. Visualization can help detach your focus from the individual data points and explore them from a different angle—to see the forest for the trees, so to speak. To keep running with this wedding photo example, Figure 1-3 uses the full wedding dataset, of which Figure 1-1 and Figure 1-2 were subsets of. Each rectangle represents a photo from our wedding album, and they are colored by the most common shade in each photo and organized by time.

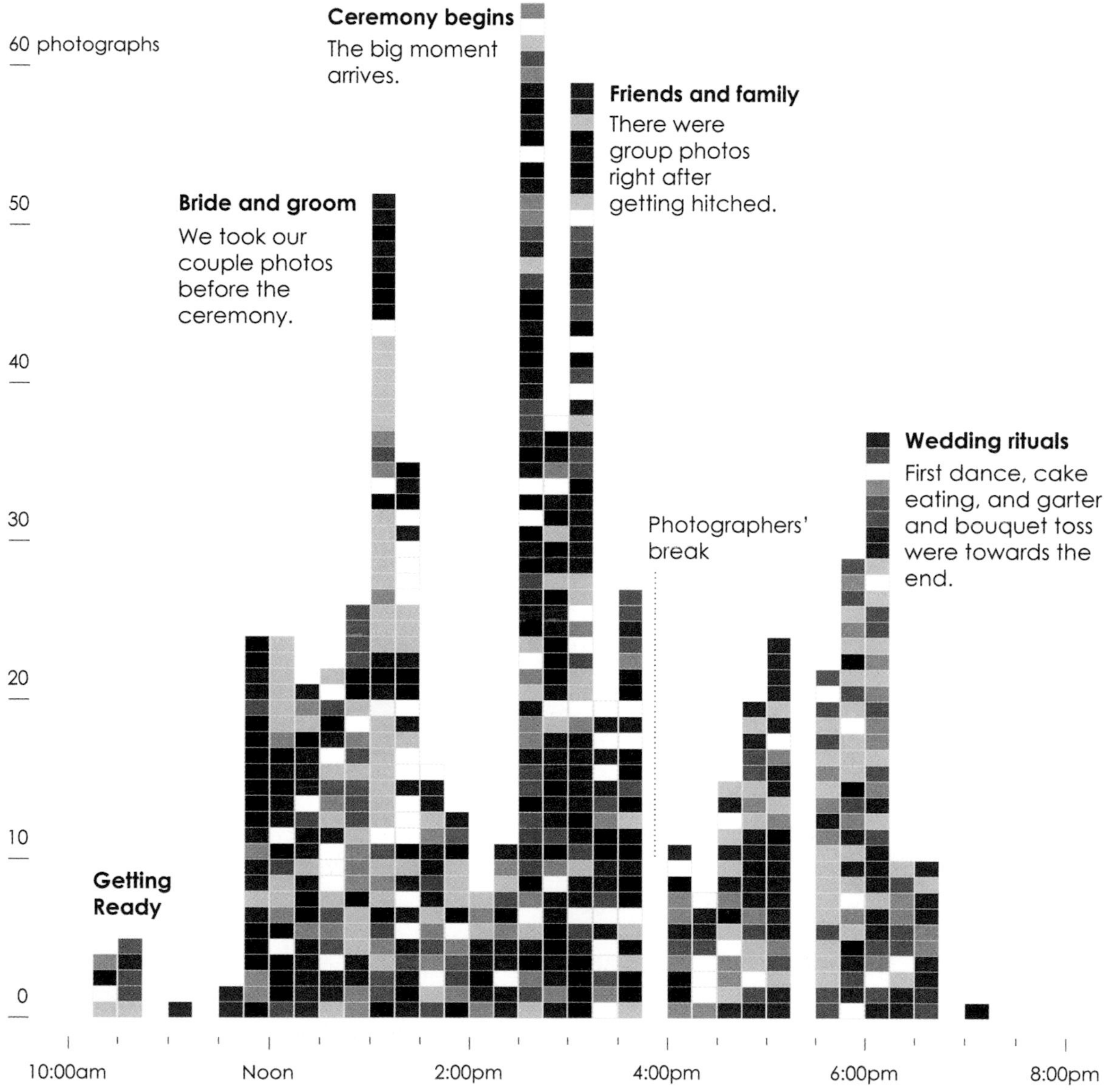

FIGURE 1-3 *Colors in the wedding*

With a time series layout, you can see the high points of the wedding, when our photographers snapped more shots, and the lulls, when only a few photos were taken. The peaks in the chart, of course, occur when there is something to take pictures of, such as when I first saw my wife in her dress or when the ceremony began. After the ceremony, we took the usual group photos with friends and family, so there was another spike at that point. Then there was food, and activity died down, especially when the photographers took a break a little before 4 o'clock. Things picked up again with typical wedding fanfare, and the day came to an end around 7 in the evening. My wife and I rode off into the sunset.

In the grid layout, you might not see this pattern because of the linear presentation. Everything seems to happen with equal spacing, when actually most pictures were taken during the exciting parts. You also get a sense of the colors in the wedding at a glance: black for the suits, white for the wedding dress, coral for the flowers and bridesmaids, and green for the trees surrounding the outdoor wedding and reception. Do you get the detail that you would from the actual photos? No. But sometimes that level isn't necessary at first. Sometimes you need to see the overall patterns before you zoom in on the details. Sometimes, you don't know that a single data point is worth a look until you see everything else and how it relates to the population.

You don't need to stop here, though. Zoom out another level to focus only on the picture-taking volumes, and disregard the colors and individual photos, as shown in Figure 1-4.

You've probably seen this layout before. It's a bar chart that shows the same highs and lows as in Figure 1-3, but it has a different feel and provides a different message. The simple bar chart emphasizes picture-taking volumes over time via 15-minute windows, whereas Figure 1-3 still carries some of the photo album's sentiment.

The main thing to note is that all four of these views show the same data, or rather, they all represent my wedding day. Each graphic just represents the day differently, focusing on various facets of the wedding. Interpretation of the data changes based on the visual form it takes on. With traditional data, you typically examine and explore from the bar chart side of the spectrum, but that doesn't mean you have to lose the sentiment of the individual data point—that single photo. Sometimes that means adding meaningful annotation that enables readers to interpret the data better, and other times the message in the numbers is clear, gleaned from the visualization itself.

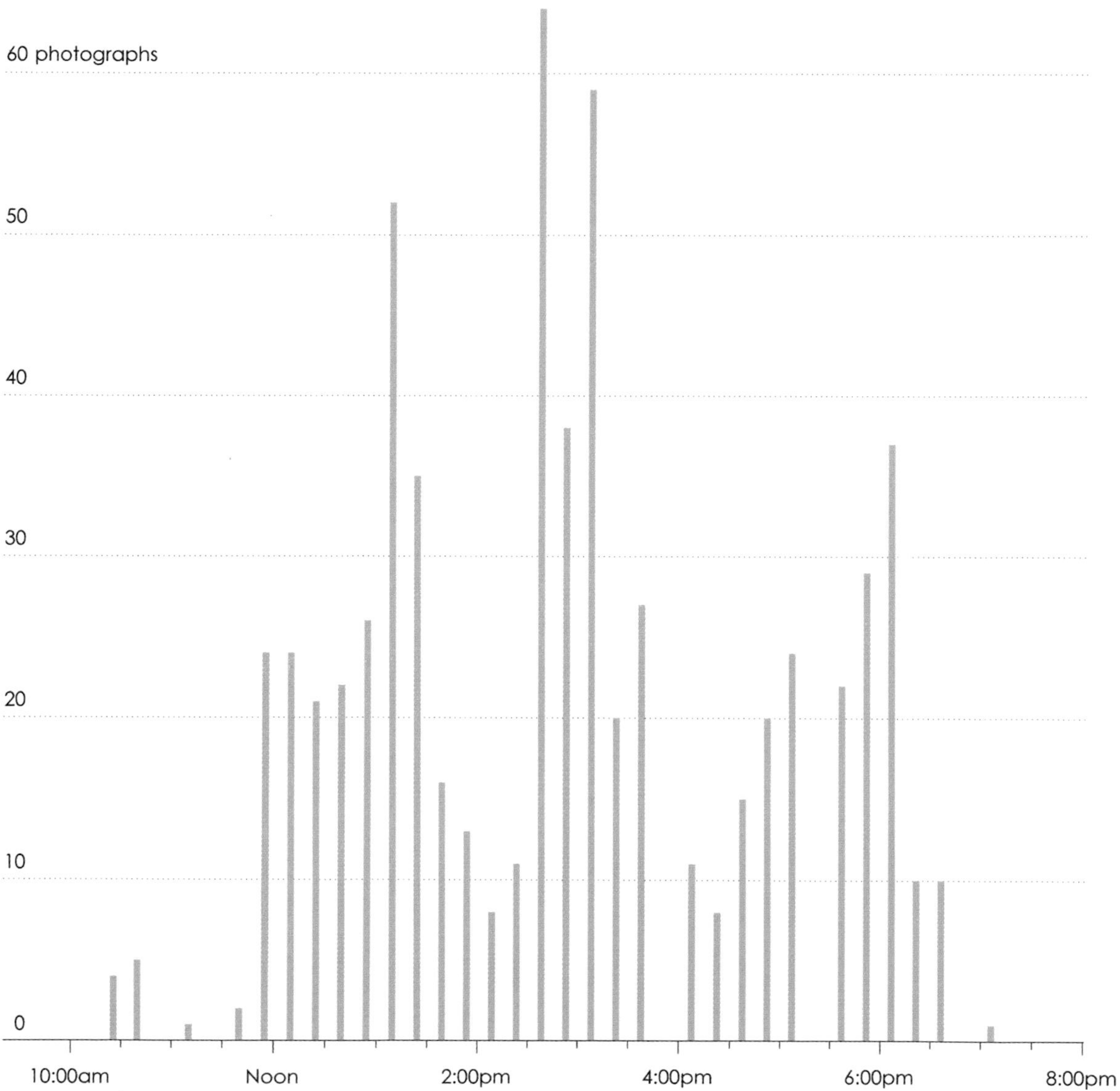
Photographs over time
Our wedding photographers snapped more pictures during the significant events with a peak of 63 during a 15-minute span.
60 photographs
50
40
30
20
10
0
10:00am
Noon
2:00pm
4:00pm
6:00pm
8:00pm

FIGURE 1-4 *Photos over time*

The connection between data and what it represents is key to visualization that means something. It is key to thoughtful data analysis. It is key to a deeper understanding of your data. Computers do a bulk of the work to turn numbers into shapes and colors, but you must make the connection between data and real life, so that you or the people you make graphics for extract something of value.

This connection is sometimes hard to see when you look at data on a large scale for thousands of strangers, but it's more obvious when you look at data for an individual. You can almost relate to that person, even if you've never met him or her. For example, Portland-based developer Aaron Parecki used his phone to collect 2.5 million GPS points over 3½ years between 2008 and 2012, about one point every 2 to 6 seconds. Figure 1-5 is a map of these points, colored by year.

As you'd expect, the map shows a grid of roads and areas where Parecki frequented that are colored more brightly than others. His housing changed a few times, and you can see his travel patterns change over the years. Between 2008 and 2010, shown in blue, travel appears more dispersed, and by 2012, in yellow, Parecki seems to stay in a couple of tighter pockets. Without more context it is hard to say anything more because all you see is location, but to Parecki the data is more personal (like the single wedding photo is to me). It's the footprint of more than 3 years in a city, and because he has access to the raw logs, which have time attached to them, he could also make better decisions based on data, like when he should leave for work.

What if there were more information attached to personal time and location data, though? What if along with where you were, you also took notes during or after about what was going on at some given time? This is what artist Tim Clark did between 2010 and 2011 for his project *Atlas of the Habitual*. Like Parecki, Clark recorded his location for 200 days with a GPS-enabled device, which spanned approximately 2,000 miles in Bennington, Vermont. Clark then looked back on his location data and labeled specific trips, people he spent time with, and broke it down by time of year.

As shown in Figure 1-6, the atlas, with clickable categorizations and time frames, shows a 200-day footprint that reads like a personal journal. Select "Running errands" and the note reads, "Doing the everyday things from running to the grocery store all the way to driving 30 miles to the only bike shop in southern Vermont opened on Sundays." The traces stay around town, with the exception of two long ones that venture out.

There is one entry titled "Reliving the breakup," and Clark writes, "A long-term girlfriend and I broke up immediately before I moved. These are the times that I had a real difficult time coming to terms that I had to move on." Two small paths, one within the city limits and one outside, appear, and the data suddenly feels incredibly personal.

FIGURE 1-5 *GPS traces collected by Aaron Parecki, http://aaronparecki.com*

This is perhaps the appeal behind the Quantified Self movement, which aims to incorporate technology to collect data about one's own activity and habits. Some people track their weight, what they eat and drink, and when they go to bed; their goal is usually to live healthier and longer. Others track a wider variety of metrics purely as a way to look in on themselves beyond what they see in the mirror; personal data collection becomes something like a journal for self-reflection at the end of the day.

200 days of GPS traces

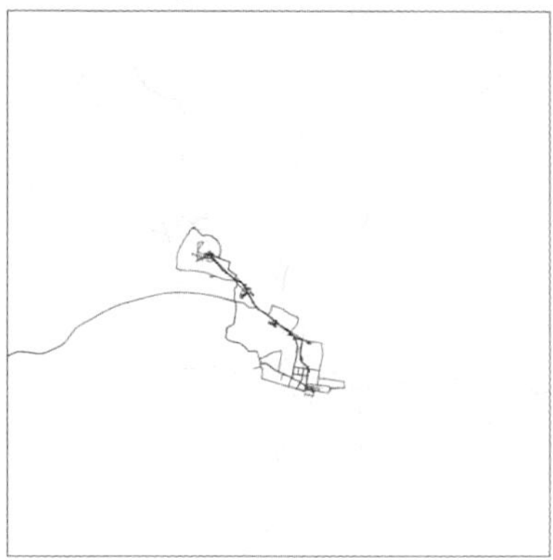

Bike riding

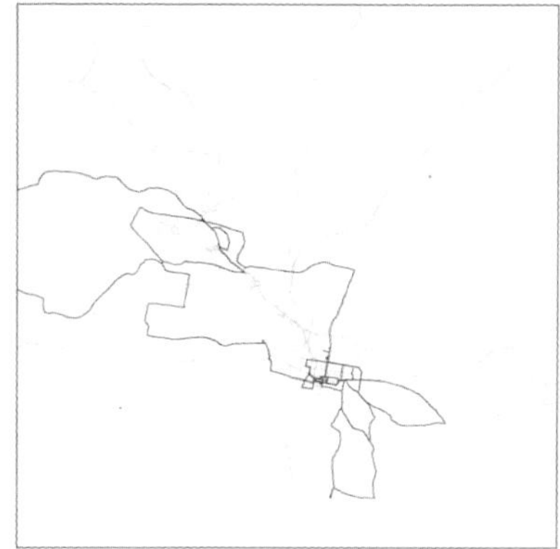

Reliving the breakup

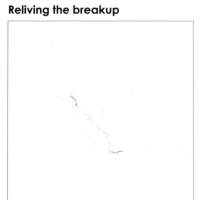

Dating

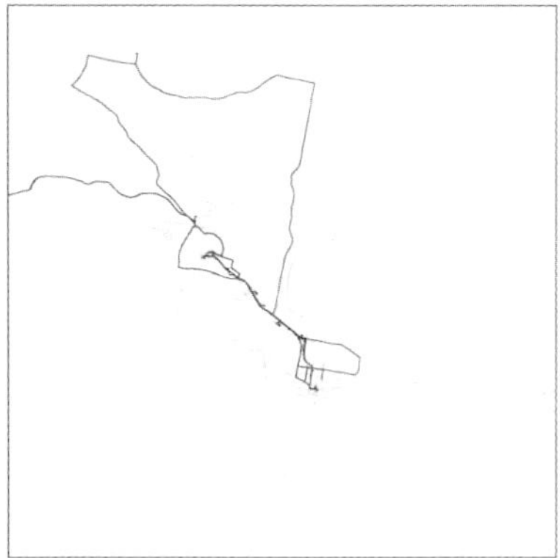

Exploring

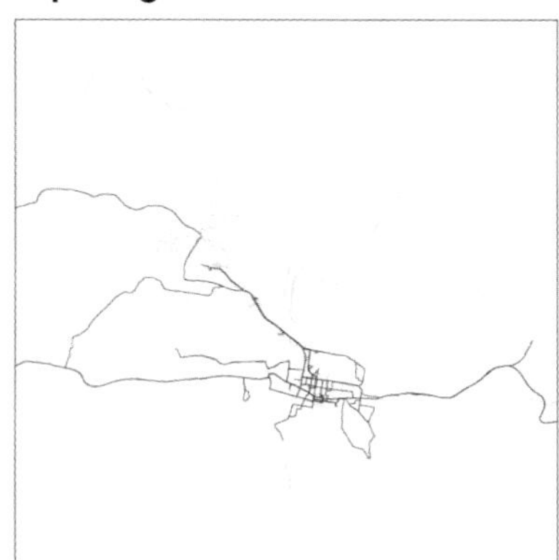

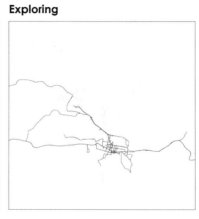

Running errands

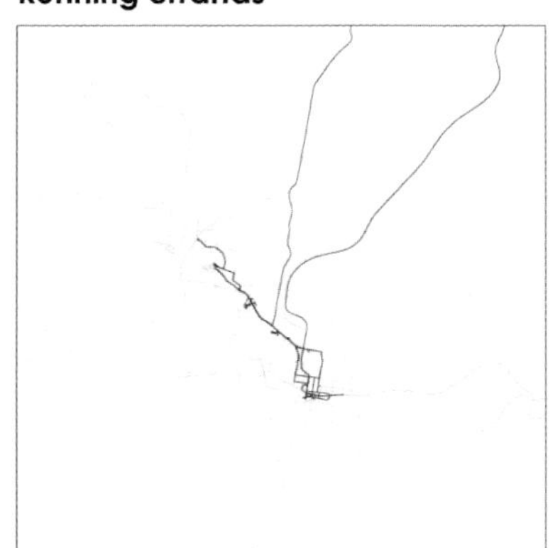

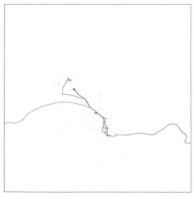

US holidays

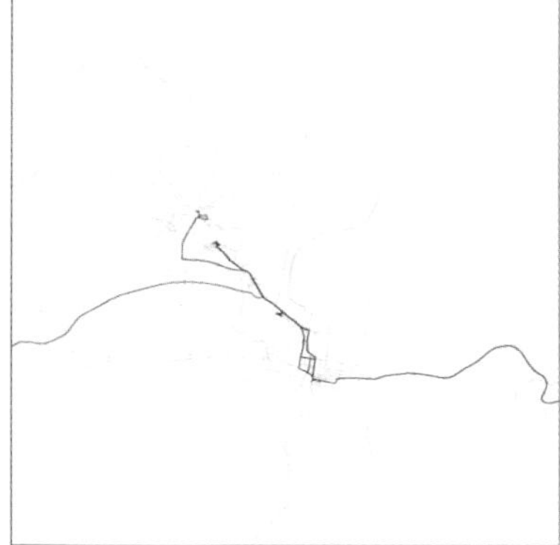

Eating

FIGURE 1-6 *Selected maps from* Atlas of the Habitual *by Tim Clark, http://www.tlclark.com/atlasofthehabitual/*

Nicholas Felton is one of the more well-known people in this area for his annual reports on himself, which highlight both his design skills and disciplined personal data collection. He keeps track of not just his location, but also who he spends time with, restaurants he eats at, movies he watches, books he reads, and an array of other things that he reveals each year. Figure 1-7 is a page out of Felton's 2010/2011 report.

Felton designed his first annual report in 2005 and has done one every year since. Individually, they are beautiful to look at and hold and satisfy an odd craving for looking in on a stranger's life. What I find most interesting, though, is the evolution of his reports into something personal and the expanding richness of data. Looking at his first report, as shown in Figure 1-8, you notice that it feels a lot like a design exercise in which there are touches of Felton's personality embedded, but it is for the most part strictly about the numbers. Each year though, the data feels less like a report and more like a diary.

This is most obvious in the *2010 Annual Report*. Felton's father passed away at the age of 81. Instead of summarizing his own year, Felton designed an annual report, as shown in Figure 1-9, that cataloged his father's life, based on calendars, slides, postcards, and other personal items. Again, although the person of focus might be a stranger, it's easy to find sentiment in the numbers.

When you see work like this, it's easy to understand the value of personal data to an individual, and maybe, just maybe, it's not so crazy to collect tidbits about yourself. The data might not be useful to you right away, but it could be a decade from now, in the same way it's useful to stumble upon an old diary from when you were just a young one. There's value in remembering. In many ways you log bits of your life already if you use social sites like Twitter, Facebook, and foursquare. A status update or a tweet is like a mini-snapshot of what you're doing at any given moment; a shared photo with a timestamp can mean a lot decades from now; and a check-in firmly places your digital bits in the physical world.

You've seen how that data can be valuable to an individual. What if you look at the data from many individuals in aggregate?

The United States Census Bureau collects the official counts of people living in the country every 10 years. The data is a valuable resource to help officials allocate funds, and from census to census, the fluctuations in population help you see how people move in the country, changing the neighborhood

FIGURE 1-7 *(following page)* *A page from* 2010/2011 Annual Report *by Nicholas Felton, http://feltron.com*

With Olga

EVERYWHERE

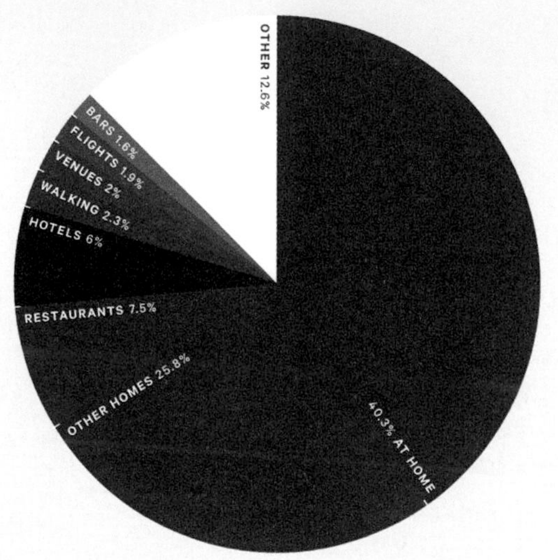

DAYS TOGETHER

191¼

315 different encounters

MOST TIME SPENT TOGETHER

MANHATTAN — 83¾ DAYS

BROOKLYN — 51½ DAYS

MILL VALLEY — 9 DAYS

ANCHORAGE — 7½ DAYS

SYDNEY — 4¾ DAYS

MOST VISITED PLACE TOGETHER

Old Apartment

194 visits

DIFFERENT CITIES VISITED TOGETHER

56

In 3 countries, 9 states and Washington D.C.

FAVORITE BEVERAGES WITH OLGA

FILTER COFFEE — 111 SERVINGS

RED WINE — 78 SERVINGS

DALE'S PALE ALE — 35 SERVINGS

CHAMPAGNE — 30 SERVINGS

LATTE — 26 SERVINGS

TIME TOGETHER

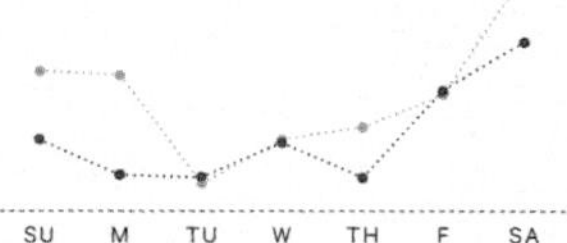

BRIEFEST MONTH TOGETHER

June 2011

40¾ hours

MOST CONSECUTIVE HOURS TOGETHER

247

Australia trip — February 2010

TIME SPENT WITH OLGA AND...

SARAH — 6¾ DAYS

MOM — 6¾ DAYS

BRIAN — 5¾ DAYS

OLGA'S MOM — 5 DAYS

RYAN — 4½ DAYS

WEDDINGS ATTENDED TOGETHER

Seven

Aaron & Jessica, Charlie & Bret, Glenn & Mariana, Lewis & Ange, Randy & Allison, Rob & Elise and Toby & Harriet

With Olga

IN THE BAY AREA

STAG'S LEAP
NAPA

DAD'S HOUSE
SAN RAFAEL

MOM'S HOUSE
MILL VALLEY

PIEDMONT PARK
PIEDMONT

SFO

FACEBOOK 1601
PALO ALTO

DAYS TOGETHER IN THE BAY AREA

13½

Approximately 7% of total time together

BAY AREA PLACES VISITED TOGETHER

77

18 stores, 13 restaurants, 10 homes, 6 outdoor places, 3 coffee shops, 3 grocery stores, 2 airport terminals, 2 bars, 2 gas stations, 2 hospitals, 2 hotels, 2 liquor stores, 2 parking garages, 2 parking lots, a cinema, a deli, a drug store, a laundromat, a library, a museum, a park and work

FAVORITE BAY AREA BOTTLESHOP

Vintage Wine & Spirits

Visited twice

FAVORITE BAY AREA BEER WITH OLGA

Lagunitas IPA

5 servings

BAY AREA MUSEUMS VISITED TOGETHER

The Exploratorium

With Marina — July 9, 2011

MOST PLAYED ARTIST TOGETHER

The Beach Boys

25 songs listened to from *Christmas with the Beach Boys*

TIME TOGETHER IN THE BAY AREA

MOST FREQUENTED CITY TOGETHER

Mill Valley

68% of time in the Bay Area

MOST VISITED BAY AREA PLACES

MOM'S HOUSE — 35 VISITS

MARIN GENERAL HOSPITAL — 6 VISITS

CHEVRON MILL VALLEY — 5 VISITS

SFO INTERNATIONAL TERMINAL — 4 VISITS

DAD'S HOUSE — 3 VISITS

CRISES INVOLVING A TICK

One

Spotted by Olga, removed by Mom

MOST VISITED RESTAURANTS TOGETHER

Le Garage, Picante and Sushi Ran

Each visited twice

With Olga

IN NEW YORK CITY

***KRAI* PERFORMANCE**
MERKIN CONCERT HALL
AT KAUFMAN CENTER

BOHEMIAN HALL
& BEER GARDEN
ASTORIA

MOMA
MIDTOWN

OFFICE
SOHO

ROB & ELISE'S
APARTMENT
JERSEY CITY

OLGA'S APARTMENT
GREENPOINT

NEW APARTMENT
WILLIAMSBURG

OLD APARTMENT
FINANCIAL DISTRICT

SARAH & BRIAN'S APARTMENT
RED HOOK

AMADOR & SARA'S APARTMENT
PROSPECT HEIGHTS

DAYS TOGETHER IN NEW YORK CITY

136¾

Approximately 72% of total time together

MOST VISITED NYC PLACES

OLD APARTMENT — 194 VISITS

OLGA'S APARTMENT — 84 VISITS

NEW APARTMENT — 67 VISITS

THE OFFICE — 35 VISITS

TAKAHACHI TRIBECA — 21 VISITS

TIME TOGETHER IN NEW YORK CITY

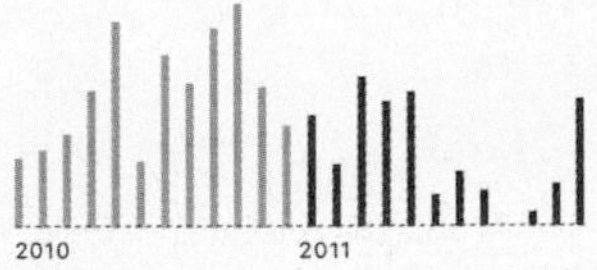

TIME IN NEW YORK SPENT WITH OLGA

31%

5% of time together spent in transit

MOST VISITED NYC RESTAURANTS

TAKAHACHI TRIBECA — 21 VISITS

LES HALLES ON JOHN STREET — 9 VISITS

DINER / ENID'S — 7 VISITS

MILLER'S TAVERN / FIVE LEAVES — 6 VISITS

RABBIT HOLE — 5 VISITS

FAVORITE NYC COCKTAIL WITH OLGA

Bloody Mary

22 servings

NYC PERFORMANCES WITH OLGA

Twenty-Eight

Bell (11), Bear in Heaven (3), Baths + How to Dress Well + Zola Jesus, Blonde Redhead + Pantha du Prince, Dexter Lake Club Band, Jason Nazary, Knights on Earth, Olga Bell *Krai*, Little Women, Nathan Fake + Four Tet, Now Ensemble + Matmos, Owen Pallett, Panda Bear, Pierre-Laurent Aimard, Sleigh Bells and *The Nose*

SIGNIFICANT NYC MISHAPS

Five

Abandoned keyboard stand, muddled dinner invitation date, missed ferry, shattered martini glass and smashed iPhone composition, and how areas grow and shrink. In short, the data paints a picture of who lives in America. However, the data, collected and maintained by the government, can show only so much about the individuals, and it's hard to grasp who the people actually are.

What are their likes and dislikes? What kind of personality do they have? Are there major differences between neighboring cities and towns?

Media artist Roger Luke DuBois took a different kind of census, via 19 million online dating profiles in *A More Perfect Union*. When you join an online dating

FIGURE 1-8 *Selected pages from* 2005 Annual Report *by Nicholas Felton, http://feltron.com*

site, you first describe yourself: who you are, where you're from, and what you're interested in. After you uncomfortably fill out that information, and perhaps choose not to share a thing or two, you describe what your ideal mate is like. In the words of DuBois, in the latter, you tell the complete truth, and in the former, you lie. So when you aggregate people's online dating profiles, you get some combination of how people see themselves and how they want to be seen.

In *A More Perfect Union*, DuBois categorized online dating profiles, digital encapsulations of hopes and dreams, by postal code, and then looked for the word that was most unique to each area. Using a tracing of a Rand McNally map, DuBois replaced each city name with the city's unique word and painted a different picture of the United States: a more recognizable and personal one.

In Figure 1-10, around southern California, where they make the talkies, words such as *acting*, *writer,* and *entertainment* appear; on the other hand, in Washington, DC, shown in Figure 1-11, words like *bureaucrat*, *partisan*, and *democratic* appear. These mostly pertain to professions, but in some areas the words describe personal attributes, favorite things, and major events.

In Louisiana, shown in Figure 1-12, *Cajun* and *curvy* pop out at you, as does *crawfish*, *bourbon*, and *gumbo*, but in New Orleans, the most unique word is *flood*, a reflection of the effects of Hurricane Katrina in 2005.

People are defined by common demographic data such as race, age, and gender, but they also identify themselves with what they like to do in their spare time, what has happened to them, and who they hang around with. The great thing about *A More Perfect Union* is that you can see that in the data on a countrywide scale.

The same sentiment—where data points are recollections and reports are portraits and diaries—is seen in Felton's reports, Clark's atlas, and Parecki's GPS traces. Statisticians and developers call this analysis. Artists and designers call this storytelling. For extracting information from data, though—to understand what's in the numbers—analysis and storytelling are one and the same.

Just like what it represents, data can be complex with variability and uncertainty, but consider it all in the right context, and it starts to make sense.

FIGURE 1-9 *(following page) Selected pages from* 2010 Annual Report *by Nicholas Felton, http://feltron.com*

2000–2010
SAN RAFAEL

THE 21ST CENTURY

138 LOCATIONS

SOCIALIZING

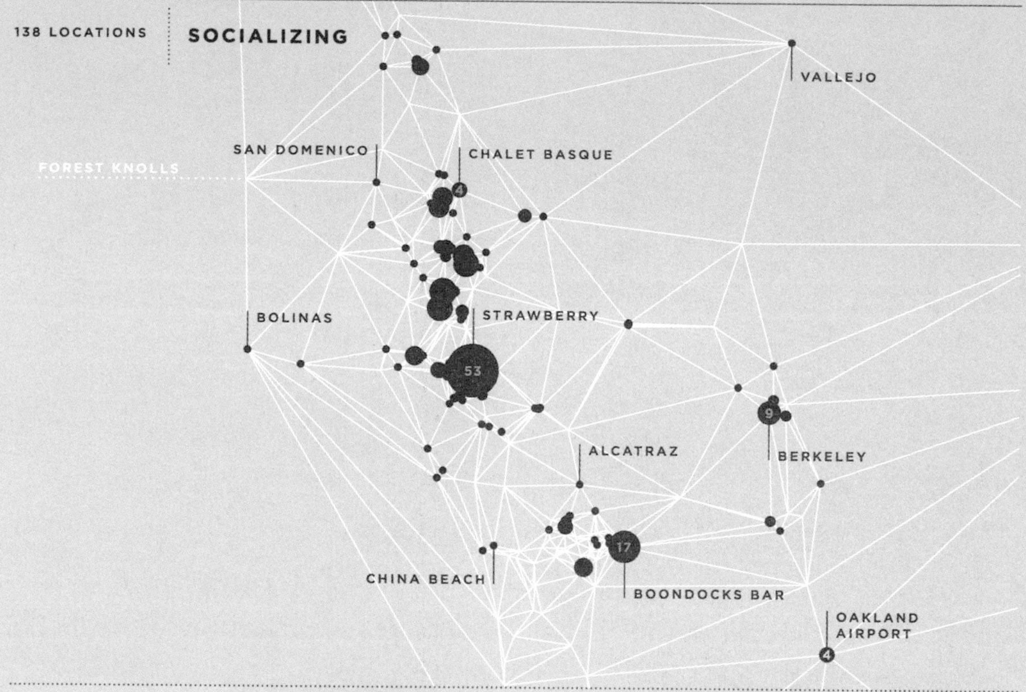

JAN 5, 2001
71 YEARS,
6 MONTHS
AND 1 DAY

PERSON SEEN THE MOST

MARINA
117 TIMES

BLACK PANTHERS MET

ONE
BOBBY SEALE

WALKS RECORDED

THIRTY-FIVE
AND 1 HIKE

2009–2010
GOLDEN GATE
TOLL BOOTH
PREFERENCE

LANE 6

14 VISITS

ENTERTAINMENT

- 123 MOVIES
- 45 MUSIC
- 38 LECTURES
- 29 DANCES
- 29 POKER
- 22 PLAYS
- 19 TELEVISION
- 8 SLIDE SHOWS
- 5 ACROBATS
- 3 MIMES

MOST WATCHED TV SHOW

THE OSCARS
8 TIMES

LAST DAY

SEP 12, 2010
81 YEARS, 2 MONTHS
AND 8 DAYS OLD

WEATHER
SEP 12, 2010
3:20 PM

49.8° F AND OVERCAST
LARKSPUR, CALIFORNIA

1848–2009

BOOKS

536 BOOKS

DATE PUBLISHED

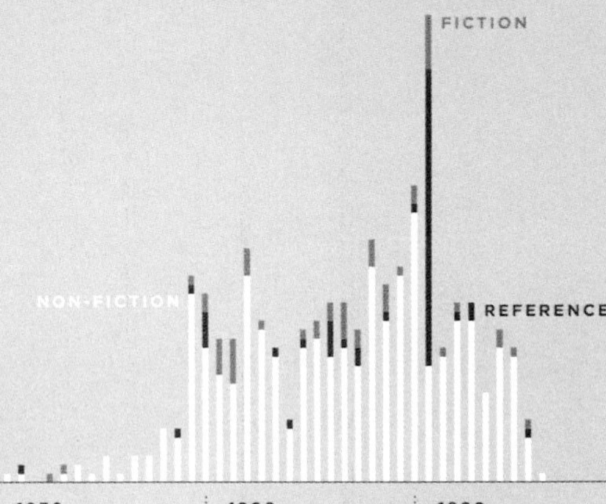

BOOKS

561

SPANNING 161 YEARS

MOST REPRESENTED AUTHOR

MARTIN GILBERT

5 BOOKS

TYPES OF BOOKS

88 TRAVEL

77 HISTORY

42 MACHINES

37 GEOGRAPHY

32 ENCYCLOPEDIA

26 RELIGION

23 HEALTH

23 NOVEL

22 SCIENCE

17 HOW-TO

BIOGRAPHIES

8

FROM ARMSTRONG TO STALIN

MEDIAN PUBLISHING DATE

1983

11 BOOKS

REGION WITH MOST TRAVEL BOOKS

RUSSIA

6 BOOKS

TRAVEL BOOKS FOR UNVISITED PLACES

SIX

AUSTRALIA, ICELAND, GREENLAND, IRAN, PAKISTAN AND VENEZUELA

COOKBOOKS

FIVE

WAR-RELATED BOOKS

51

35 BOOKS ABOUT WORLD WAR 2

ELEVATOR BOOKS

TWELVE

1941–1991

HOW-TO TOPICS

FOURTEEN

BICYCLES, CLEANING, CROSS COUNTRY SKIING, DOING EVERYTHING RIGHT, HANDICRAFT, HOME REPAIR, PEST CONTROL, PHOTOGRAPHY, PREVENTING AND SURVIVING FIRES, SAILING, SURVIVAL AND TAI CHI

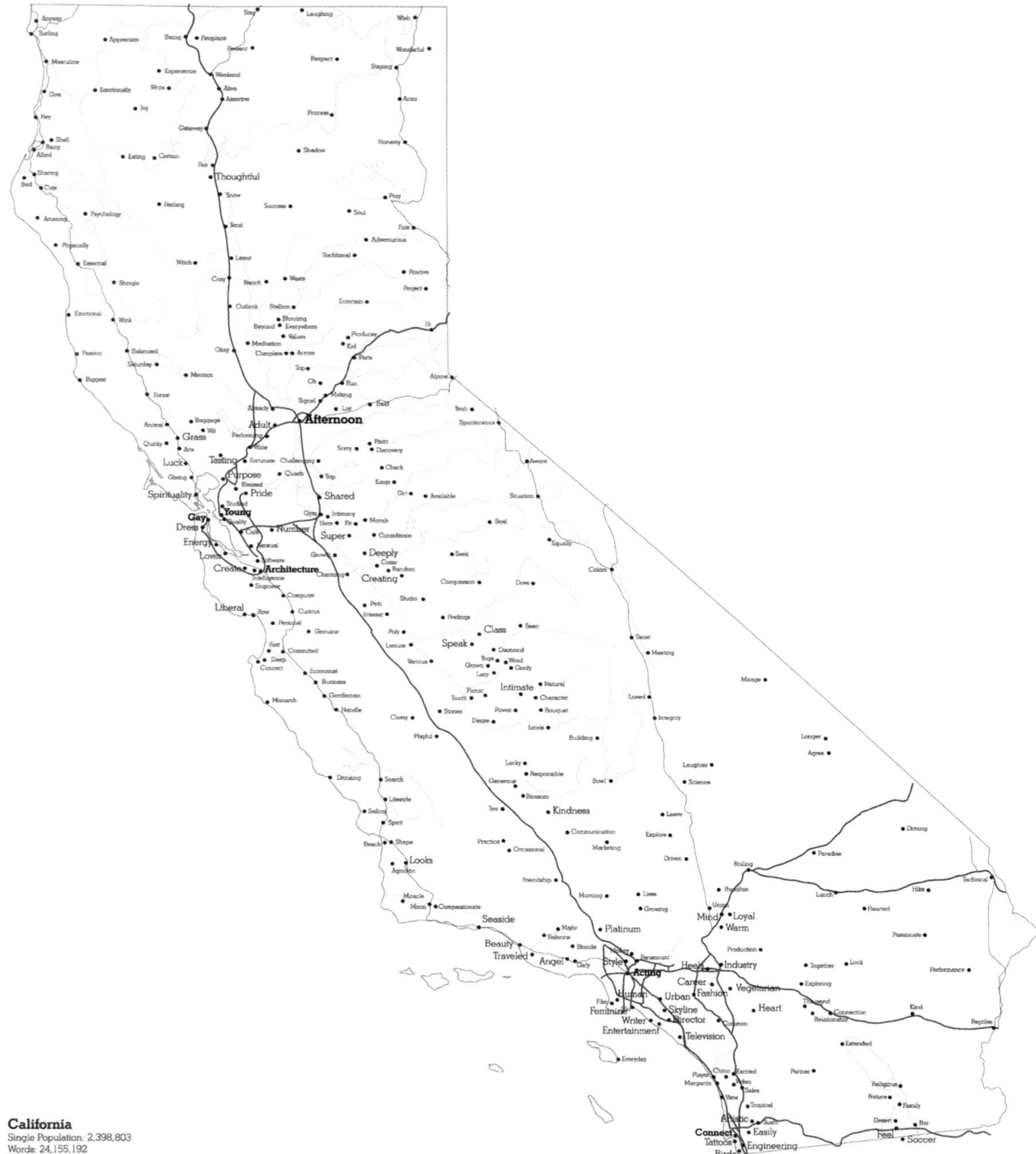

FIGURE 1-10 *California map from* A More Perfect Union *(2011) by R. Luke DuBois, courtesy of the artist and bitforms gallery, New York City, http://perfect.lukedubois.com*

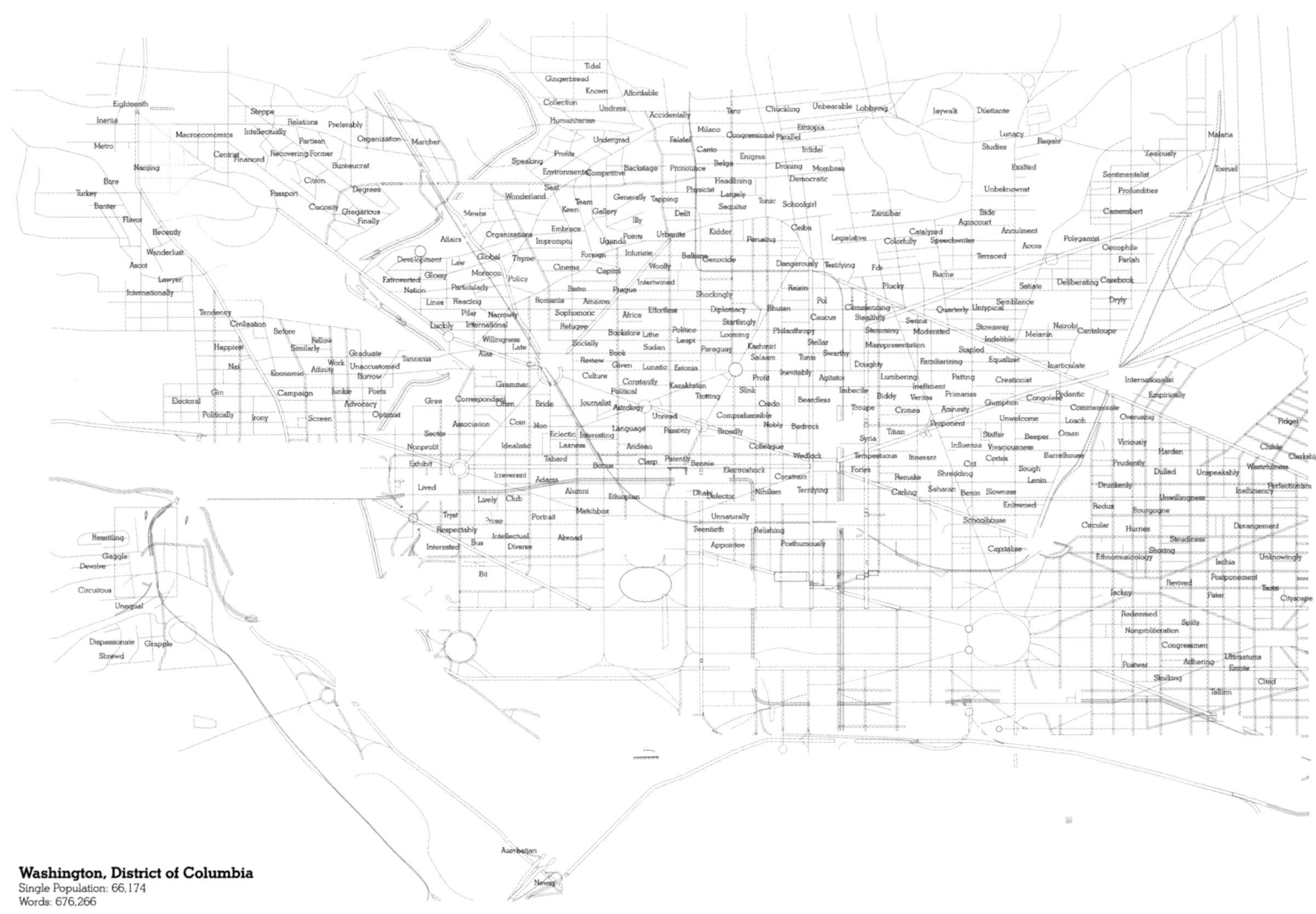

FIGURE 1-11 *Washington, DC map from* A More Perfect Union *(2011)*

FIGURE 1-12: *Louisiana map from* A More Perfect Union *(2011)*

VARIABILITY

In a small town in Germany, amateur photographer and full-time physicist Kristian Cvecek heads out into the forest at night with his camera. Using long-exposure photography, Cvecek captures the movements of fireflies as they prance between the trees. The insect, as shown in Figure 1-13, is tiny and barely noticeable during the day, but in the dark, it's hard to look elsewhere.

FIGURE 1-13 *A firefly in the night by Kristian Cvecek, http://quit007.deviantart.com/*

Although each moment in flight seems like a random point in space to an observer, a pattern emerges in Cvecek's photos, as shown in Figure 1-14. It's as if the fireflies move along the walking path and circle around the trees with a predetermined destination.

There is randomness, though. You can guess where a firefly goes next based on its flight path, but how sure are you? A firefly can bolt left, right, up, and down at any moment, and that variability, which makes each flight unique, is what makes fireflies so fun to watch and the picture so beautiful. The path is what you care about. The end point, start point, and average position don't mean nearly as much.

With data, you can find patterns, trends, and cycles, but it's not always (rarely, actually) a smooth path from point A to point B. Total counts, means, and other aggregate measurements can be interesting, but they're only part of the story, whereas the fluctuations in the data might be the most interesting and important part.

Between 2001 and 2010, according to the National Highway Traffic Safety Administration, there were 363,839 fatal automobile crashes in the United States. No doubt this total count, over one-third of a million, carries weight because it represents the lost lives of even more than that. Place all focus on the one number, as in Figure 1-15, and it makes you think or maybe even reflect on your own life.

However, is there anything you can learn from the data, other than that you should drive safely? The NHTSA provides data down to individual accidents, which includes when and where each occurred, so you can look closer.

In Figure 1-16, every fatal crash in the contiguous United States between 2001 and 2010 is mapped. Each dot represents a crash. As you might expect, there is a higher concentration of accidents in large cities and major highways; there are fewer accidents where there are fewer people and roads.

FIGURE 1-14 Path of a firefly *by Kristian Cvecek, http://quit007.deviantart.com/*

FIGURE 1-15 *(facing page) One aggregate*

Fatal Crashes, 2001–2010

363,839

Source: National Highway Traffic Safety Administration

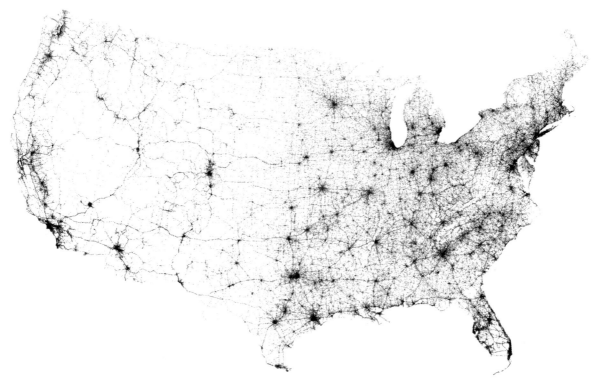

FIGURE 1-16 *Everything mapped at once*

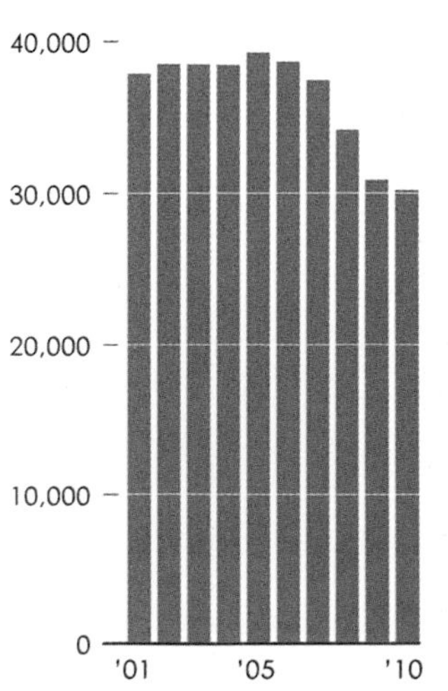

FIGURE 1-17 *Annual fatal accidents*

Again, although not to be taken lightly, the map tells you more about the country's road network than it does the accidents.

A look at crashes over time shifts focus to the events themselves. For example, Figure 1-17 shows the number of accidents per year, which tells a different story than the total in Figure 1-15. Accidents still occurred in the tens of thousands annually, but there was a significant decline from 2006 through 2010, and fatalities per 100 million vehicle miles traveled (not shown) also decreased.

Seasonal cycles become obvious at month-by-month granularity, as shown in Figure 1-18. Incidents peak during the summer months when people go on vacation and spend more time outside, whereas during the winter, fewer people drive, so there are fewer crashes. This happens every year. At the same time, you can still see the annual decline overall between 2006 and 2010.

However, there's variability when you compare specific months over the years. For example, in 2001, the most crashes occurred in August, and there was a small, relative drop the following month. The same thing happened in 2002 through 2004. However, in 2005 through 2007, July had the most accidents. Then it was back to August in 2008 through 2010.

On the other hand, February, the month with the fewest days had the least accidents every year, with the exception of 2008. So there are seasonal variations and variation within the seasons.

Go down another level to daily crashes, as shown in Figure 1-19, and you see even higher variability, but it's not all noise. There still appears to be a pattern of peaks and valleys. Although it's harder to make out the seasonal patterns, you can see a weekly cycle with more accidents during the weekends than during the middle of the week. The peak day each week fluctuates between Friday, Saturday, and Sunday.

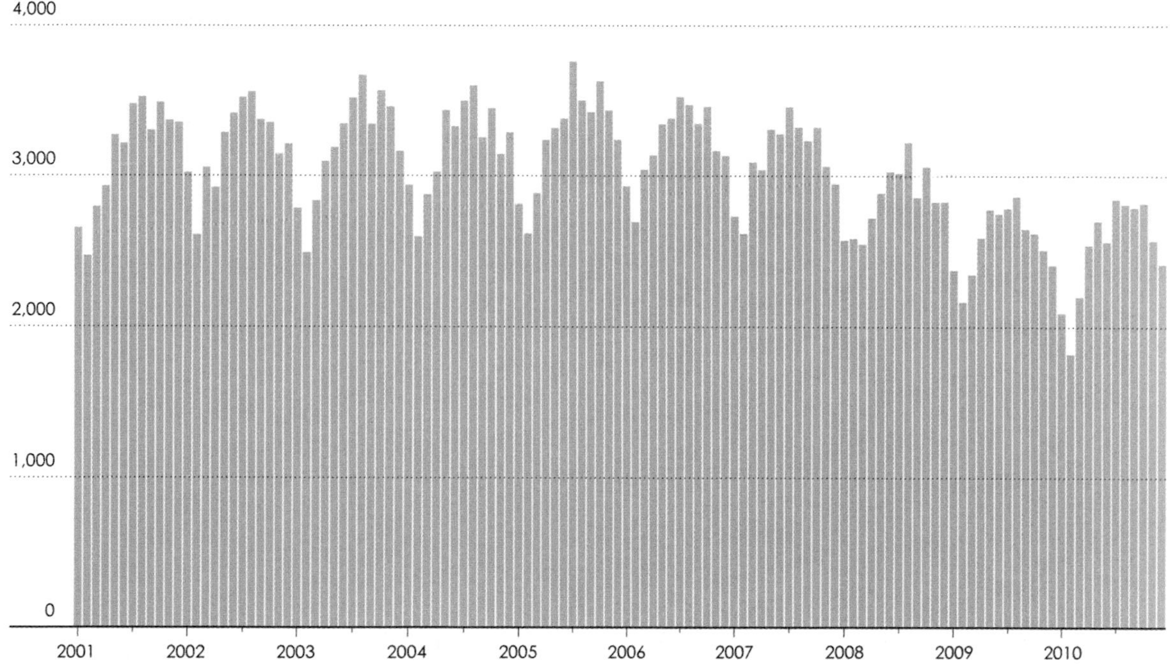

FIGURE 1-18 *Monthly fatal accidents*

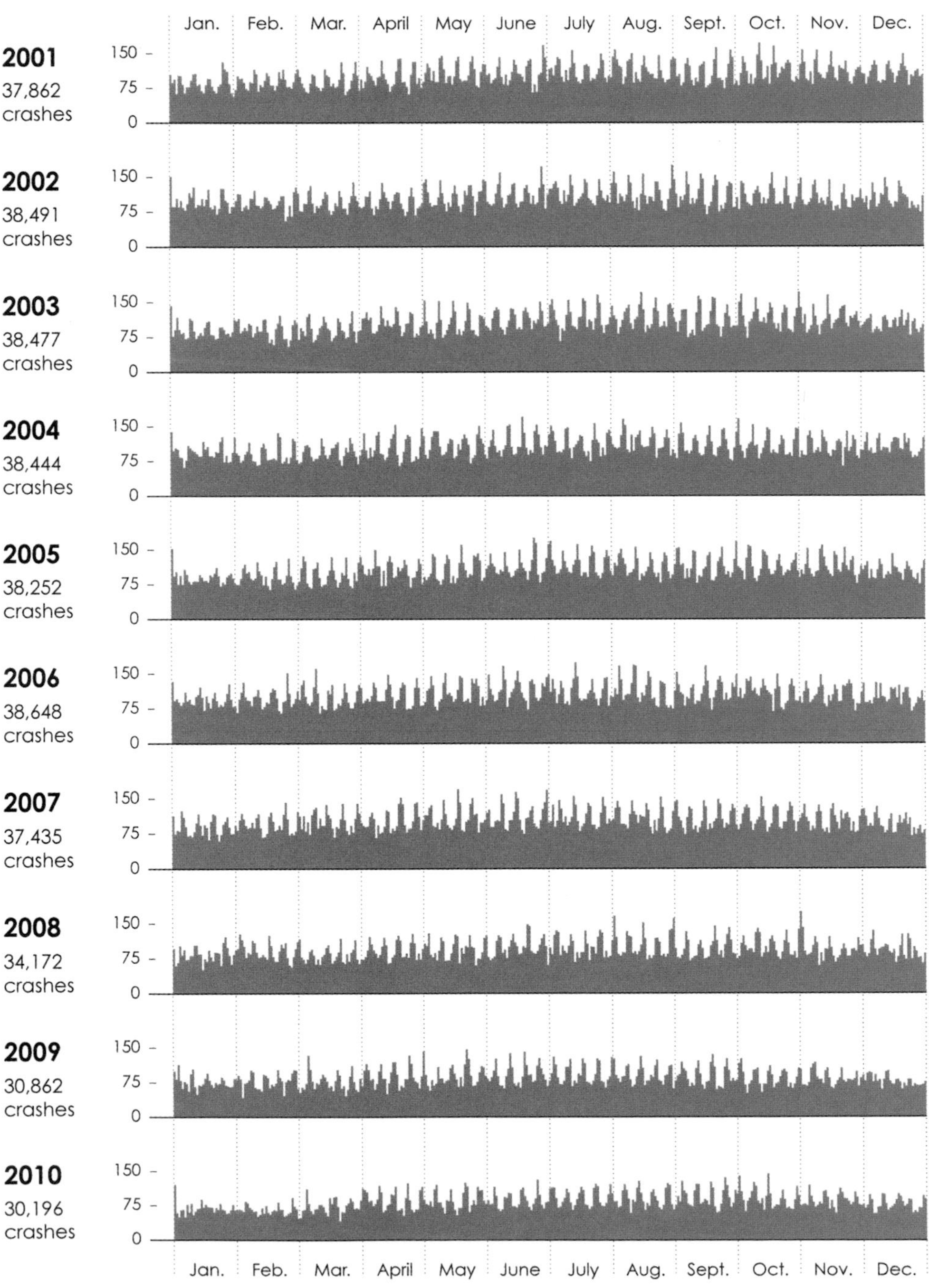

FIGURE 1-19 *Daily fatal accidents*

But guess what: You can increase granularity to crashes by the hour. Figure 1-20 breaks it down. Each row represents a year, so each cell in the grid shows an hourly time series for the corresponding month.

With the exception of a new year's spike during the midnight hour, it's hard to make out patterns at this level because of the variability. Actually, the monthly chart is hard to interpret, too, if you don't know what you're looking for. There are clear patterns, though, if you aggregate, as shown in Figure 1-21. Instead of showing values at every hour, day, or month, you can aggregate on specific time segments to explore the distributions.

What was hard to discern, or looked like noise before, is easy to see here. There's a small bump in the morning when people commute to work, but most fatal crashes occur in the evening after work. As you saw in Figure 1-19, there are more crashes during the weekend, but summed up, it's more obvious. Finally, you can see the seasonal patterns, but more clearly, with a greater number of accidents during the summer than in the winter.

The main point is that there's value in looking at the data beyond the mean, median, or total because those measurements tell you only a small part of the story. A lot of the time, aggregates or values that just tell you where the middle of a distribution is hide the interesting details that you should actually focus on, for both decision making and storytelling.

An outlier that stands out from the crowd could be something that you need to fix or pay special attention to. Maybe the changes over time are a signal that something good (or bad) is happening in your system. Cycles or regular occurrences could help you prepare for the future. However, sometimes it isn't helpful to see so much variability; in which case you can dial back the granularity for generalizations and distributions.

You lose this information—the juicy bits—when you step too far away from the data.

Think of it this way: When you look back on your life, would you rather just remember what your days were like on average, or is it the highs and the lows that are most important? I bet it's some combination of the two.

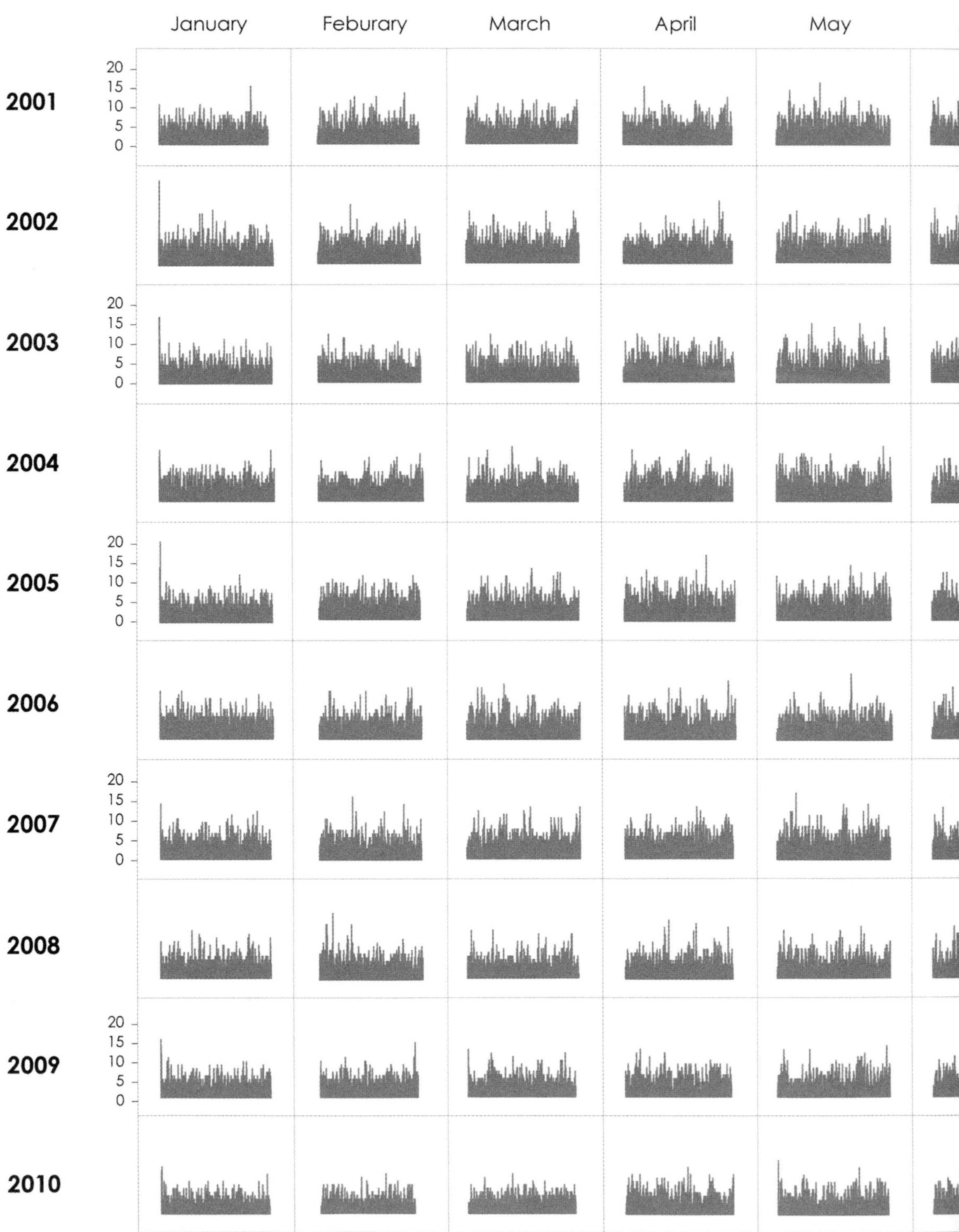

FIGURE 1-20 *Hourly fatal accidents*

July	August	September	October	November	December

- 20
- 15
- 10
- 5
- 0

- 20
- 15
- 10
- 5
- 0

- 20
- 15
- 10
- 5
- 0

- 20
- 15
- 10
- 5
- 0

- 20
- 15
- 10
- 5
- 0

2001–2010

Fatal crashes by...

Time of day

Most in the evening and least early morning

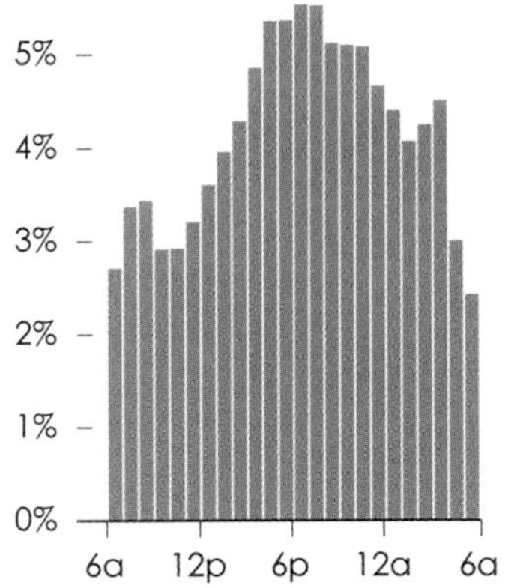

Day of the week

Most on weekends and least middle week

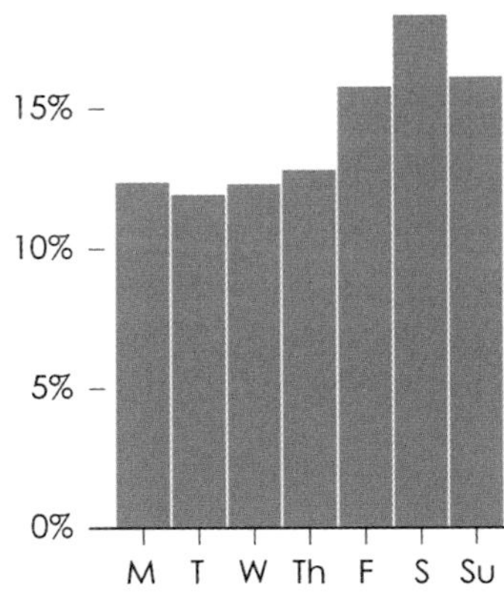

Month

Most in the summer and least during the winter

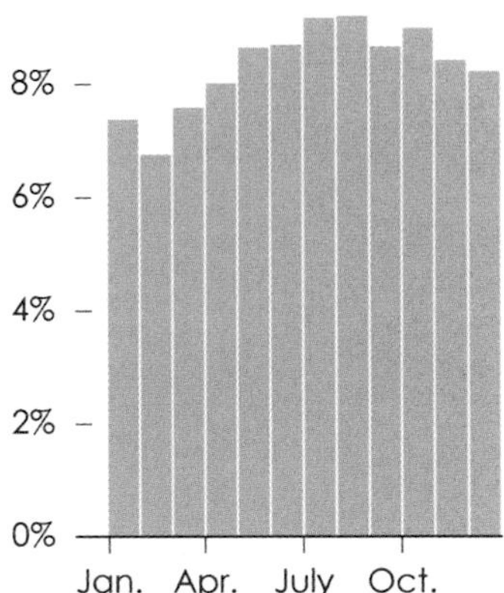

FIGURE 1-21 *Accident distributions over time*

UNCERTAINTY

A lot of data is estimates rather than absolute counts. An analyst considers the evidence (such as a sample), and makes an education guess about a full population. That educated guess has uncertainty attached to it. You do this all the time in your day-to-day. You make a guess based on what you know, read, or what someone told you, and you can say with some (possibly rough) certainty that you're right. Are you absolutely positive or are you basically clueless? It works the same with data.

Note: It's tempting to look at data as absolute truth, because we associate numbers with fact, but more often than not, data is an educated guess. Your goal is to use data that doesn't have large levels of uncertainty attached.

When I was a young lad, a recent engineering graduate with a statistics minor, I had a 9-month gap in between college and graduate school. I took a few temporary jobs that paid a little more than minimum wage, and they were mind-numbingly boring, so naturally my mind wandered to more engaging things.

One day I thought to myself, "Hey, I have some statistics and probability know-how and a deck of cards. I'm going to become an expert blackjack player like those kids from MIT. Forget this stupid job. I'm gonna be rich!" And my 1-month obsession with blackjack began. (To save you the suspense, I didn't get rich, and it's not nearly as exciting as they make it look in the movies.)

In case you're unfamiliar with the game, here's a quick rundown. There's a dealer and a player. The dealer deals two cards to each (one of his is face down), and the goal is to get a card total as close to 21 as possible, without going over. You can choose to take additional cards (called a *hit*) or not (*stay*). In some cases, you can also *split* your hand of two cards, as if you're playing two separate hands; you can also *double down*, which means to double your bet. The more you bet, the more you can win. If you go over 21, to *bust,* you automatically lose, and if not, the dealer hits or stays, and whoever is closer to 21 wins.

By design, the dealer has the advantage, but if you hit and stay when you're supposed to, you can decrease that advantage. These rules are based on averages, but as anyone who has played blackjack can tell you, there is uncertainty in each hand of cards. You can still lose even when you make the right move. For example, imagine you are dealt a 5 and a 6 for a total of 11, and the dealer shows a 6. The right move is to double your bet because it's impossible for you to bust with an additional card, and there's a decent chance of getting 21. There's also a good chance the dealer will bust with a 6 showing.

So you double down, and you get a 3, for a total of 14. Ouch. That's not good. Your only hope is for a dealer bust. So he flips his hidden card, and it's a 10 for a total of 16. By rule, he has to hit, and it's a 5. Dealer total: 21. You lose.

Had you not double downed, you would have lost only half the money that you did playing the right way. But if it were that easy to win, the casino wouldn't bother putting the game on the floor.

There's uncertainty in each hand because you are playing against distributions, or rather, you know only the approximate probabilities of drawing cards. You might have an idea of what cards are in the deck, but you can make only an educated guess about what card comes next.

Of course, uncertainty applies to things outside of cards, and it comes in a variety of forms. Take the weather for example. How many times have you looked up the forecast for the

Note: If you count cards, or keep track of what's left in the deck, the probabilities change as you modify your bet based on your advantage, but uncertainty remains.

next day or for the next week as you pack for a trip, only to find, when the time comes, that the weather isn't how you expected it to be?

What about the meter in cars that tells you how much farther you can drive with the current amount of fuel in your tank? I was running errands with my wife, and the meter said I could drive an estimated 16 more miles, but home was about 18 miles away. Dilemma. Instead of stopping at the nearest gas station, I drove toward the one nearest home, and the meter said I had zero miles left for about 2 miles, but we made it. (Good thing because someone kept insisting that I would be the one to push the car.)

Weigh yourself more than once, and you might get different readings; typically though, breathing for a few seconds does not lead to weight loss or gain. The estimated battery life on your laptop can jump around by hour increments when only minutes have passed. The subway announcement says a train will arrive in 10 minutes, but it comes in 11, or a delivery is estimated to arrive on Monday, but it comes on Wednesday instead.

When you have data that is a series of means and medians or a collection of estimates based on a sample population, you should always wonder about the uncertainty.

Note: Numbers seem concrete and absolute, but estimates carry uncertainity with them. Data is an abstraction of what it represents, and the level of exactness varies.

This is especially important when people base major decisions, which affect millions, on estimates, such as with national and global demographics. Program creation and funding is often based on these numbers, so even a small margin of error can make a big difference.

The United States Census Bureau releases data about the country on topics such as migration, poverty, and housing, which are estimates based on samples from the population. (This is different from the decennial census, which aims to count every person in the United States.) A margin of error is provided with each estimate, which means that the actual count or percentage is likely within a given range. For example, Figure 1-22 shows estimates about housing. The margin of error for total households is almost one-quarter of a million.

To put it differently, imagine you have a jar of gumballs that you can't see into, and you want to guess how many of each color there are. (Why do you care about gumball distribution? I don't know. Use your imagination. You're a gumball connoisseur who works for a gumball factory, and you bet your snotty statistician friend that every jar on your watch is uniformly distributed, so it's a matter of pride and cash.)

If you were to pour all the gumballs on to the table and count every one, you wouldn't have to guess because you would get the full tally.

But say you can grab only a handful, and you have to guess the contents of the entire jar, based on what you have in your hand. A larger handful would make it easier to guess because it's more likely a better representation of the entire jar. On the other side of the spectrum, you could take just one gumball out, and it'd be much harder to guess what else is in the jar.

With one gumball, your margin of error would be high; with a large handful of gumballs, your margin of error would be lower; and if you counted all the gumballs, you would have zero margin of error.

Apply that to millions of gumballs in thousands of differently sized jars, with different distributions and big and small handfuls, and estimation

	Estimate	Margin of error
Total households	114,235,996	± 248,114
Total families	76,254,318	± 230,785
Average family size	3.17	± 0.01
Married-couple family households	56,655,412	± 293,638
Married, 15 and over	50.2%	± 0.2
Divorced, 15 and over	10.5%	± 0.1

Source: 2010 American Community Survey

FIGURE 1-22 *Household estimates in 2010*

Jar of gumballs

Population

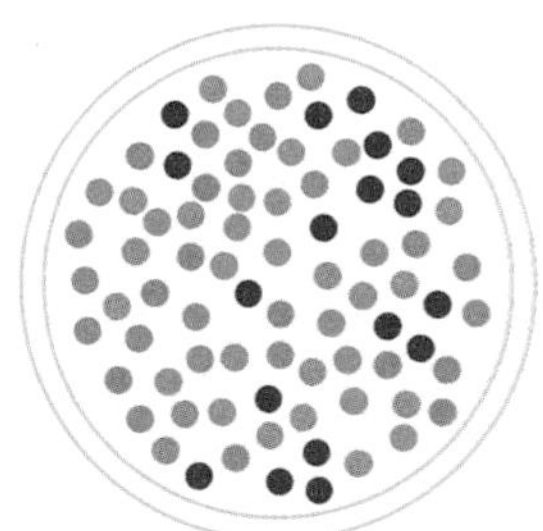

Guess what's in the jar based on...

Tiny sample **Small** **Medium** **Large**

Larger margin of error → Smaller margin of error

FIGURE 1-23 *Gumballs and margin of error*

grows more complex. Then substitute the gumballs for people, the jars for towns, cities, and counties, and the handfuls for randomly distributed surveys, and a mean with a margin of error carries more weight.

According to Gallup, 48 percent of Americans disapproved of the job Barack Obama was doing from June 11 through 13 in 2012. However, there was a 3 percent margin of error, which means the difference between more than half and less than half of the country disapproving. Similarly, during election season, polls estimate which candidates lead, and if the margin of error is wide, the results can put more than one person in front, which kind of defeats the purpose of the poll.

Estimates get tricky when you rank people, places, and things, especially when you combine measurements (and create statistical models with multiple variables).

Take education evaluation, for example, which is under constant scrutiny. Cities, schools, and teachers are often compared against one another, but what defines a good education or makes an entire city smart? Is it the percentage of high school students who graduate? The percentage of students who go to college? Is it the number of universities, libraries, and museums per capita? If it's all of this, is one count more important than the other, or do you give all of them equal weight? Answers change depending on who you ask, as do ratings.

Note: My hometown was ranked the "dumbest" city in America by a publication that shall go unnamed. The rankings were estimates, which were based on estimates with questionable uncertainty.

In 2011, the New York City Department of Education released Teacher Data Reports that tried to measure teaching quality. The reports were originally given only to schools and teachers but were later made publicly available in early 2012. The estimates took several factors into account, but one of the main ones was the change in test percentiles from the seventh to eighth grade.

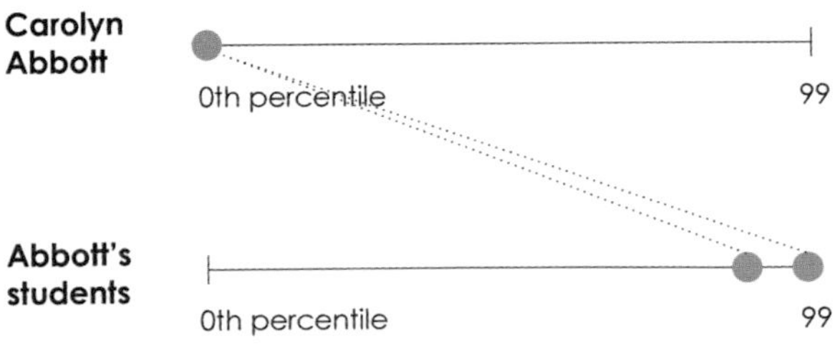

FIGURE 1-24 *Carolyn Abbott's rating compared to her students'*

This is how seventh- and eighth-grade math teacher Carolyn Abbott became known as the worst math teacher in the city, placed in the 0th percentile. However, her seventh-grade students scored in the 98th percentile. What?

Those students were predicted to score in the 97th percentile in the eighth grade, but they instead scored in the 89th percentile, which according to the statistical model, was not progress. Most would agree that students wouldn't earn the scores they did with a poor teacher. The challenge is that there's uncertainty and variability within teacher ratings. A rating represents a distribution of teachers, who are ranked based on estimates with uncertainty attached, but the ratings are treated as absolute. A general audience won't understand that concept, so it's your responsibility to and communicate it clearly.

When you don't consider what your data truly represents, it's easy to accidently misinterpret. Always take uncertainty and variability into account. This is also when context comes into play.

CONTEXT

Look up at the night sky, and the stars look like dots on a flat surface. The lack of visual depth makes the translation from sky to paper fairly straightforward, which makes it easier to imagine constellations. Just connect the dots. However, although you perceive stars to be the same distance away from you, they are actually varying light years away.

If you could fly out beyond the stars, what would the constellations look like? This is what Santiago Ortiz wondered as he visualized stars from a different perspective, as shown in Figure 1-25.

The initial view places the stars in a global layout, the way you see them. You look at Earth beyond the stars, but as if they were an equal distance away from the planet.

Zoom in, and you can see constellations how you would from the ground, bundled in a sleeping bag in the mountains, staring up at a clear sky.

The perceived view is fun to see, but flip the switch to show actual distance, and it gets interesting. Stars transition, and the easy-to-distinguish constellations are practically unrecognizable. The data looks different from this new angle.

This is what context can do. It can completely change your perspective on a dataset, and it can help you decide what the numbers represent and how to interpret them. After you do know what the data is about, your understanding helps you find the fascinating bits, which leads to worthwhile visualization.

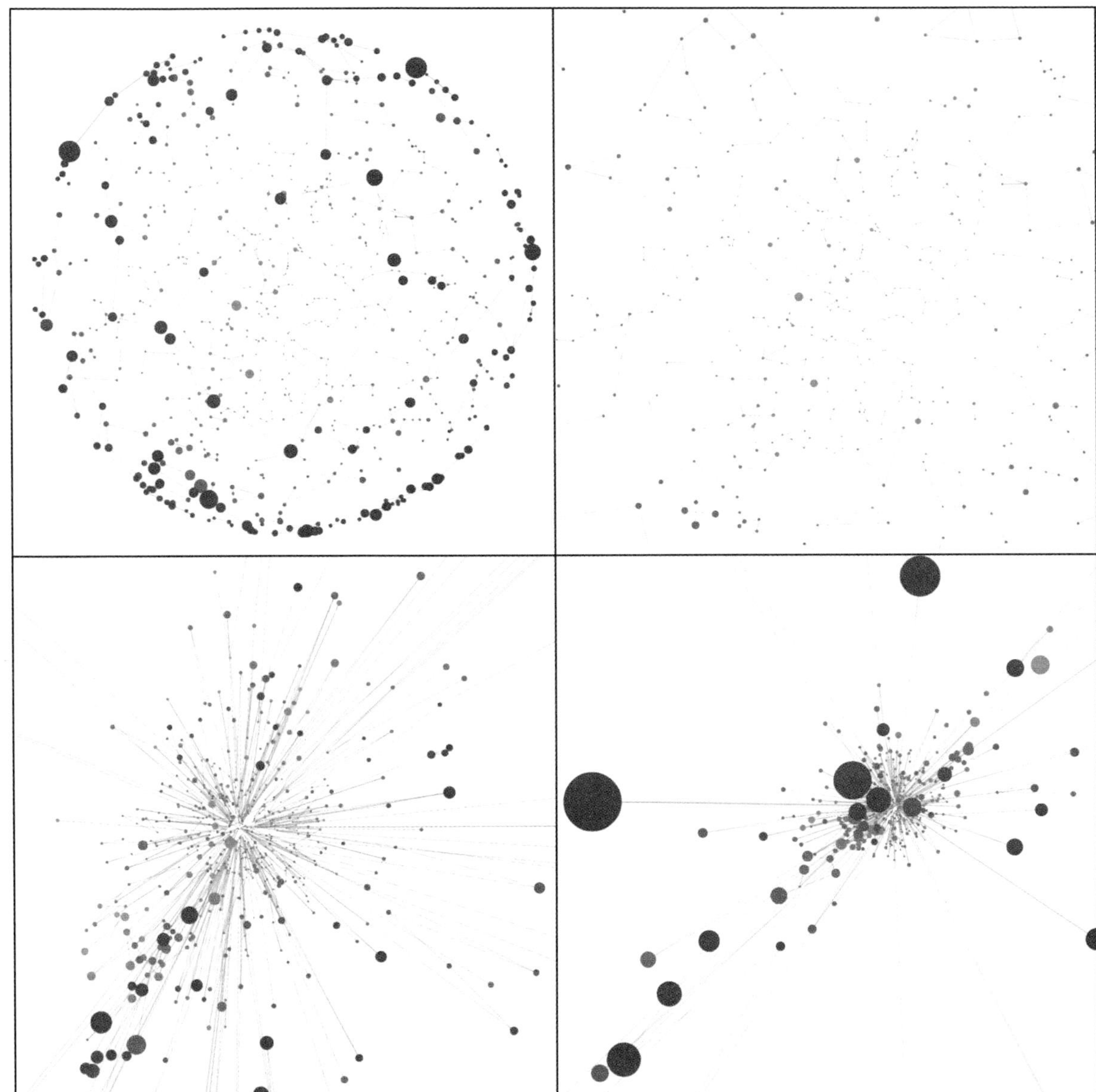

FIGURE 1-25 View of the Sky *by Santiago Ortiz, http://moebio.com/exomap/viewsofthesky/2/*

Without context, data is useless, and any visualization you create with it will also be useless. Using data without knowing anything about it, other than the values themselves, is like hearing an abridged quote secondhand and then citing it as a main discussion point in an essay. It might be okay, but you risk finding out later that the speaker meant the opposite of what you thought.

You have to know the who, what, when, where, why, and how—the metadata, or the data about the data—before you can know what the numbers are actually about.

Who: A quote in a major newspaper carries more weight than one from a celebrity gossip site that has a reputation for stretching the truth. Similarly, data from a reputable source typically implies better accuracy than a random online poll.

For example, Gallup, which has measured public opinion since the 1930s, is more reliable than say, someone (for example, me) experimenting with a small, one-off Twitter sample late at night during a short period of time. Whereas the former works to create samples representative of a region, there are unknowns with the latter.

Speaking of which, in addition to who collected the data, who the data is about is also important. Going back to the gumballs, it's often not financially feasible to collect data about everyone or everything in a population. Most people don't have time to count and categorize a thousand gumballs, much less a million, so they sample. The key is to sample evenly across the population so that it is representative of the whole. Did the data collectors do that?

How: People often skip methodology because it tends to be complex and for a technical audience, but it's worth getting to know the gist of how the data of interest was collected.

If you're the one who collected the data, then you're good to go, but when you grab a dataset online, provided by someone you've never met, how will you know if it's any good? Do you trust it right away, or do you investigate? You don't have to know the exact statistical model behind every dataset, but look out for small samples, high margins of error, and unfit assumptions about the subjects, such as indices or rankings that incorporate spotty or unrelated information.

Sometimes people generate indices to measure the quality of life in countries, and a metric like literacy is used as a factor. However, a country might not have up-to-date information on literacy, so the data gatherer simply uses an estimate from a decade earlier. That's going to cause problems because then the index works only under the assumption that the literacy rate one decade earlier is comparable to the present, which might not be (and probably isn't) the case.

What: Ultimately, you want to know what your data is about, but before you can do that, you should know what surrounds the numbers. Talk to subject experts, read papers, and study accompanying documentation.

In introduction statistics courses, you typically learn about analysis methods, such as hypothesis testing, regression, and modeling, in a vacuum, because the goal is to learn the math and concepts. But when you get to real-world data, the goal shifts to information gathering. You shift from, "What is in the numbers?" to "What does the data represent in the world; does it make sense; and how does this relate to other data?"

A major mistake is to treat every dataset the same and use the same canned methods and tools. Don't do that.

When: Most data is linked to time in some way in that it might be a time series, or it's a snapshot from a specific period. In both cases, you have to know when the data was collected. An estimate made decades ago does not equate to one in the present. This seems obvious, but it's a common mistake to take old data and pass it off as new because it's what's available. Things change, people change, and places change, and so naturally, data changes.

Where: Things can change across cities, states, and countries just as they do over time. For example, it's best to avoid global generalizations when the data comes from only a few countries. The same logic applies to digital locations. Data from websites, such as Twitter or Facebook, encapsulates the behavior of its users and doesn't necessarily translate to the physical world.

Although the gap between digital and physical continues to shrink, the space between is still evident. For example, an animated map that represented the "history of the world" based on geotagged Wikipedia, showed popping dots for each entry, in a geographic space. The end of the video is shown in Figure 1-26.

The result is impressive, and there is a correlation to the real-life timeline for sure, but it's clear that because Wikipedia content is more prominent in English-speaking countries the map shows more in those areas than anywhere else.

Why: Finally, you must know the reason data was collected, mostly as a sanity check for bias. Sometimes data is collected, or even fabricated, to serve an agenda, and you should be wary of these cases. Government and elections might be the first thing that come to mind, but so-called information graphics around the web, filled with keywords and published by sites trying to grab Google juice, have also grown up to be a common culprit. (I fell for these a couple of times in my early days of blogging for FlowingData, but I learned my lesson.)

Learn all you can about your data before anything else, and your analysis and visualization will be better for it. You can then pass what you know on to readers.

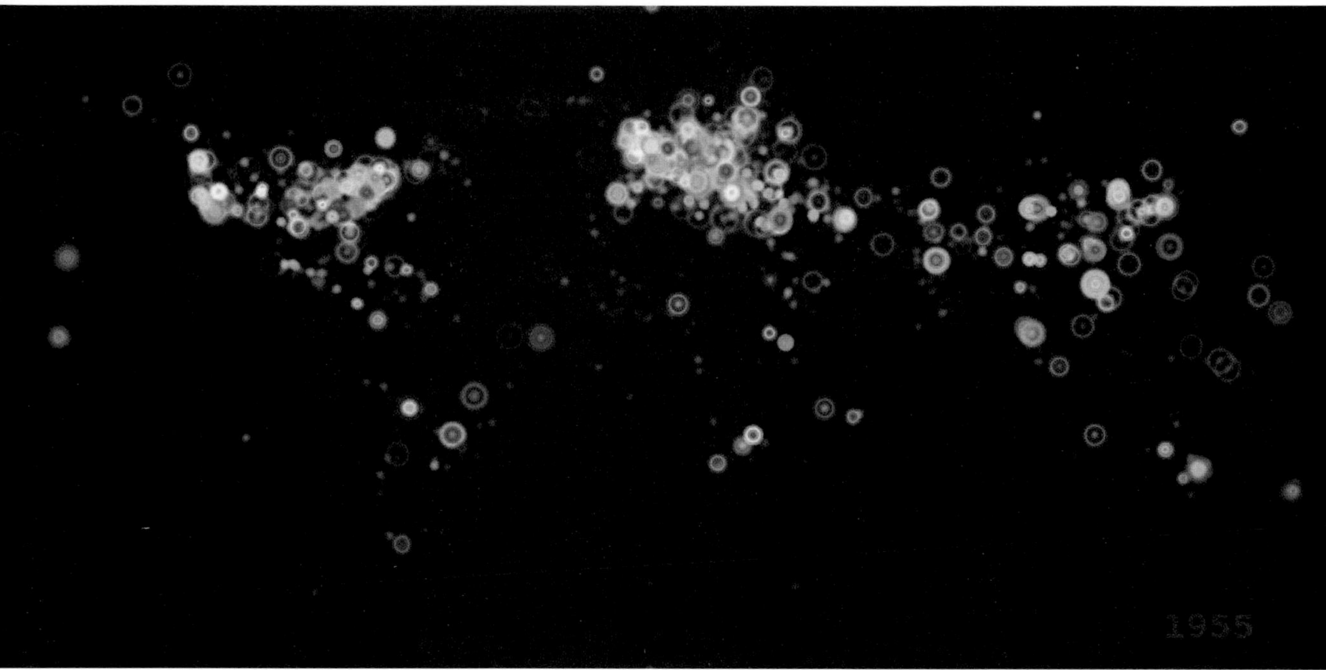

FIGURE 1-26 A History of the World in 100 Seconds *by Gareth Lloyd, http://datafl.ws/24a*

However, just because you have data doesn't mean you should make a graphic and share it with the world. Context can help you add a dimension—a layer of information—to your data graphics, but sometimes it means it's better to hold back because it's the right thing to do.

In 2010, Gawker Media, which runs large blogs like Lifehacker and Gizmodo, was hacked, and 1.3 million usernames and passwords were leaked. They were downloadable via BitTorrent. The passwords were encrypted, but the hackers cracked about 188,000 of them, which exposed more than 91,000 unique passwords. What would you do with that kind of data?

The mean thing to do would be to highlight usernames with common (read that poor) passwords, or you could go so far as to create an application that guessed passwords, given a username.

A different route might be to highlight just the common passwords, as shown in Figure 1-27. This offers some insight into the data without making it too easy to log in with someone else's account. It might also serve as a warning to others to change their passwords to something less obvious. You know, something with at least two symbols, a digit, and a mix of lowercase and uppercase letters. Password rules are ridiculous these days. But I digress.

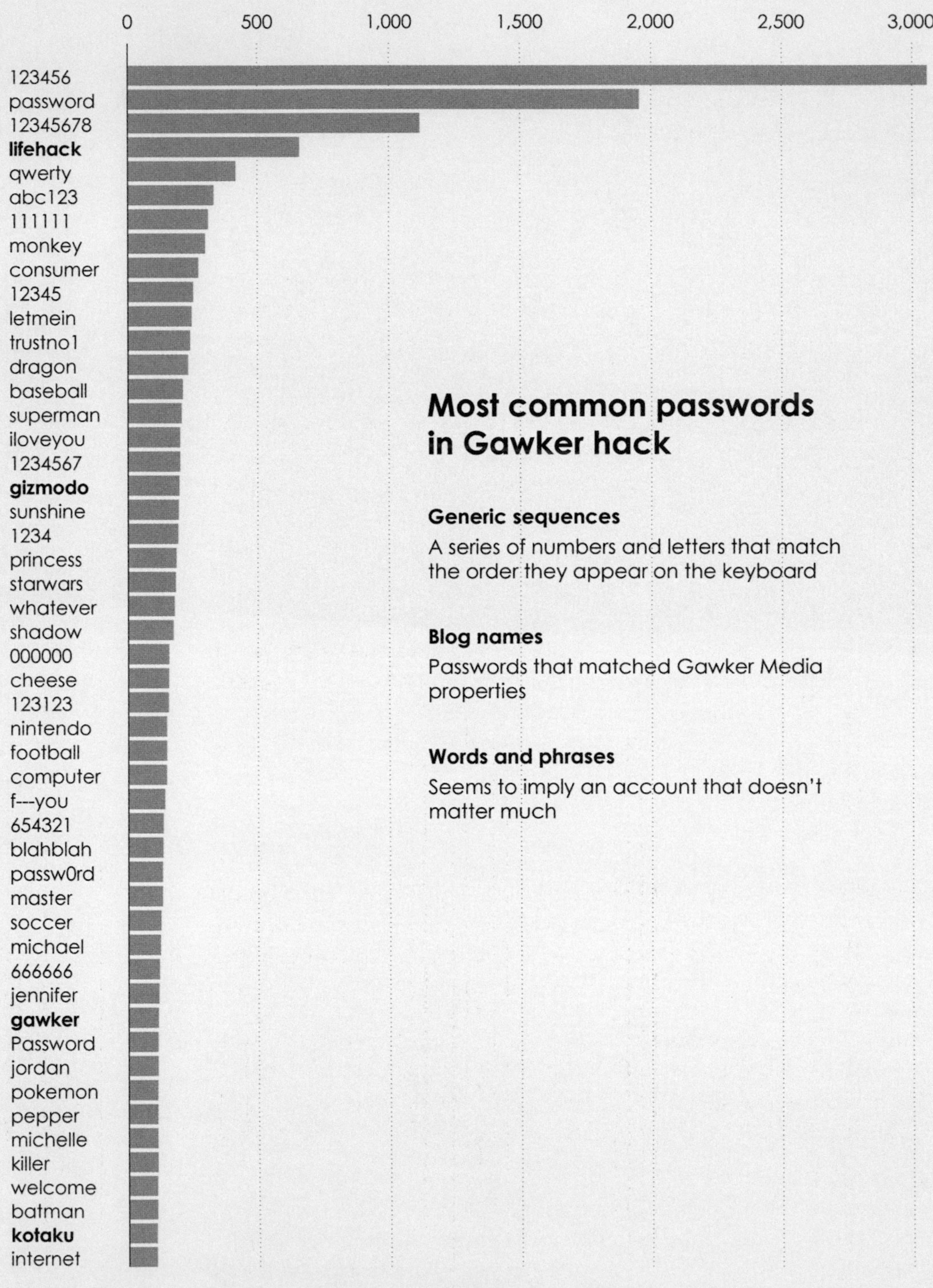

0
500
1,000
1,500
2,000
2,500
3,000
123456
password
12345678
lifehack
qwerty
abc123
111111
monkey
consumer
12345
letmein
trustno1
dragon
baseball
superman
iloveyou
1234567
gizmodo
sunshine
1234
princess
starwars
whatever
shadow
000000
cheese
123123
nintendo
football
computer
f---you
654321
blahblah
passw0rd
master
soccer
michael
666666
jennifer
gawker
Password
jordan
pokemon
pepper
michelle
killer
welcome
batman
kotaku
internet
Most common passwords in Gawker hack
Generic sequences
A series of numbers and letters that match the order they appear on the keyboard
Blog names
Passwords that matched Gawker Media properties
Words and phrases
Seems to imply an account that doesn't matter much

FIGURE 1-27 *Commonly used passwords in Gawker hack*

With data like the Gawker set, a deep analysis might be interesting, but it could also do more harm than good. In this case, data privacy is more important, so it's better to limit what you show and look at.

Whether you should use data is not always clear-cut though. Sometimes, the split between what's right and wrong can be gray, so it's up to you to make the call. For example, on October 22, 2010, Wikileaks, an online organization that releases private documents and media from anonymous sources, released 391,832 United States Army field reports, now known as the Iraq War Logs. The reports recorded 66,081 civilian deaths out of 109,000 recorded deaths, between 2004 and 2009.

The leak exposed incidents of abuse and erroneous reporting, such as civilian deaths classified as "enemy killed in action." On the other hand, it can seem unjustified to publish findings about classified data obtained through less than savory means.

Maybe there should be a golden rule for data: Treat others' data the way you would want your data treated.

In the end, it comes back to what data represents. Data is an abstraction of real life, and real life can be complicated, but if you gather enough context, you can at least put forth a solid effort to make sense of it.

WRAPPING UP

Visualization is often thought of as an exercise in graphic design or a brute-force computer science problem, but the best work is always rooted in data. To visualize data, you must understand what it is, what it represents in the real world, and in what context you should interpret it in.

Data comes in different shapes and sizes, at various granularities, and with uncertainty attached, which means totals, averages, and medians are only a small part of what a data point is about. It twists. It turns. It fluctuates. It can be personal, and even poetic. As a result, you can find visualization in many forms.

Visualization: The Medium

Visualization has been around for centuries, but it is relatively new as a field of study, and experts in the area haven't even decided what exactly visualization is yet. Should it be used for only analysis? Is visualization specifically for quantitative insights, or can you use it to evoke emotions? At what point does visualization—a field deeply rooted in, well, visual things—become art?

The answers to these questions vary between who you ask. The questions have created heated debates within and in between subject areas, and this is just amongst the academics and practitioners.

I was on a consulting gig at a large, data-centric organization because it wanted to inject more visualization into its work. It wanted the public to know what it was doing and wanted to improve its existing work in reports and data summaries, along with tools within the organization.

So I was in a meeting with about 40 people, which was a diverse group of marketers, developers, and statisticians. The group worked on a variety of projects, from quick, made-for-blog graphics to interactive data exploration tools. We were discussing an online application, and part of the group felt there should be more editorial content on what the data was about, whereas another part insisted that any interpretation should be left to the users. A few others leaned toward graphics that looked like abstract paintings. The ideas for visualization were all over the place, and a long argument ensued.

They were all right. Everyone argued in support of visualization for a specific purpose and insisted that others' visualization had to fit the same criteria, even if the others designed applications for different reasons and with a different audience in mind. They approached visualization as if it were a monolithic thing that had a defined set of rules. This might have been true a century ago (or not), but visualization has grown into more than just a tool. Visualization is a medium: a way to explore, present, and express meaning in data.

Rather than disjoint categories that work independently from others, you can think of visualization as a continuous spectrum that stretches from statistical graphics to data art. There is visualization that is clearly one or the other, but there are many works that are a blend of both and can't be put in a bin. Where statistics, design, and aesthetics find a balance is where a lot of the best work comes from.

This is not to say that the blend is always best, nor are statistical graphics better than data art, or vice versa. They all serve their own purposes and should be judged by how well they achieve their goals. You don't critique a documentary in the same way you might judge a slapstick comedy because you go in with different expectations and a different mindset. Similarly, you don't expect a romance novel from a textbook or complain about how unfunny a television crime drama is.

A series of comical pie charts shouldn't be put under the same microscope as visualization research, unless those pie charts happen to be part of research on how people react to comical pie charts. If so, I would like to read that paper, because I'm sure it's hilarious.

Note: There are rules and design suggestions for visualization. These are fine, but you can't just blindly follow them. Consider goals and applications.

Again, this is not to say that you should be less critical of funny graphs or data art than you are of exploratory visualization. People examine comedy and art all the time. Just know what you're critiquing.

ANALYSIS AND EXPLORATION

William Playfair is credited with inventing many of the traditional chart types used today: the line graph, the bar chart, and the pie chart. In 1786, Playfair published the first bar charts in *The Commercial and Political Atlas*, which showed progress via indicators such as imports and exports, as shown in Figure 2-1. Figure 2-2 shows one of the first pie charts. These charts were handmade on paper, of course.

It's hard to believe that as late as the 1970s this was how people looked at data—by hand. In John Tukey's seminal *Exploratory Data Analysis*, published in 1977, he described how to darken the shade of lines by using a pen instead of a pencil. Such a technique seems ancient now. The good news is that technology advanced, and Tukey continued to innovate with what became available.

In line with the improvements of technology, the volume of data and availability has increased dramatically, too, which in turn gave people something new to visualize (and new jobs and fields of study). Because remember: There is no visualization without the data.

In 2001, Wikipedia (the collaborative, online encyclopedia) launched, and as of this writing has 35 million registered users. Anyone can edit Wikipedia entries, so when someone starts an article, it can grow and shrink as others add and delete information. This creates a dynamic within every article, especially as individuals argue over what should and shouldn't be written.

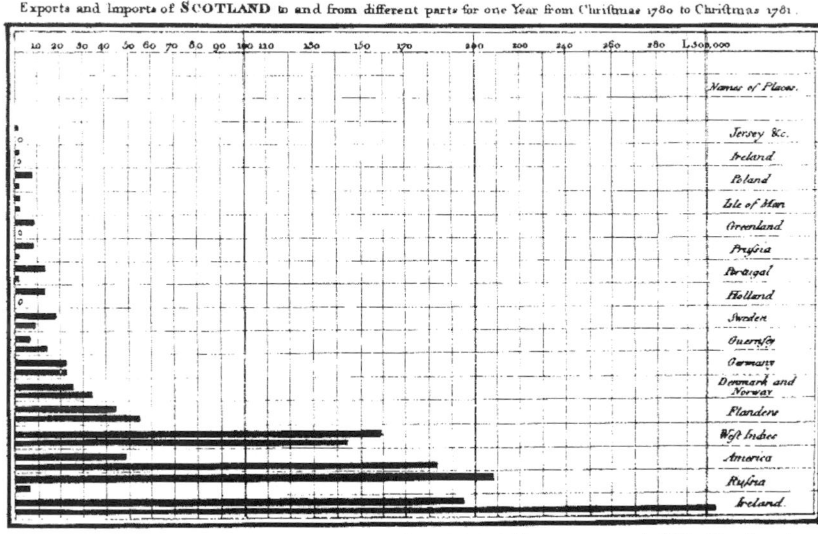

FIGURE 2-1 *Bar chart from* The Commercial and Political Atlas *by William Playfair*

The fun starts when you look at the event history for articles, which is freely available via the site. Fernanda Viégas and Martin Wattenberg explored this concept in 2003 with *History Flow*, a tool that enables you to explore the history of any Wikipedia entry over time.

As shown in Figure 2-3, the visualization looks like an inverted stacked area chart, where each layer represents a body of text. As time passes, new layers (encoded with different colors) are added (or removed), and you can see the change in overall size via the total vertical height of the full stack.

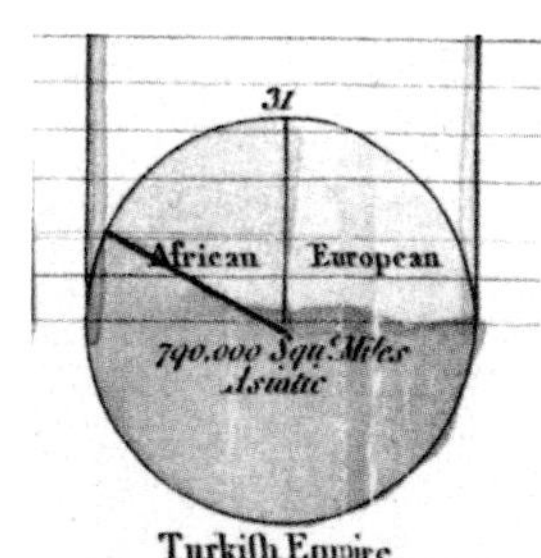

FIGURE 2-2 *William Playfair created the pie chart for* Statistical Breviary *in 1801*

Notice the zigzag pattern and the seemingly random sections of black? The former shows debate between users, and the latter is when someone deleted a portion of the article, either because of a disagreement or a user is just in it for the lulz.

The most interesting aspect of *History Flow* is the changes over time for each article. When events occur in real life, it's hard to see the big picture because you're so focused on a single event. As a Wikipedia user in a heated debate, your main concern is what the opposition just did and then you figure out how to react, but when you take a step back to see the overall changes after the fact, it's likely you'll see something interesting.

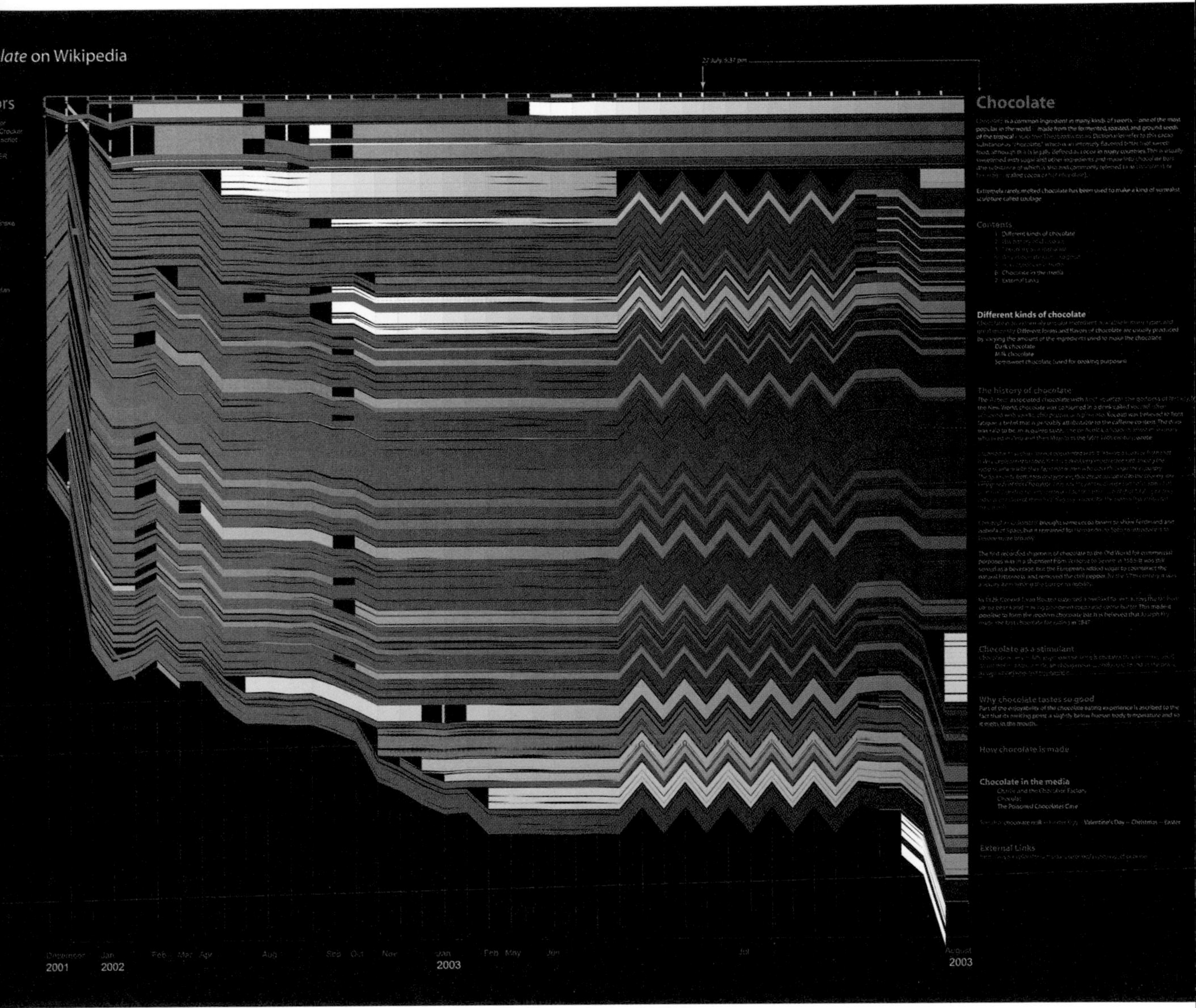

FIGURE 2-3 History Flow *(2003) by Fernanda Viégas and Martin Wattenberg, http://hint.fm*

Note: Although Wikipedia is an encyclopedia, because it's always changing, you can also easily relate activity to current events, such as times of unrest or shifts in political power.

The World Bank provides countrywide data in an easy-to-download format to help you gain an understanding on the progress of the world. Figure 2-4 (an interactive I made to look at life expectancy over the years for different countries) shows an overall increase for most regions; but at the same time, big dips indicate wars and times of struggle in some places. You can, for example, see the Bangladesh Liberation War in the 1970s, the Iran-Iraq War in the 1980s, and the Rwanda Civil War in the 1990s. There's a smaller dip for Iraq in early 2000. Selectable regions and countries enable you to highlight specifics.

From a methodology point of view, both *History Flow* and the life expectancy charts are a modified stacked area chart and multiple time series, respectively. The data makes them interesting, but pre-Internet, these numbers would have been harder to come by, if they existed at all.

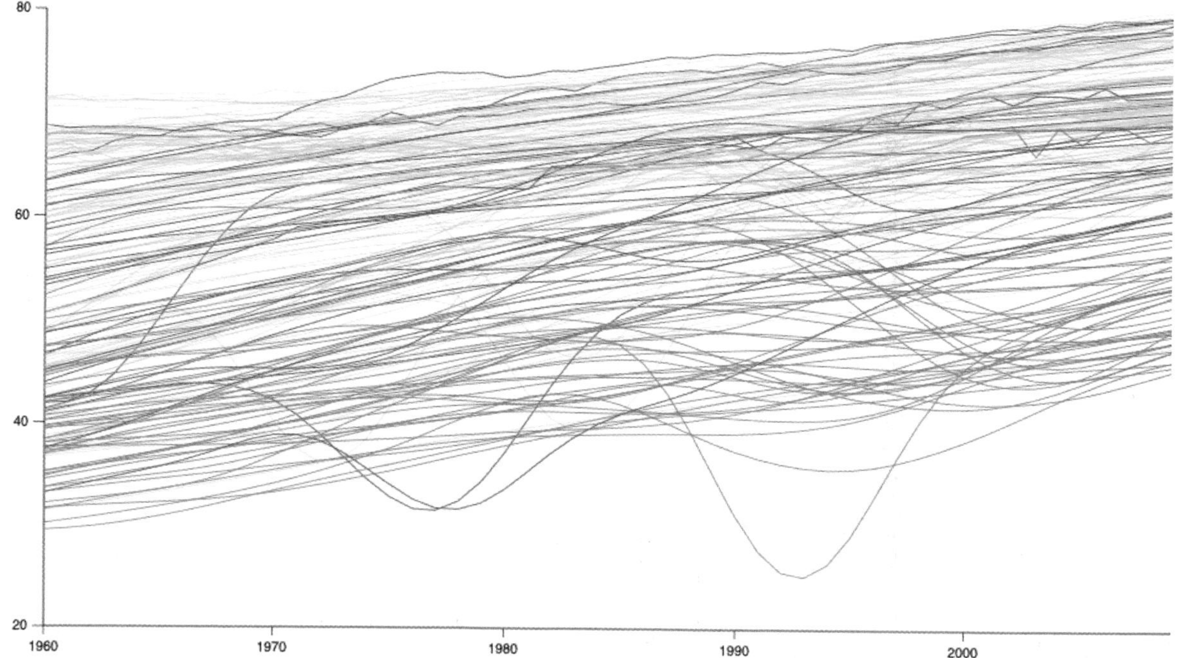

FIGURE 2-4 *Life expectancy around the world, http://datafl.ws/24w*

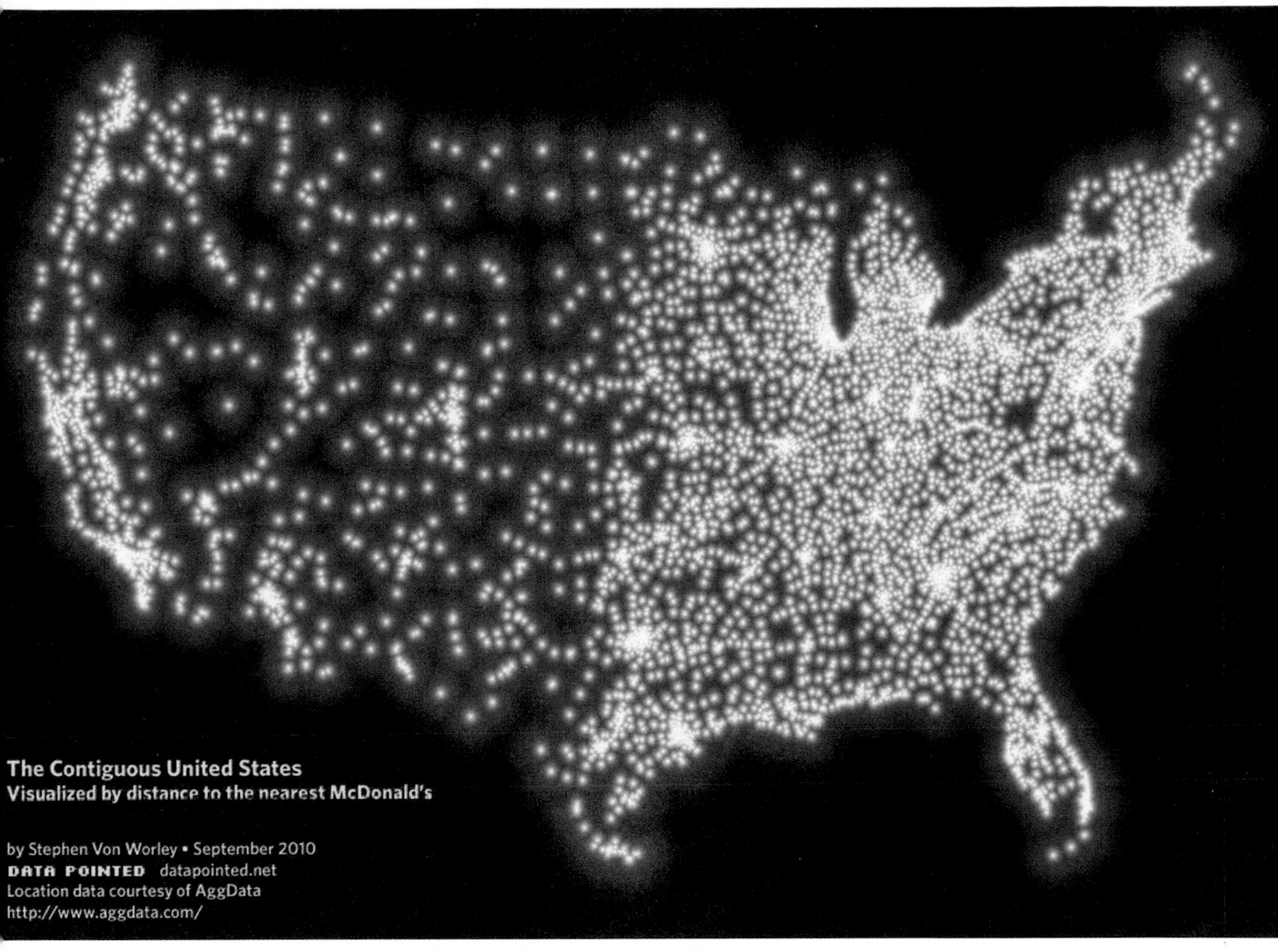

The Contiguous United States
Visualized by distance to the nearest McDonald's
by Stephen Von Worley • September 2010
DATA POINTED datapointed.net
Location data courtesy of AggData
http://www.aggdata.com/

FIGURE 2-5 Distance to McDonald's *(2010) by Stephen Von Worley, http://datafl.ws/24y*

Now it seems like you can find data on almost anything if you look hard enough. Using a ready-to-use, comma-delimited file, Stephen Von Worley calculated the distance to the nearest McDonald's everywhere in the contiguous United States and mapped it. As shown in Figure 2-5, the brighter an area, the less time it'll take to grab a Big Mac.

Even newer, social media sites, such as Twitter and Facebook, continue to flourish, which provide a new source of information on what people are talking about and concerned with. Data is readily accessible via the application programming interfaces (API). Photo-sharing site Flickr also has an accessible API. Eric Fischer combined data from Twitter and Flickr in his series of maps titled *See Something or Say Something*, shown in Figure 2-6.

FIGURE 2-6 *(following page)* See Something or Say Something *(2011) by Eric Fischer, http://datafl.ws/2ba*

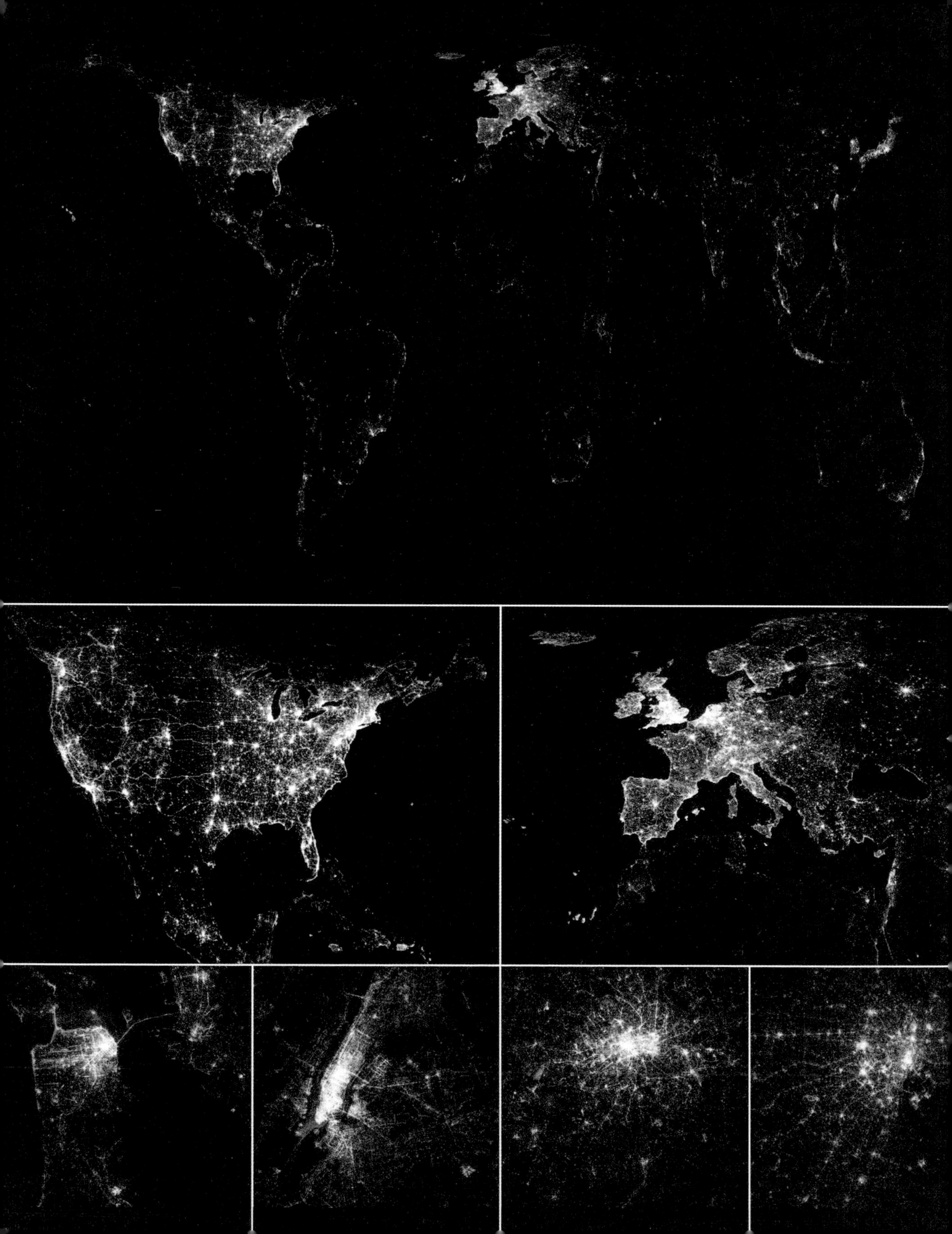

Blue dots are locations where people tweeted, red dots represent places where people took Flickr photos, and white dots represent where people did both. So you can see where people usually tweet (say something) or take a picture (see something). It's a simple idea with great execution and beautiful results.

From an even wider point of view—from space—NASA uses satellite data to monitor activity on Earth. For example, Figure 2-7 is a snapshot from an animation that shows components of the water cycle: evaporation, water vapor, and precipitation. The data feeds into atmospheric models, which allows you to see significant changes over time for the planet.

Perpetual Ocean, also by NASA, uses similar data and models to estimate ocean currents, as shown in Figure 2-8. It might remind you of Vincent van Gogh's painting *The Starry Night.*

How amazing is that? Large amounts of data make that possible. Of course, new data types in growing volumes demands new tools to explore beyond pen and paper.

Note: See also the animated world map from NASA that shows a decade of fires: http://datafl.ws/2bb.

TOOLS

The introduction of computers changed how you can analyze and explore data. You can make a lot of charts in a few seconds, view data from many angles, and sift through more complex datasets than those who had to chart manually. There are also more data exploration tools than ever before. Microsoft Excel is still the software of choice in many offices, which can work for a lot of jobs, but the methods that people want to use and the depth they want to explore is changing.

Tableau Software is one of the more popular desktop programs that enables you to visually analyze your data. Everything is done via a click interface, so no programming skills are required, and it can handle a healthy amount of data at once, so you're free to roam. Tableau Public enables you to build visualization dashboards and share them online.

There's also desktop software to visualize specific types of data. For example, ImagePlot by Software Studies Lab at California Institute for Telecommunication and Information Technology (Calit2) enables you to explore images. You can do this with other software, but ImagePlot specializes in handling millions of images at once and can place them in a two-dimensional space to analyze aspects of a collection, such as color or volume.

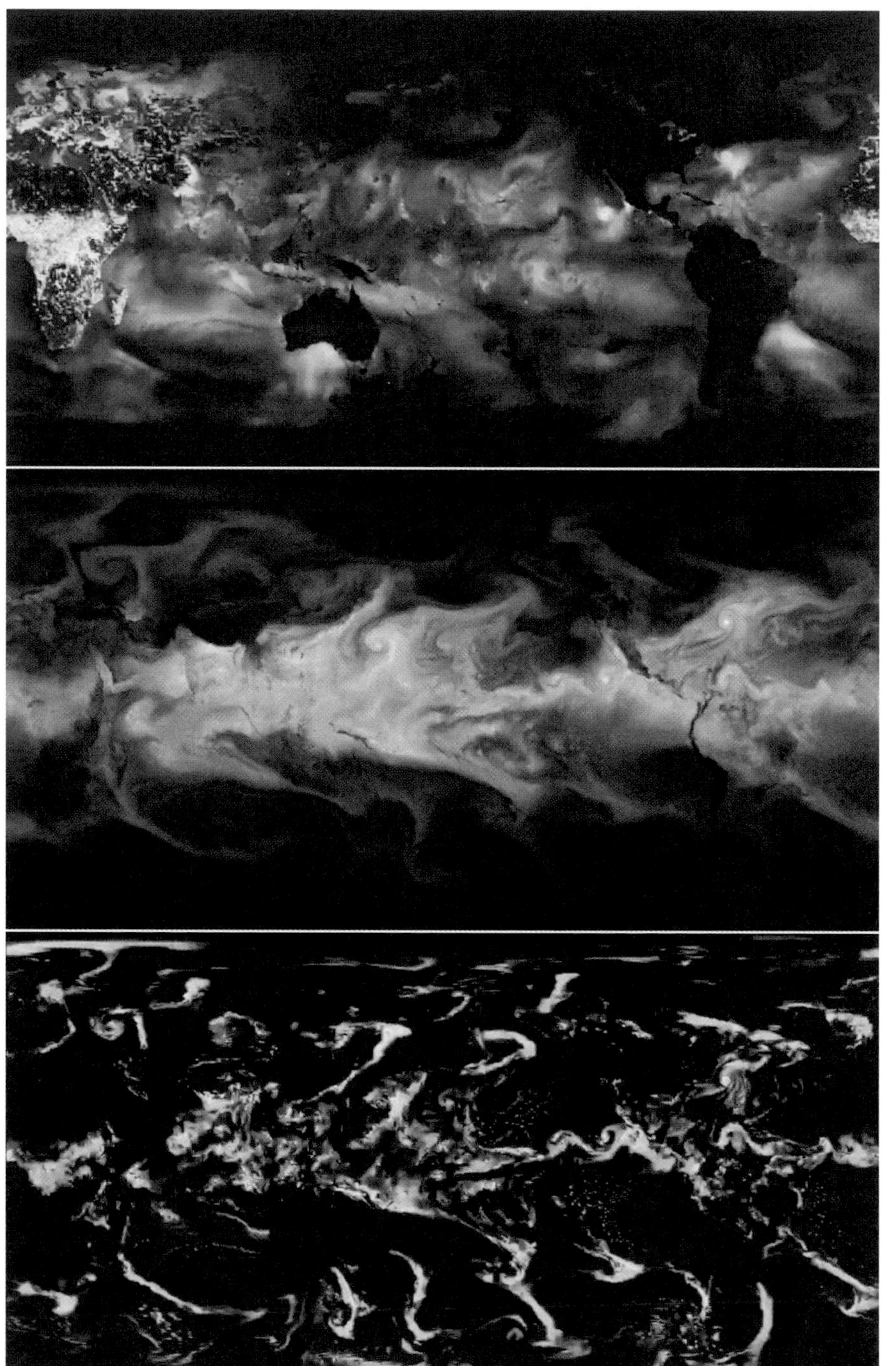

FIGURE 2-7 *Components of the* Water Cycle on a Flat Map *(2011) by NASA/Goddard Space Flight Center Scientific Visualization Studio, http://svs.gsfc.nasa.gov/goto?3811*

This specialization is a current theme in visualization development. It's easier and more efficient to build tools to handle specific data types than try to develop software that handles everything imaginable.

Gephi is the go-to open source software to visualize networks and systems. It's "like Photoshop for graphs." Whenever you see a static graph with a lot of nodes and edges, it was most likely created with this software. On the desktop, you can easily explore and interact by clicking and dragging, and you can export images when you find something interesting.

Treemap, developed by the Human-Computer Interaction Lab at the University of Maryland, enables you to explore hierarchical data via said treemaps. Originally created by Ben Shneiderman in 1990 to visualize the contents in a hard drive, the software, shown in Figure 2-11, is now more flexible, interactive, and free to use for noncommercial use.

For static, statistical graphics, my personal favorite is R in combination with Adobe Illustrator. R is the statistical computing language of choice, which has recently gained steam in the data community, and Illustrator is a program that a lot of designers use.

As you venture out to visualization for the web, the programming skill requirements seem to increase, but there are a lot of packages to help ease you into the area. It's not drag-and-drop simple, but developers have learned that providing a lot of examples get people to use their software, which is great for everyone.

Note: See Chapter 7, "Where to Go from Here," for more on tools and programming for visualization.

A few years ago, online visualization was almost all in Flash, but that has since faded. It's all about JavaScript and HTML5 these days. Again, there are a lot of libraries, but Data-Driven Documents (D3) by Mike Bostock, Raphaël by Dmitry Baranovskiy, and the JavaScript InfoVis Toolkit by Nicolas Garcia Belmonte are your best bets for starting. You can of course always upload static images online, but loading visualization native in the browser brings the added benefit of graphics that can update based on current data. Programming in JavaScript also allows you to incorporate interaction and animation, which can add another dimension to data exploration and presentation.

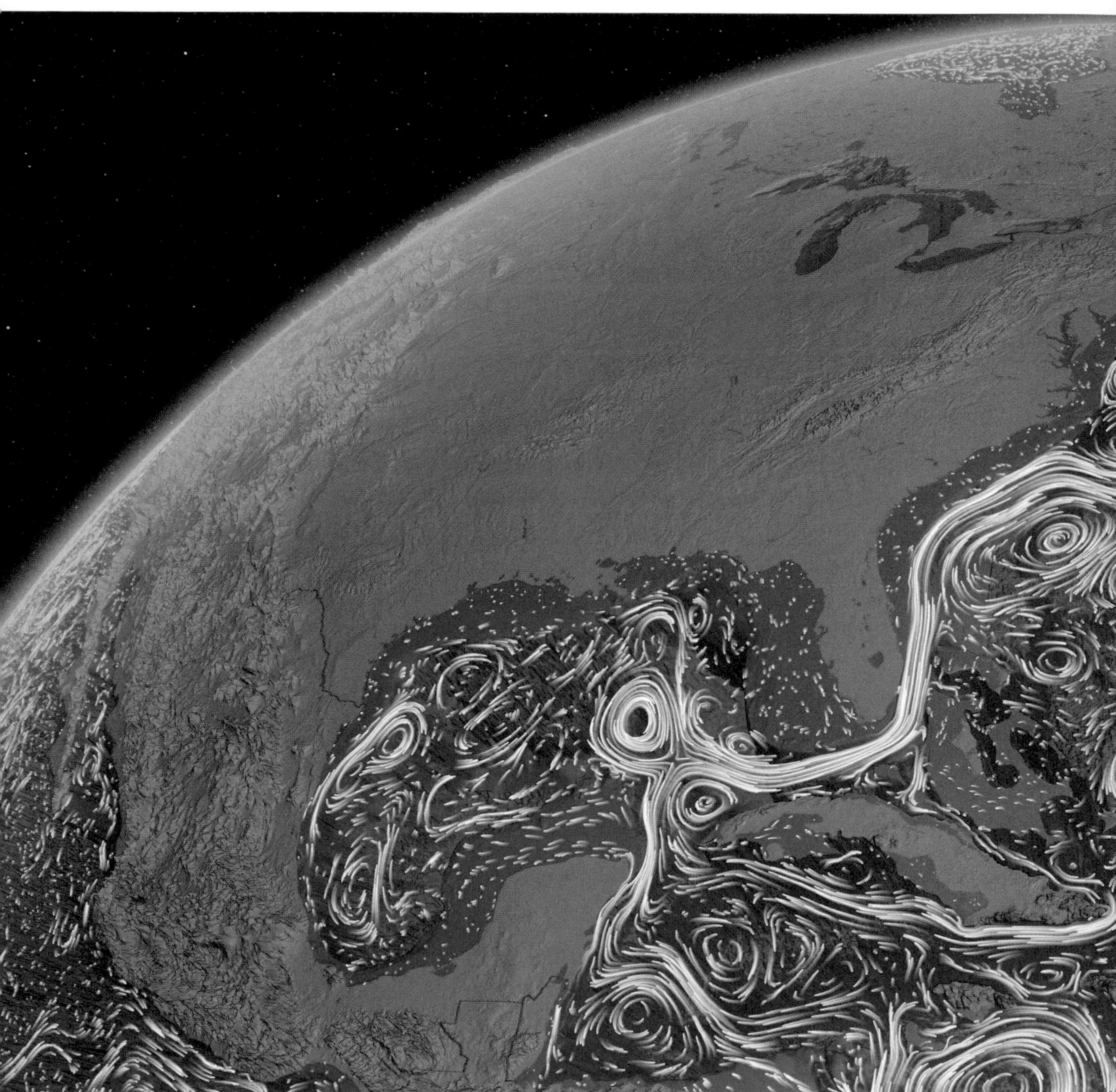

FIGURE 2-8 Perpetual Ocean *(2012) by NASA/Goddard Space Flight Center Scientific Visualization Studio, http://datafl.ws/2bc*

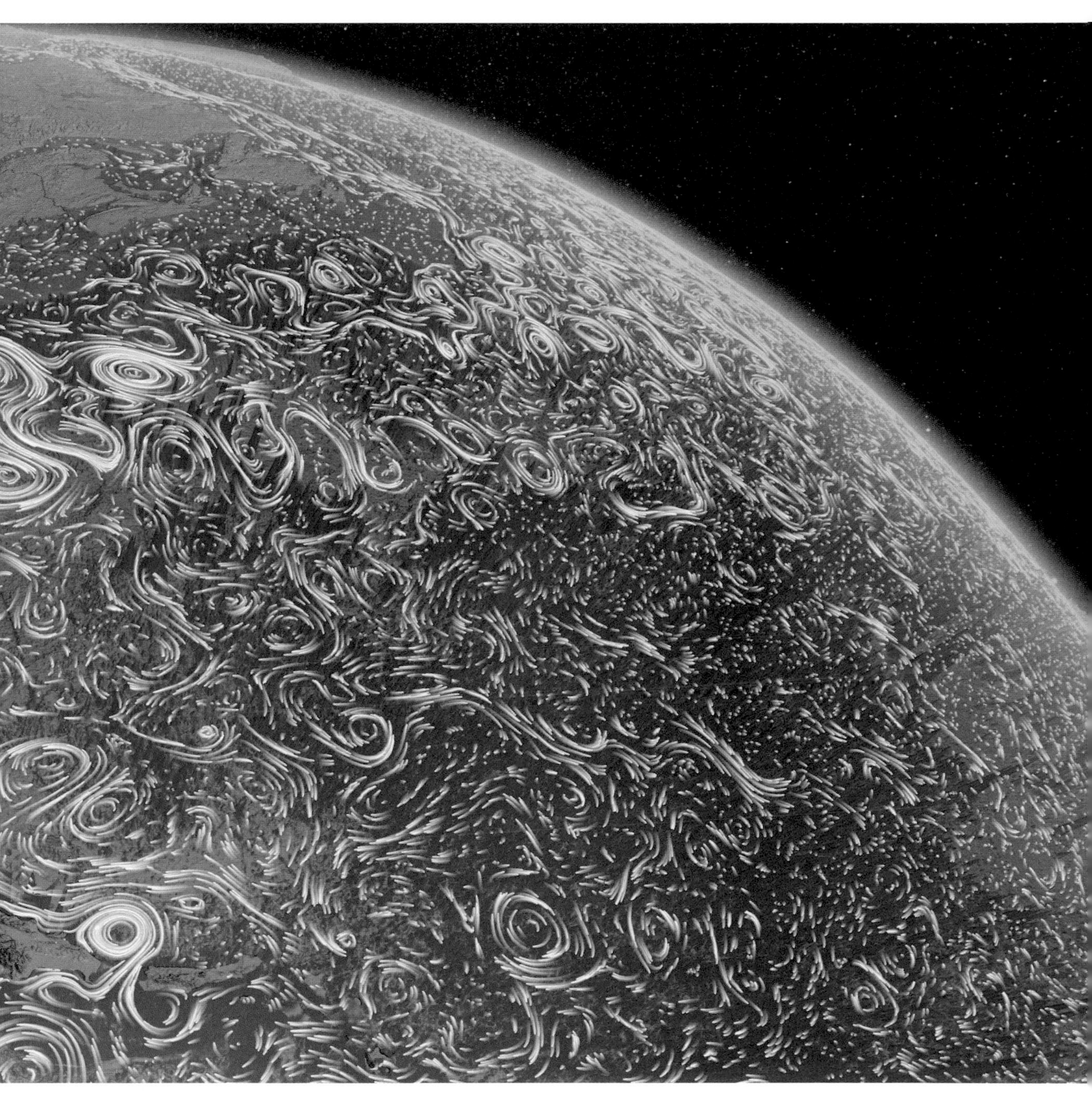

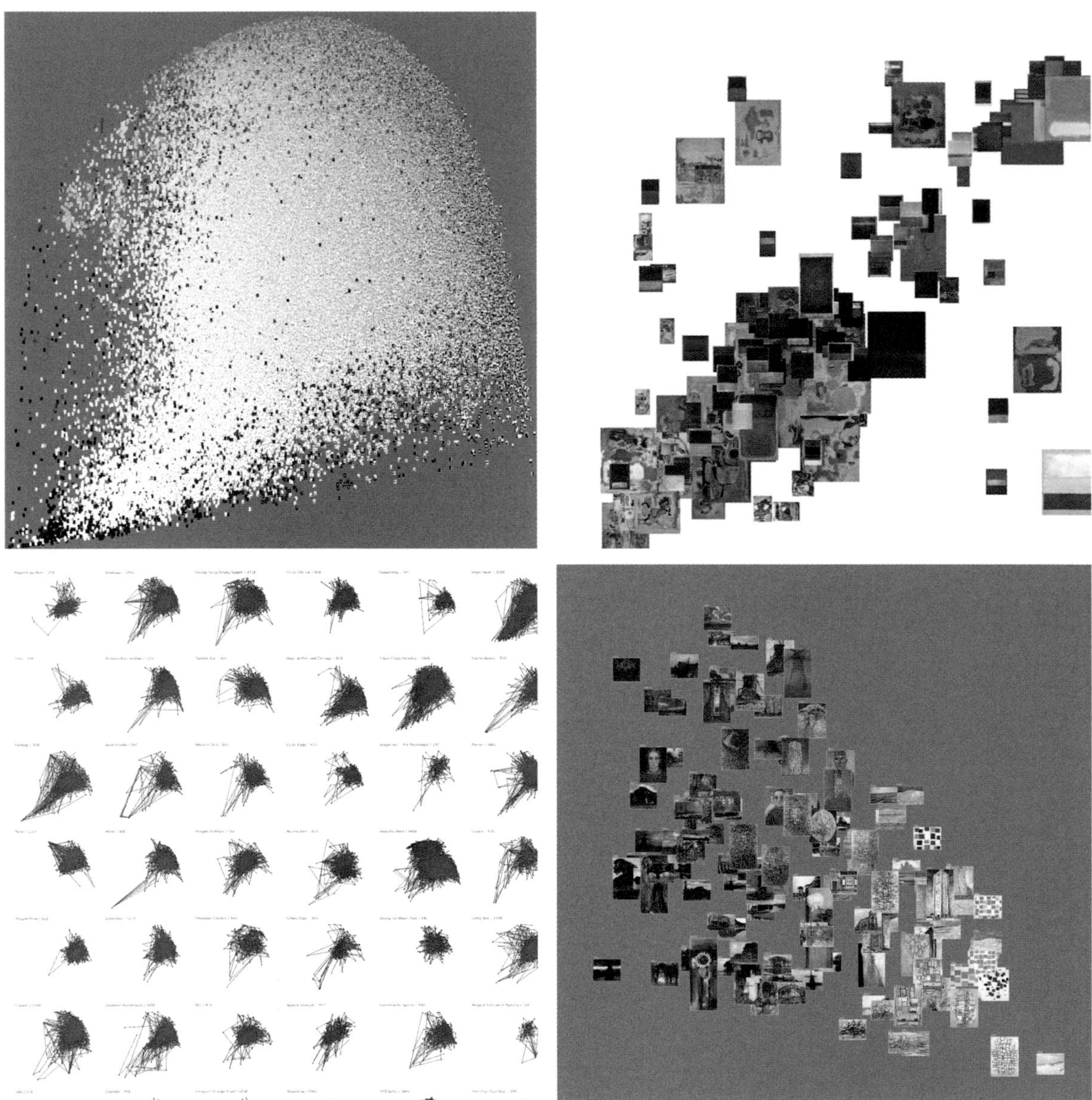

FIGURE 2-9 *ImagePlot by Software Studies Lab at California Institute for Telecommunication and Information Technology, http://datafl.ws/24x*

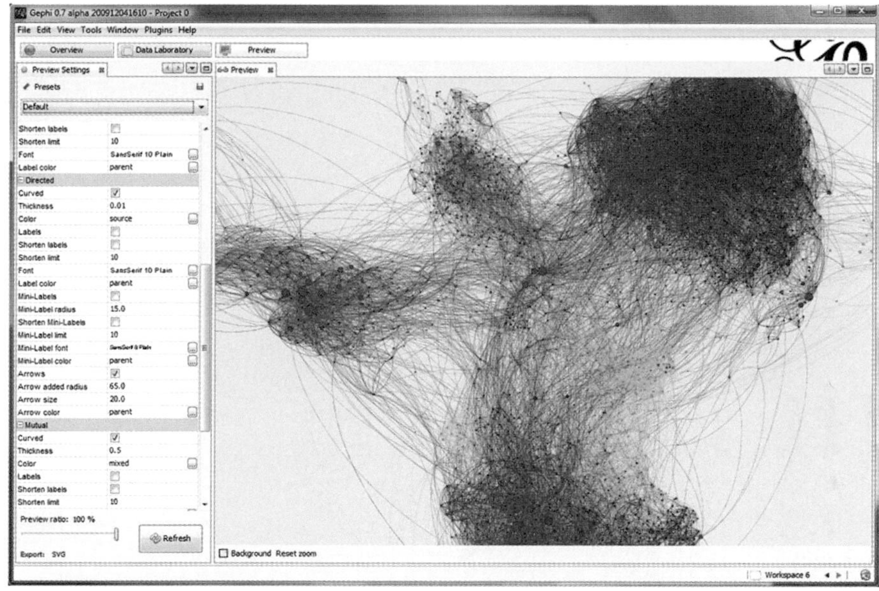

FIGURE 2-10 *Gephi, http://gephi.org*

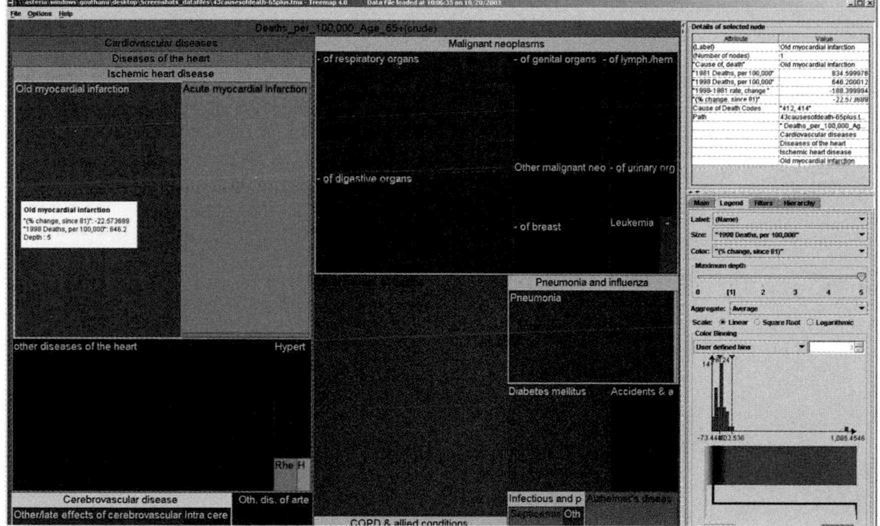

FIGURE 2-11 *Treemap by the Human-Computer Interaction Lab at the University of Maryland, http://www.cs.umd.edu/hcil/treemap/*

INFORMATION GRAPHICS AND PRESENTATION

When you explore your data, you gather your own insights, so there's no need to explain interesting facets of the data to yourself. But when your audience of one increases to more than that, you must provide guidance and context for what the data is about.

Often, this has little to do with accompanying graphics with long, detailed essays (or dissertations) and more to do with carefully placed labels, titles, and text to set up readers for what they're about to see. The visualization itself—the shapes, colors, and sizes—represent the data, whereas words can make your graphics easier to read and understand. Attention to typography, contextual elements, and logical layouts also lend to an additional layer of information on top of raw, statistical output.

A common saying in visualization design is to "let the data speak." It means to visualize your data (or information) and then get out of the way, which works great when the data is familiar and the patterns are obvious.

For example, Patrick Smith used a minimalist approach to describe mental disorders, such as obsessive-compulsive disorder, depression, and narcolepsy, as shown in Figure 2-12. He used basic shapes that are relatively small compared to the space available on each poster, but the isolation lends to the seriousness of the conditions.

Coffee Drinks Illustrated, shown in Figure 2-13, by Lokesh Dhakar is a nice example of how small enhancements to basic charts can provide readers a connection. Stacked bars for each drink form the core of this graphic, and labels tell you what each bar represents. Dhakar also includes the name of each coffee drink, making the content simple to read. The coffee mug and steam illustration around each bar graph sets the context immediately.

The True Size of Africa by Kai Krause communicates its point by rotating countries away from their geographic orientation to fit inside Africa, which is explained in the lead-in, as shown in Figure 2-14. You typically view Africa to be a relatively smaller continent, based on the Mercator projection used in online maps. However, in reality, Africa is much larger by area. The title makes this obvious, and smaller maps and tables provide details.

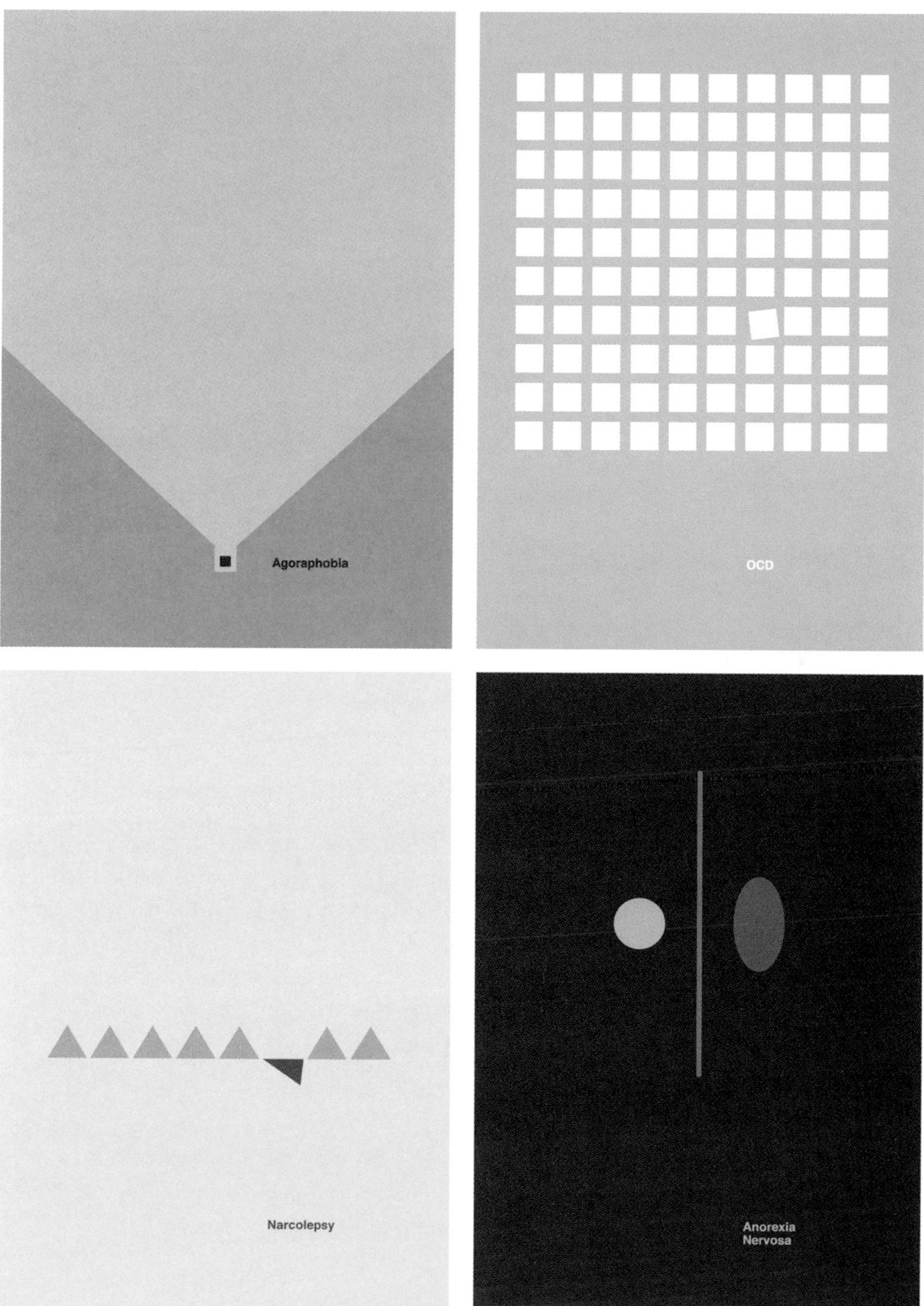

FIGURE 2-12 Mental disorder posters *(2010) by Patrick Smith, http://datafl.ws/259*

FIGURE 2-13 Coffee Drinks Illustrated *(2007) by Lokesh Dhakar, http://lokeshdhakar.com/coffee-drinks-illustrated/*

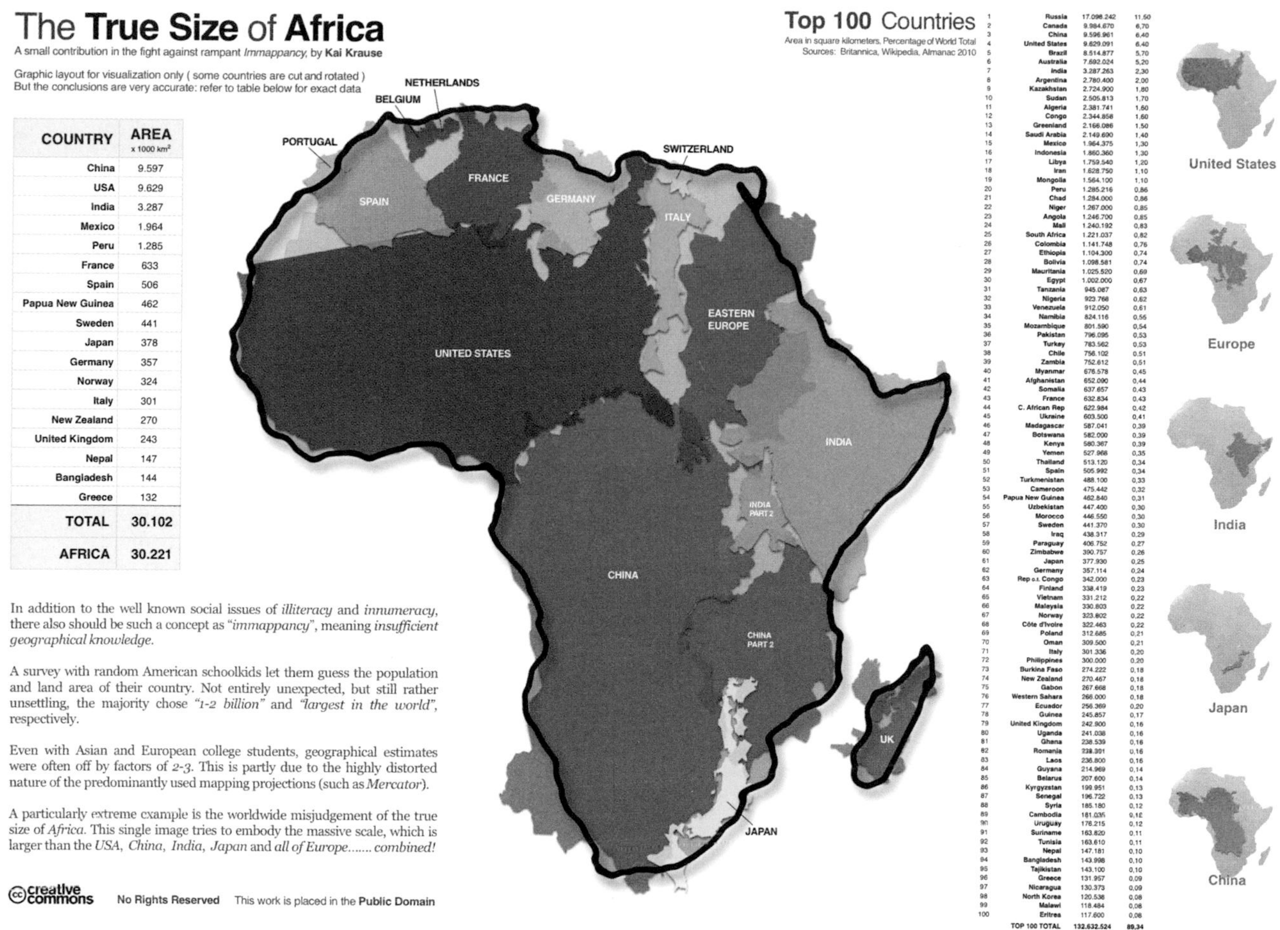

COUNTRY	AREA x 1000 km²
China	9.597
USA	9.629
India	3.287
Mexico	1.964
Peru	1.285
France	633
Spain	506
Papua New Guinea	462
Sweden	441
Japan	378
Germany	357
Norway	324
Italy	301
New Zealand	270
United Kingdom	243
Nepal	147
Bangladesh	144
Greece	132
TOTAL	**30.102**
AFRICA	**30.221**

Top 100 Countries
Area in square kilometers, Percentage of World Total
Sources: Britannica, Wikipedia, Almanac 2010

1	Russia	17.098.242	11,50
2	Canada	9.984.670	6,70
3	China	9.596.961	6,40
4	United States	9.629.091	6,40
5	Brazil	8.514.877	5,70
6	Australia	7.692.024	5,20
7	India	3.287.263	2,30
8	Argentina	2.780.400	2,00
9	Kazakhstan	2.724.900	1,80
10	Sudan	2.505.813	1,70
11	Algeria	2.381.741	1,60
12	Congo	2.344.858	1,60
13	Greenland	2.166.086	1,50
14	Saudi Arabia	2.149.690	1,40
15	Mexico	1.964.375	1,30
16	Indonesia	1.860.360	1,30
17	Libya	1.759.540	1,20
18	Iran	1.628.750	1,10
19	Mongolia	1.564.100	1,10
20	Peru	1.285.216	0,86
21	Chad	1.284.000	0,86
22	Niger	1.267.000	0,85
23	Angola	1.246.700	0,85
24	Mali	1.240.192	0,83
25	South Africa	1.221.037	0,82
26	Colombia	1.141.748	0,76
27	Ethiopia	1.104.300	0,74
28	Bolivia	1.098.581	0,74
29	Mauritania	1.025.520	0,69
30	Egypt	1.002.000	0,67
31	Tanzania	945.087	0,63
32	Nigeria	923.768	0,62
33	Venezuela	912.050	0,61
34	Namibia	824.116	0,55
35	Mozambique	801.590	0,54
36	Pakistan	796.095	0,53
37	Turkey	783.562	0,53
38	Chile	756.102	0,51
39	Zambia	752.612	0,51
40	Myanmar	676.578	0,45
41	Afghanistan	652.090	0,44
42	Somalia	637.657	0,43
43	France	632.834	0,43
44	C. African Rep	622.984	0,42
45	Ukraine	603.500	0,41
46	Madagascar	587.041	0,39
47	Botswana	582.000	0,39
48	Kenya	580.367	0,39
49	Yemen	527.968	0,35
50	Thailand	513.120	0,34
51	Spain	505.992	0,34
52	Turkmenistan	488.100	0,33
53	Cameroon	475.442	0,32
54	Papua New Guinea	462.840	0,31
55	Uzbekistan	447.400	0,30
56	Morocco	446.550	0,30
57	Sweden	441.370	0,30
58	Iraq	438.317	0,29
59	Paraguay	406.752	0,27
60	Zimbabwe	390.757	0,26
61	Japan	377.930	0,25
62	Germany	357.114	0,24
63	Rep o.t. Congo	342.000	0,23
64	Finland	338.419	0,23
65	Vietnam	331.212	0,22
66	Maleysia	330.803	0,22
67	Norway	323.802	0,22
68	Côte d'Ivoire	322.463	0,22
69	Poland	312.685	0,21
70	Oman	309.500	0,21
71	Italy	301.336	0,20
72	Philippines	300.000	0,20
73	Burkina Faso	274.222	0,18
74	New Zealand	270.467	0,18
75	Gabon	267.668	0,18
76	Western Sahara	266.000	0,18
77	Ecuador	256.369	0,20
78	Guinea	245.857	0,17
79	United Kingdom	242.900	0,16
80	Uganda	241.038	0,16
81	Ghana	238.539	0,16
82	Romania	238.391	0,16
83	Laos	236.800	0,16
84	Guyana	214.969	0,14
85	Belarus	207.600	0,14
86	Kyrgyzstan	199.951	0,13
87	Senegal	196.722	0,13
88	Syria	185.180	0,12
89	Cambodia	181.035	0,12
90	Uruguay	176.215	0,12
91	Suriname	163.820	0,11
92	Tunisia	163.610	0,11
93	Nepal	147.181	0,10
94	Bangladesh	143.998	0,10
95	Tajikistan	143.100	0,10
96	Greece	131.957	0,09
97	Nicaragua	130.373	0,09
98	North Korea	120.538	0,08
99	Malawi	118.484	0,08
100	Eritrea	117.600	0,08
	TOP 100 TOTAL	**132.632.524**	**89,34**

FIGURE 2-14 The True Size of Africa *(2010) by Kai Krause, http://datafl.ws/12t*

There's an easy takeaway from each of these figures because you're familiar with the data and the message is clear, but more often, the data is unfamiliar and the patterns are only obvious after someone points them out to you. That's when storytelling and data narratives come into the picture.

STORYTELLING

As a medium, visualization has developed into a great way to tell stories. News organizations are learning to do this well in a budding field known as data journalism. (This is perhaps where Tufte's chartjunk and data-ink ratio suggestions are most applicable.) Actually though, it's just good analysis and reporting.

For example, when the Deepwater Horizon oil rig exploded in the Gulf of Mexico April 2010, which led to gallons of oil, in the hundreds of millions, spilled into the ocean, *The New York Times* graphically reported several facets of the 3-month accident. This provided context for how the oil spill was playing out, what was affected, and why the spill happened. Looking back on the interactive series now, well after the initial spill, the graphics are still informative and will be for years.

Note: See the in-depth interactive on the oil spill by *The New York Times* at http://datafl.ws/254. You can also find more of their work at http://datafl.ws/2bd.

Digital Narratives, a project by Microsoft Research that demonstrates their Rich Interactive Narratives (RIN) technology, experiments with combining various types of media—video, audio, and text—with visualization that users can interact with. The great thing about the project is that an author can string media together and add a vocal component so that a piece plays as a continuous narrative. A user can pause a narrative at any time to interact with the visualization on the screen.

For example, as shown in Figure 2-15, an author might verbally describe a visualization, and users can pause to interact with the visualization within the

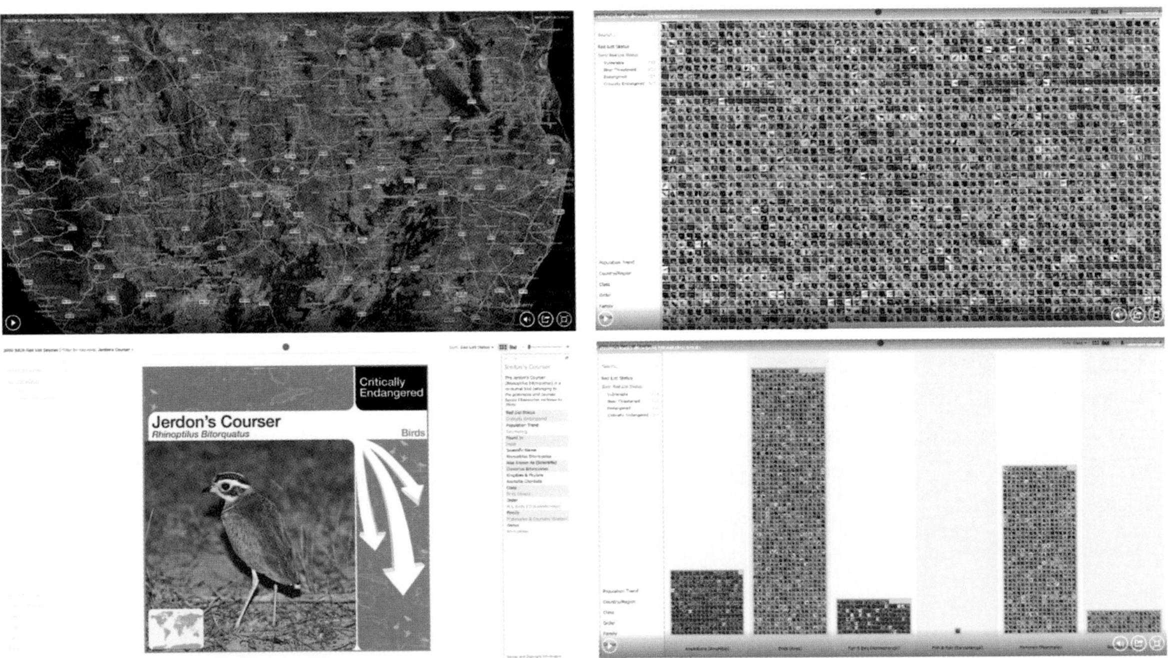

FIGURE 2-15 Digital Narratives *(2011) by Microsoft Research, http://www.digitalnarratives.net/ and http://datafl.ws/2be*

media and explore the data themselves (and resume the story when they're done interacting).

Visualization not only lends itself well to narratives but also communicating and clarifying ideas. Maybe you just want to get a quick point across—data vignettes, so to speak. After all, one of the main selling points of visualization is that it helps you digest a lot of information at once.

Flowcharts, for example, are straightforward ways to communicate process and decision making. You start in a state and then move to adjacent states as you answer questions. Eventually, you end up in a state that helps you make a decision. For example, *So You Need a Typeface* by Julian Hansen, as shown in Figure 2-16 helps you choose the right typography based on task and preference.

Sometimes you want to see an entire process, such as Michael Niggel's flowchart that maps all possible outcomes to *Choose Your Own Adventure #2: Journey Under the Sea*. If you're unfamiliar, the *Choose Your Adventure* books are divided into sections, and at the end of each section, the author provides choices for where to go next. The goal is to stay alive, so it's kind of like a game. As a whole, the flowchart, shown in Figure 2-17, represents the book's complete storyline. FYI: Most likely you'll die.

Information graphics can also cover topics close to the heart. As shown in Figure 2-18, *What Love Looks Like* by Louise Ma is a chart series that conceptualizes what love looks like. Love is a complicated feeling that can be hard to describe in words, but Ma's charts are beautifully poetic in describing the emotion's many facets, focusing both inward and outward.

Notice that Ma doesn't use actual data. Rather, she uses abstract trends and patterns to illustrate miniature stories. Matthew Might uses this to great effect in *The Illustrated Guide to a Ph.D.* shown in Figure 2-19. It was made directed toward graduate students (and of course struck a chord with me right away), but it applies to everyone who's learning and wants to push their field forward, regardless of whether you're in an academic setting, office, or at home.

FIGURE 2-16 *(following page)* So You Need a Typeface *(2010) by Julian Hansen, http://julianhansen.com*

SO YOU NEED A TYPEFACE is a project by Julian Hansen. It's an alternative way on how to choose fonts (or just be inspired) for a specific project, not just by browsing through the pages of FontBook. The list is (very loosely) based on the top 50 of **Die 100 Besten Schriften** by Font Shop.

© 2010 Julian Hansen www.julianhansen.com

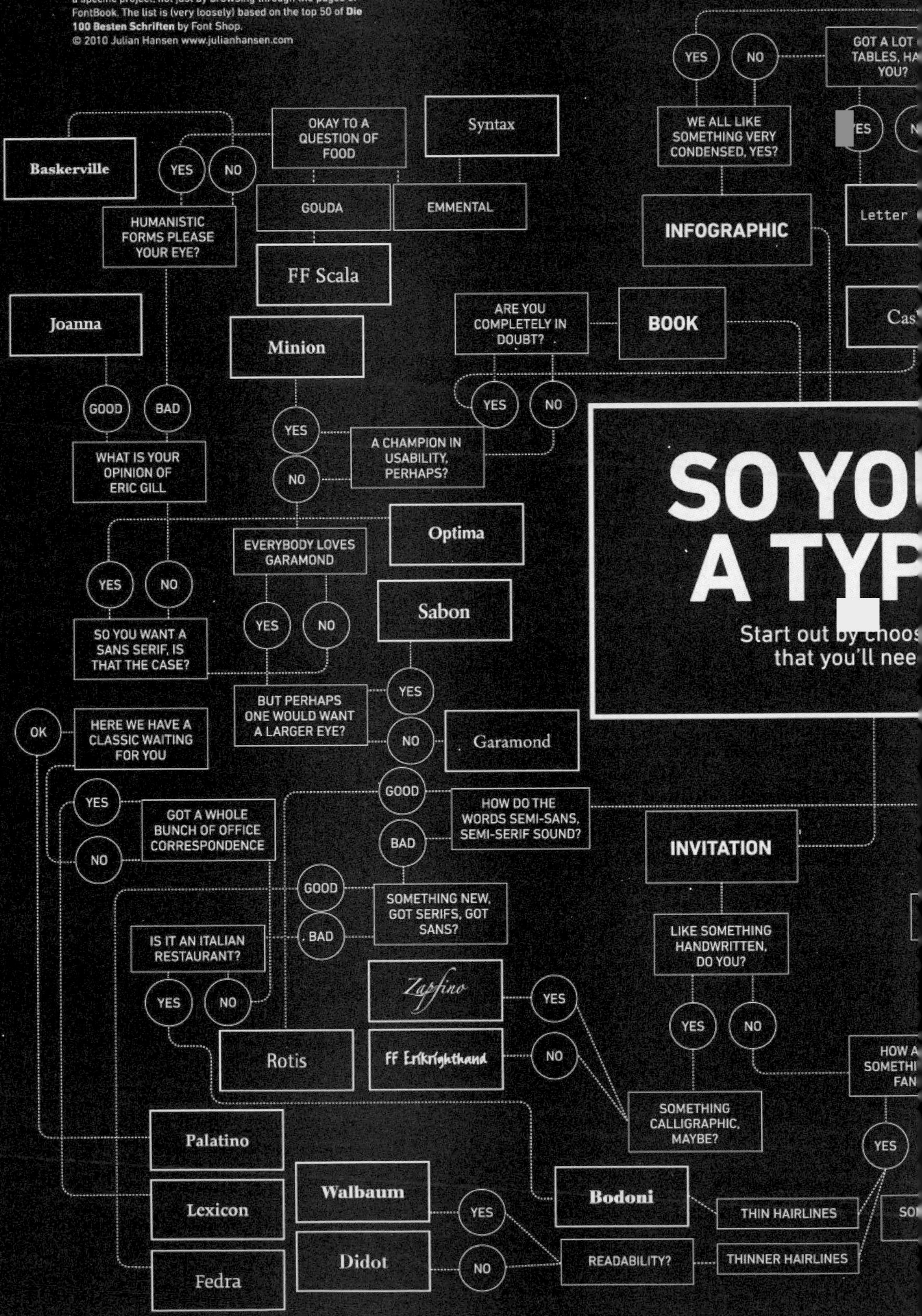

SO YOU
A TYP

Start out by choos
that you'll nee

Baskerville
YES
NO
OKAY TO A QUESTION OF FOOD
Syntax
GOUDA
EMMENTAL
HUMANISTIC FORMS PLEASE YOUR EYE?
FF Scala
Joanna
Minion
GOOD
BAD
YES
A CHAMPION IN USABILITY, PERHAPS?
WHAT IS YOUR OPINION OF ERIC GILL
NO
Optima
EVERYBODY LOVES GARAMOND
YES
NO
YES
NO
Sabon
SO YOU WANT A SANS SERIF, IS THAT THE CASE?
BUT PERHAPS ONE WOULD WANT A LARGER EYE?
YES
NO
Garamond
OK
HERE WE HAVE A CLASSIC WAITING FOR YOU
GOOD
HOW DO THE WORDS SEMI-SANS, SEMI-SERIF SOUND?
BAD
YES
GOT A WHOLE BUNCH OF OFFICE CORRESPONDENCE
NO
GOOD
SOMETHING NEW, GOT SERIFS, GOT SANS?
BAD
IS IT AN ITALIAN RESTAURANT?
YES
NO
Zapfino
YES
Rotis
FF Erikrighthand
NO
SOMETHING CALLIGRAPHIC, MAYBE?
Palatino
Lexicon
Walbaum
Bodoni
YES
THIN HAIRLINES
Fedra
Didot
NO
READABILITY?
THINNER HAIRLINES

YES
NO
GOT A LOT
TABLES, HA
YOU?
WE ALL LIKE SOMETHING VERY CONDENSED, YES?
YES
INFOGRAPHIC
Letter
ARE YOU COMPLETELY IN DOUBT?
BOOK
Cas
YES
NO
INVITATION
LIKE SOMETHING HANDWRITTEN, DO YOU?
YES
NO
HOW A
SOMETHI
FAN
YES
SO

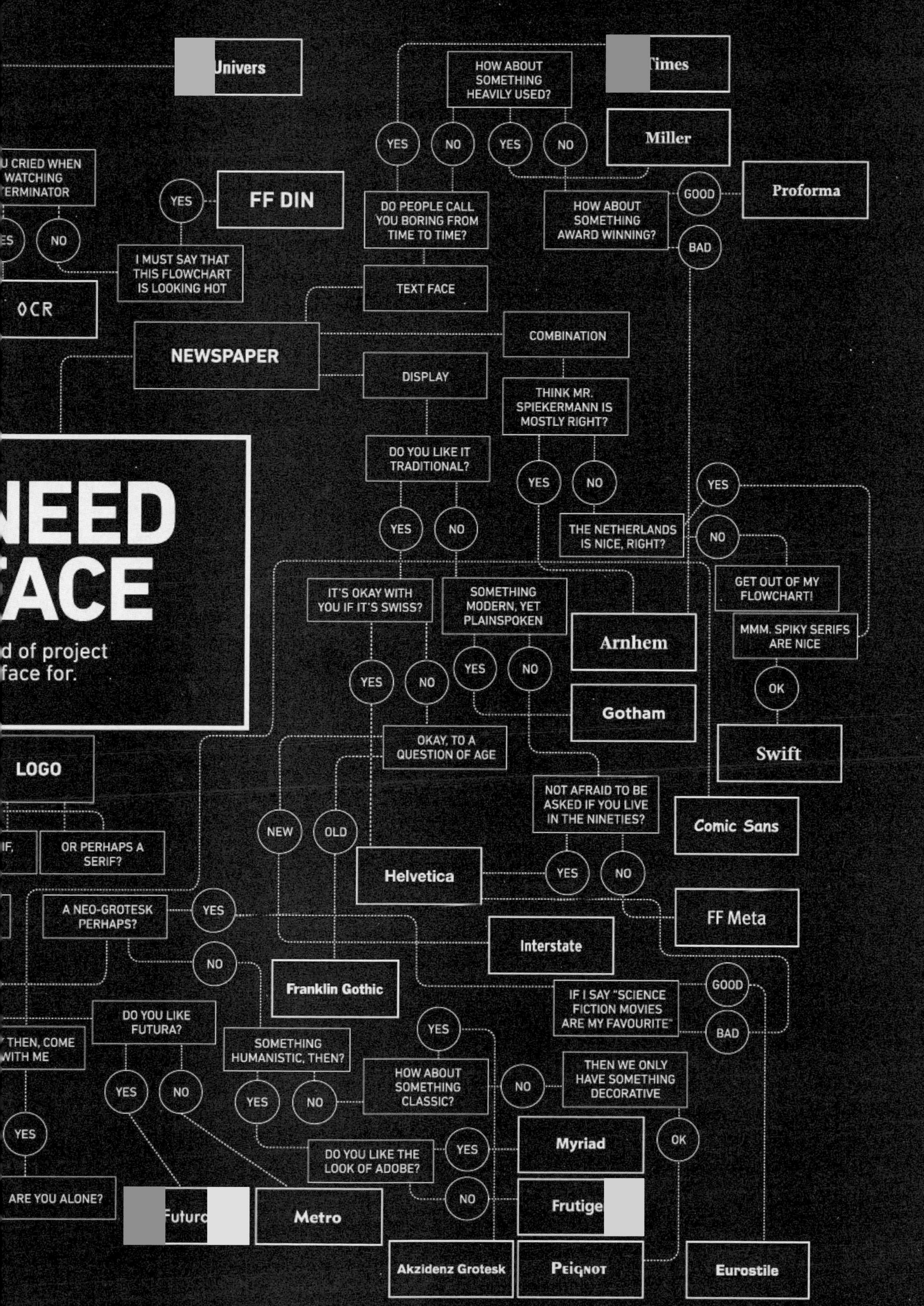

Univers
HOW ABOUT SOMETHING HEAVILY USED?
Times
YES
NO
YES
NO
Miller
U CRIED WHEN WATCHING ERMINATOR
YES
FF DIN
DO PEOPLE CALL YOU BORING FROM TIME TO TIME?
HOW ABOUT SOMETHING AWARD WINNING?
GOOD
Proforma
BAD
ES
NO
I MUST SAY THAT THIS FLOWCHART IS LOOKING HOT
TEXT FACE
OCR
NEWSPAPER
COMBINATION
DISPLAY
THINK MR. SPIEKERMANN IS MOSTLY RIGHT?
DO YOU LIKE IT TRADITIONAL?
NEED
ACE
d of project
face for.
YES
NO
YES
YES
NO
THE NETHERLANDS IS NICE, RIGHT?
NO
GET OUT OF MY FLOWCHART!
IT'S OKAY WITH YOU IF IT'S SWISS?
SOMETHING MODERN, YET PLAINSPOKEN
Arnhem
MMM. SPIKY SERIFS ARE NICE
YES
NO
YES
NO
OK
Gotham
OKAY, TO A QUESTION OF AGE
Swift
LOGO
NOT AFRAID TO BE ASKED IF YOU LIVE IN THE NINETIES?
Comic Sans
NEW
OLD
IF,
OR PERHAPS A SERIF?
Helvetica
YES
NO
A NEO-GROTESK PERHAPS?
YES
FF Meta
Interstate
NO
Franklin Gothic
GOOD
IF I SAY "SCIENCE FICTION MOVIES ARE MY FAVOURITE"
DO YOU LIKE FUTURA?
YES
BAD
THEN, COME WITH ME
SOMETHING HUMANISTIC, THEN?
HOW ABOUT SOMETHING CLASSIC?
NO
THEN WE ONLY HAVE SOMETHING DECORATIVE
YES
NO
YES
NO
OK
YES
DO YOU LIKE THE LOOK OF ADOBE?
YES
Myriad
ARE YOU ALONE?
Futur
Metro
NO
Frutige
Akzidenz Grotesk
Peignot
Eurostile

Choose Your Own Adventure #2: Journey Under the Sea

Analysis of paths and outcomes

Michael Niggel—2009-07-31

Page Sequence

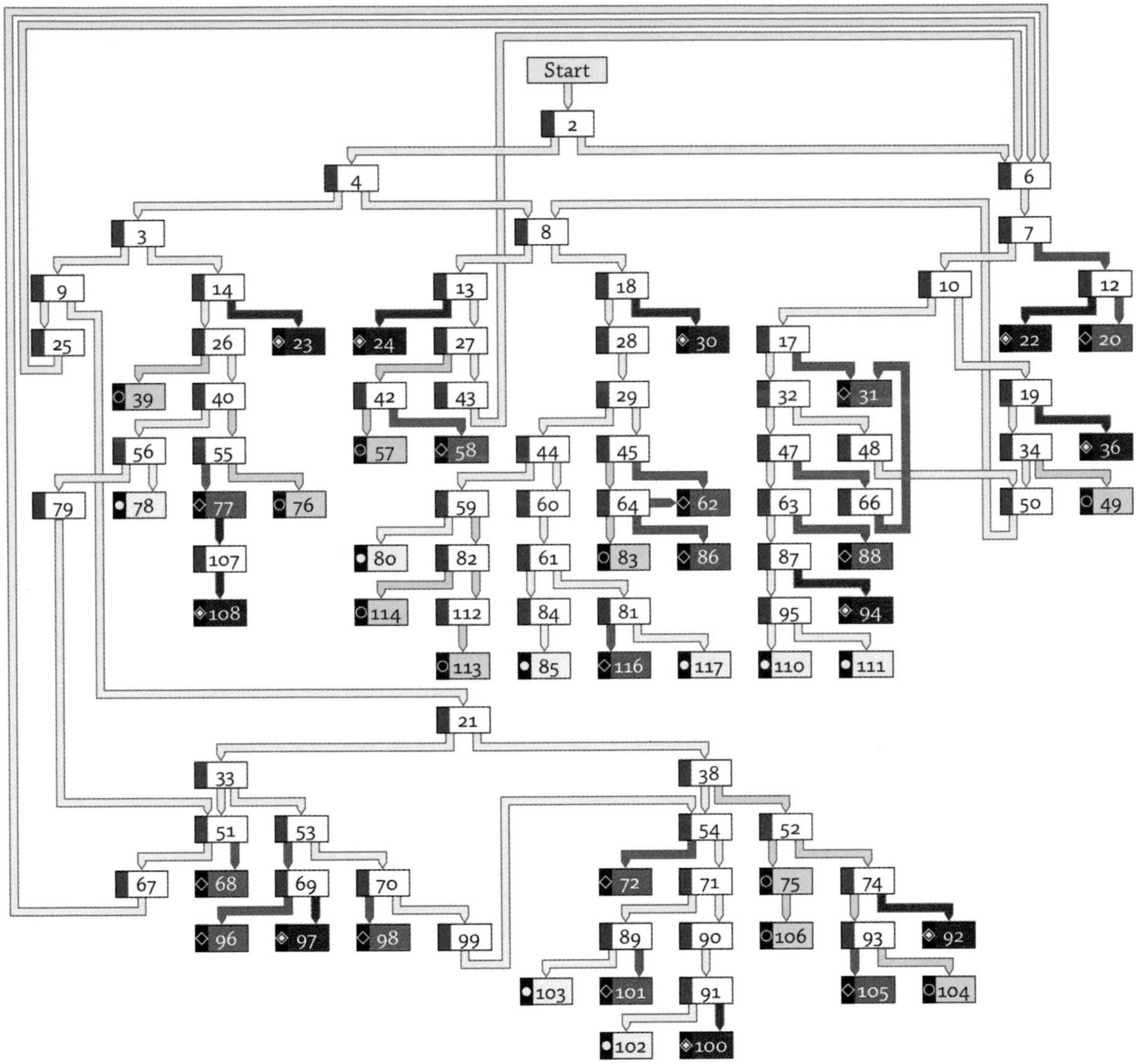

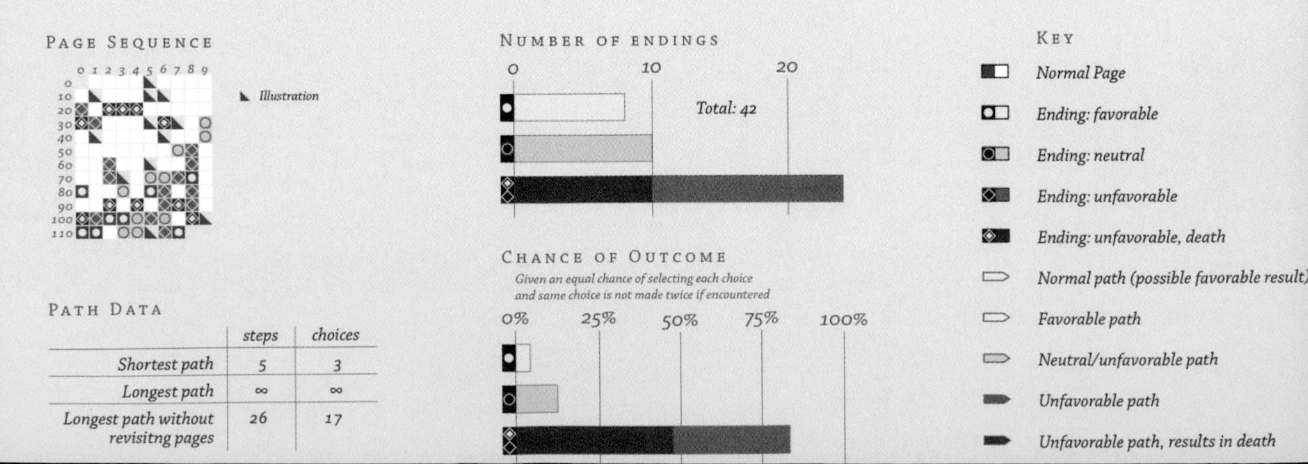

Path Data

	steps	choices
Shortest path	5	3
Longest path	∞	∞
Longest path without revisitng pages	26	17

Number of Endings

Chance of Outcome

Given an equal chance of selecting each choice and same choice is not made twice if encountered

Key

- *Normal Page*
- *Ending: favorable*
- *Ending: neutral*
- *Ending: unfavorable*
- *Ending: unfavorable, death*
- *Normal path (possible favorable result)*
- *Favorable path*
- *Neutral/unfavorable path*
- *Unfavorable path*
- *Unfavorable path, results in death*

Although the illustrations aren't flashy, the idea drives the series forward, which shows you don't need a lot of bells and whistles to get people to look. The same goes for data. Worthwhile data makes graphics worth looking at. It drives data stories forward.

FIGURE 2-17 *(facing page)* Analysis of paths and outcomes *(2009) to* Choose Your Own Adventure #2 *by Michael Niggel, http://datafl.ws/6p*

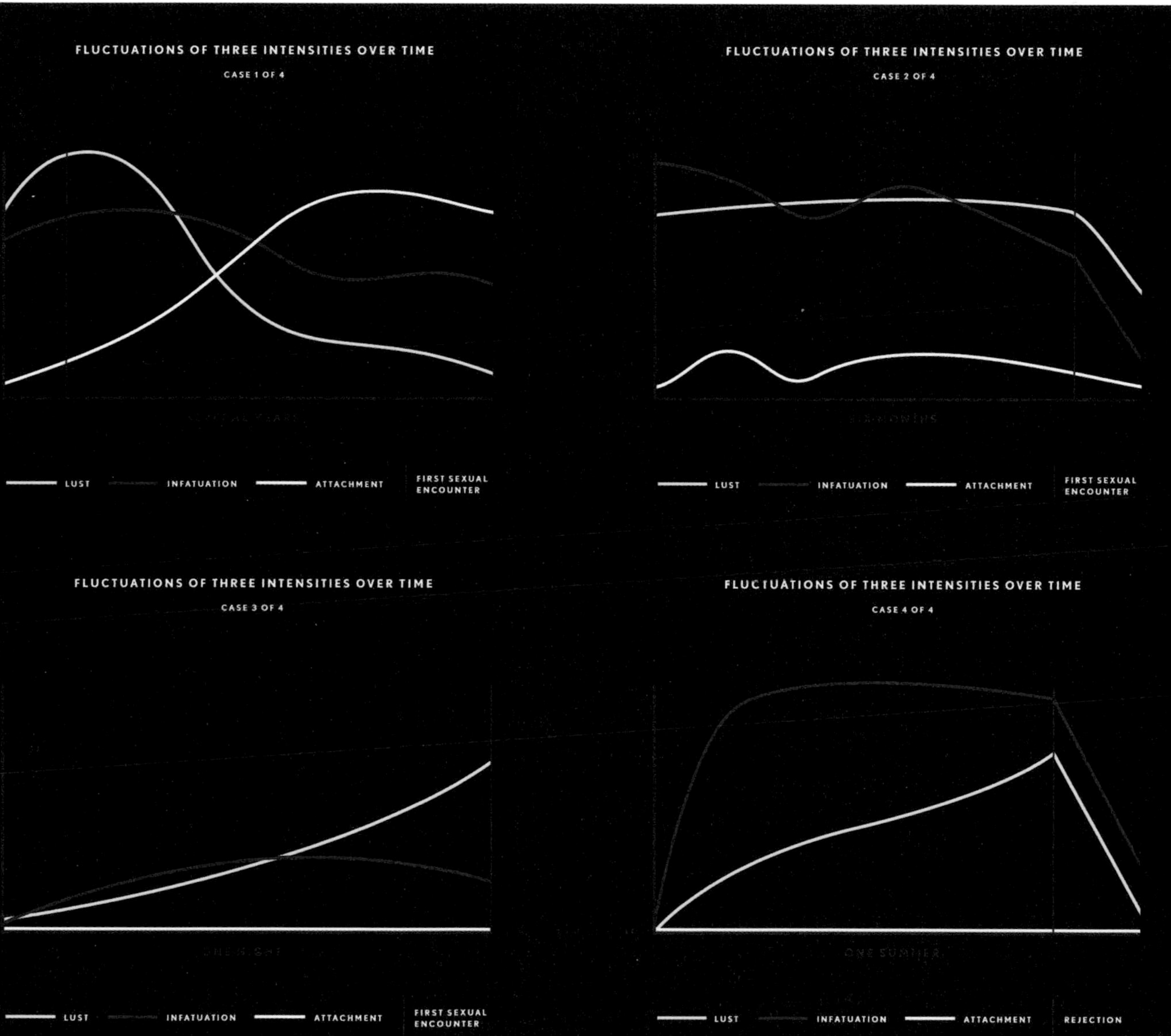

FIGURE 2-18 What Love Looks Like *(2012) by Louise Ma, http://love.seebytouch.com*

Imagine a circle that contains all of human knowledge:

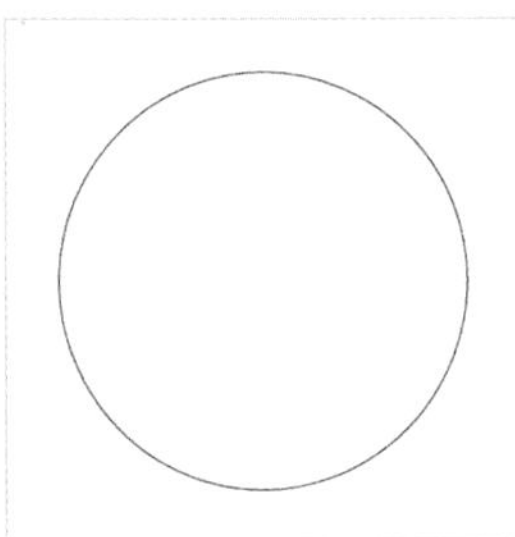

By the time you finish elementary school, you know a little:

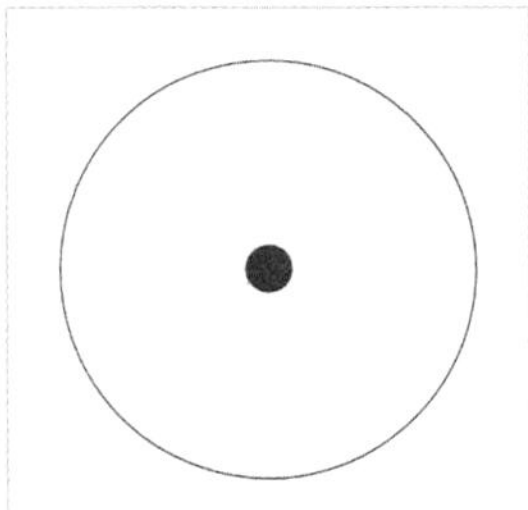

By the time you finish high school, you know a bit more:

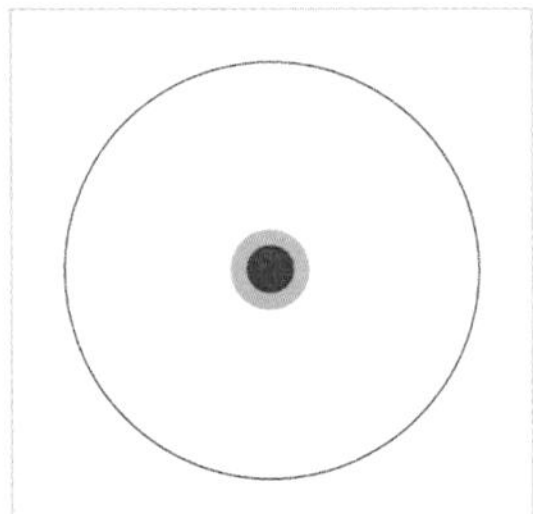

With a bachelor's degree, you gain a specialty:

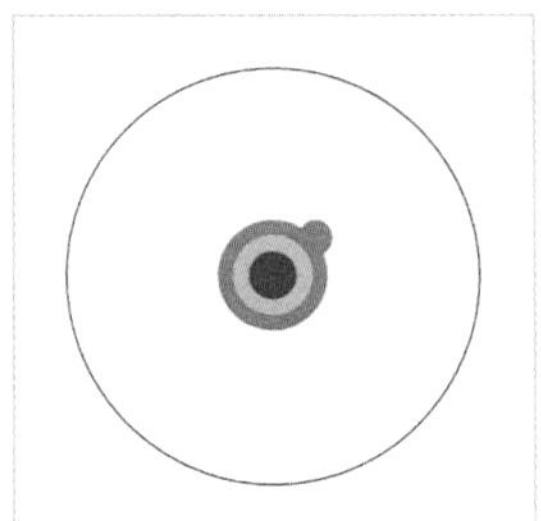

A master's degree deepens that specialty:

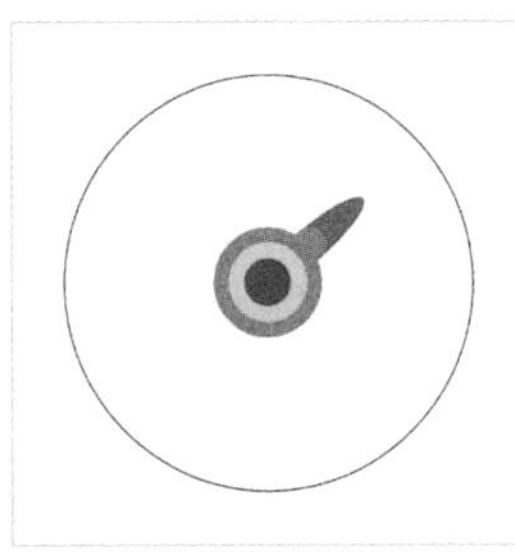

Reading research papers takes you to the edge of human knowledge:

Once you're at the boundary, you focus:

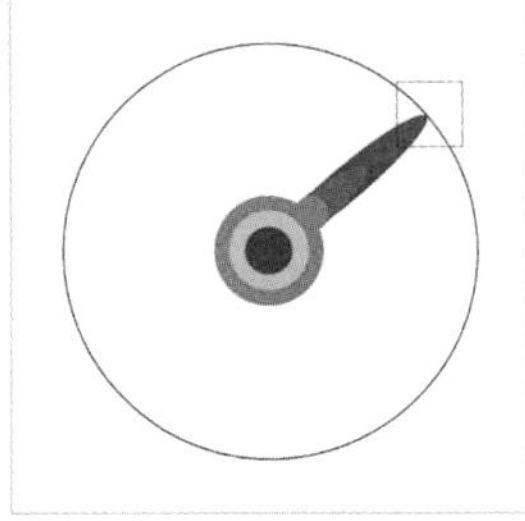

You push at the boundary for a few years:

Until one day, the boundary gives way:

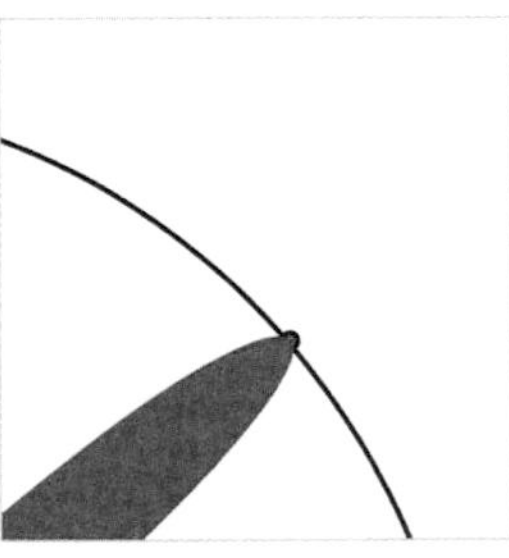

And, that dent you've made is called a Ph.D.:

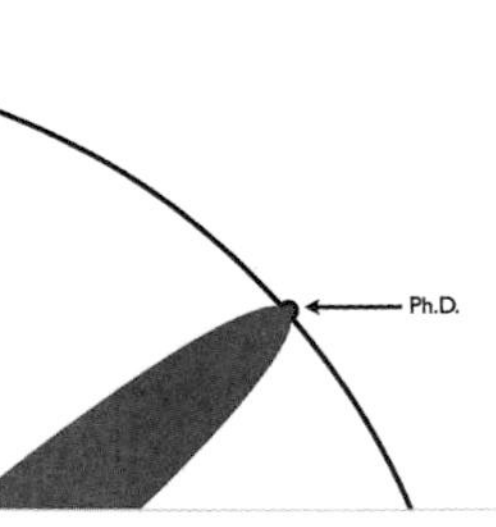

Of course, the world looks different to you now:

So, don't forget the bigger picture:

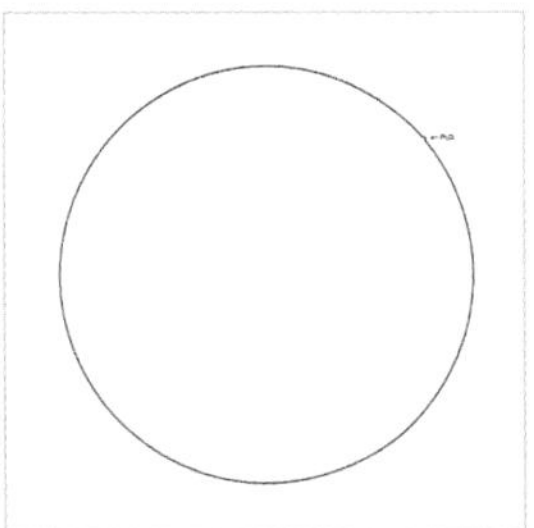

Keep pushing.

FIGURE 2-19 Illustrated Guide to a Ph.D. *(2010) by Matthew Might, http://datafl.ws/25c*

ENTERTAINMENT

Approaching the middle of the visualization spectrum, I start to lose people rooted in the analysis side. I might have already lost you at the end of the last section. This is when reader attention, engagement, and happiness tend to grow more important and useful for the task than minimizing chartjunk and increasing data-ink ratios. Although the latter is still important, people tend to relate with the former more readily.

Some might feel antsy or scoff at the work that follows (I do still call myself a statistician, so I can relate.), but there is value in visualization that isn't a traditional, just-the-facts chart. There is value in entertaining, putting a smile on someone's face, and making people feel something, as much as there is in optimized presentation. Obviously, you don't embed a comic within a business dashboard, but an entertainment-based publication? That's not so crazy.

> *"All of the great chartmakers make me feel something: alarm, wonder, surprise, joy ... something. Even, I think you might argue in the case of something like dashboard design, calm."*
>
> *—Amanda Cox*

The definition split has a lot to do with use of the visualization label. In research and academia, visualization is typically a data exploration tool that requires precision and visual efficiency. You look at data, understand what you can, and then quickly move on to another part of the data. Visualization researchers hope to generalize their results to be used with similar data types and situations.

Practitioners, on the other hand, tend to design and create on a case-by-case basis. They certainly draw from past work and experience, but often the goal is to design a tool, interactive, or single graphic tailored to a dataset.

Because this dataset-tailored visualization is more visible, and academic research can feel out of reach, the general public thinks of visualization mostly of the former and less of the latter. The general public considers visualization to be anything that places numbers in a graphical context, and you're either someone who completely resists this, or you embrace it. There's no in between, and I have yet to meet anyone who shifted to the other side.

I embrace it.

Taxonomies and frameworks are important to advance research, if only to make it easier to discuss topics, but from a practical standpoint, the definition of visualization or what you call your work has little to do with what you make. Even while working with potential clients, a quick glance at a portfolio makes it easy to see what you do.

Who knows what visualization will be in 10 years? After all, a web search for visualization only 10 years ago returned results for a mental exercise to set goals and calm your nerves.

HUMOR

Along the same lines as Ma's charts about love, a genre of graphs have popped up in recent years used to tell jokes. They seem to stem from everyone's love (that is, hate) of PowerPoint presentations, starting as satire and morphing into a chart subclass.

Jessica Hagy was one of the first to do this online with her blog *Indexed*, which she started in 2006. As shown in Figure 2-20, Hagy uses line charts and Venn diagrams to communicate observations and ideas. Even after several years of regular updates, Hagy's cards never fail to make me smile. Sometimes she explores complex ideas and other times simple observations, but the hand-drawn sketches seem to lend a dose of clarity that only charts can provide.

Although *Indexed* is in its own genre that's some hybrid of reports and satire, charts have found their way into comics, too. For example, *Doghouse Diaries* often uses charts to make people smile. In Bed Cartography, shown in Figure 2-21, the odd sleeping zones of one's significant other are described. Dog and cat areas are not shown, nor is the child scared by nightmare.

Manu Cornet caricaturized organizational charts for major technology companies in his comic *Bonkers World*, as shown in Figure 2-22. They range from the strict, top-down structure at Amazon.com to Facebook's seemingly self-managed, small groups.

Taking it down to the human level, don't forget *The Trustworthiness of Beards* by Matt McInerney, who mapped beard types against a range of trustworthy to disastrous, as shown in Figure 2-23. Never trust a werewolf, no matter how hunky or shirtless he might be.

A B C D E F G

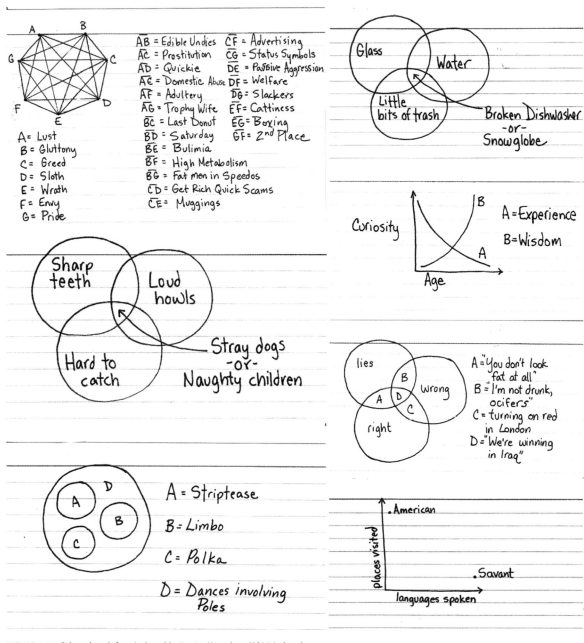

A = Lust
B = Gluttony
C = Greed
D = Sloth
E = Wrath
F = Envy
G = Pride

$\overline{AB}$ = Edible Undies
$\overline{AC}$ = Prostitution
$\overline{AD}$ = Quickie
$\overline{AE}$ = Domestic Abuse
$\overline{AF}$ = Adultery
$\overline{AG}$ = Trophy Wife
$\overline{BC}$ = Last Donut
$\overline{BD}$ = Saturday
$\overline{BE}$ = Bulimia
$\overline{BF}$ = High Metabolism
$\overline{BG}$ = Fat men in Speedos
$\overline{CD}$ = Get Rich Quick Scams
$\overline{CE}$ = Muggings
$\overline{CF}$ = Advertising
$\overline{CG}$ = Status Symbols
$\overline{DE}$ = Passive Aggression
$\overline{DF}$ = Welfare
$\overline{DG}$ = Slackers
$\overline{EF}$ = Cattiness
$\overline{EG}$ = Boxing
$\overline{GF}$ = 2nd Place

Sharp teeth
Loud howls
Hard to catch
Stray dogs
-or-
Naughty children

A
B
C
D

A = Striptease
B = Limbo
C = Polka
D = Dances involving Poles

Glass
Water
Little bits of trash
Broken Dishwasher
-or-
Snowglobe

Curiosity
Age
A
B
A = Experience
B = Wisdom

lies
wrong
right
A B C D
A = "You don't look fat at all"
B = "I'm not drunk, ocifers"
C = turning on red in London
D = "We're winning in Iraq"

.American
.Savant
places visited
languages spoken

FIGURE 2-20 *Selected cards from* Indexed *by Jessica Hagy, http://thisisindexed.com*

Bed Cartography
Her pillow
Apparently also her pillow
Where she's supposed to sleep
Where I'm supposed to sleep
Peacekeeping buffer zone
Where she actually sleeps
Where I actually sleep
Uncharted territory

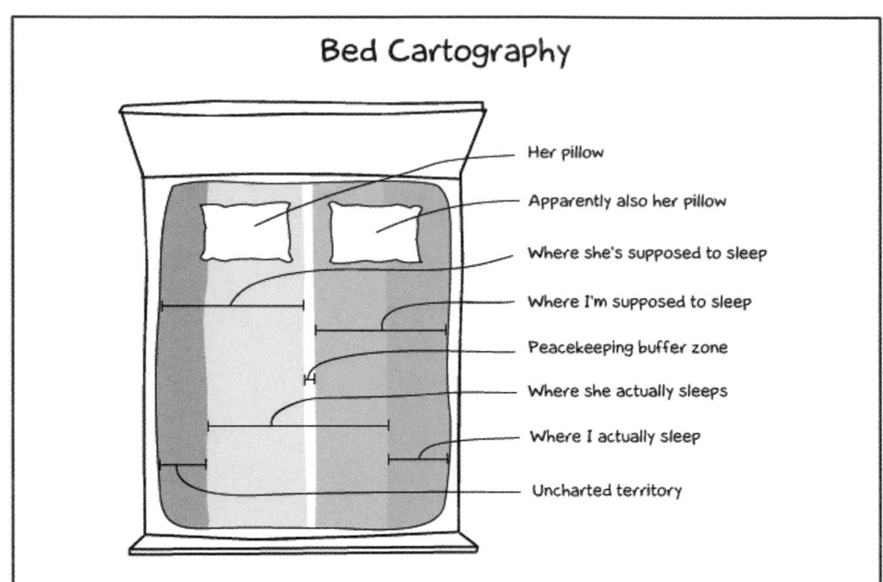

FIGURE 2-21 Bed Cartography *(2012) from Doghouse Diaries, http://thedoghousediaries .com/3586*

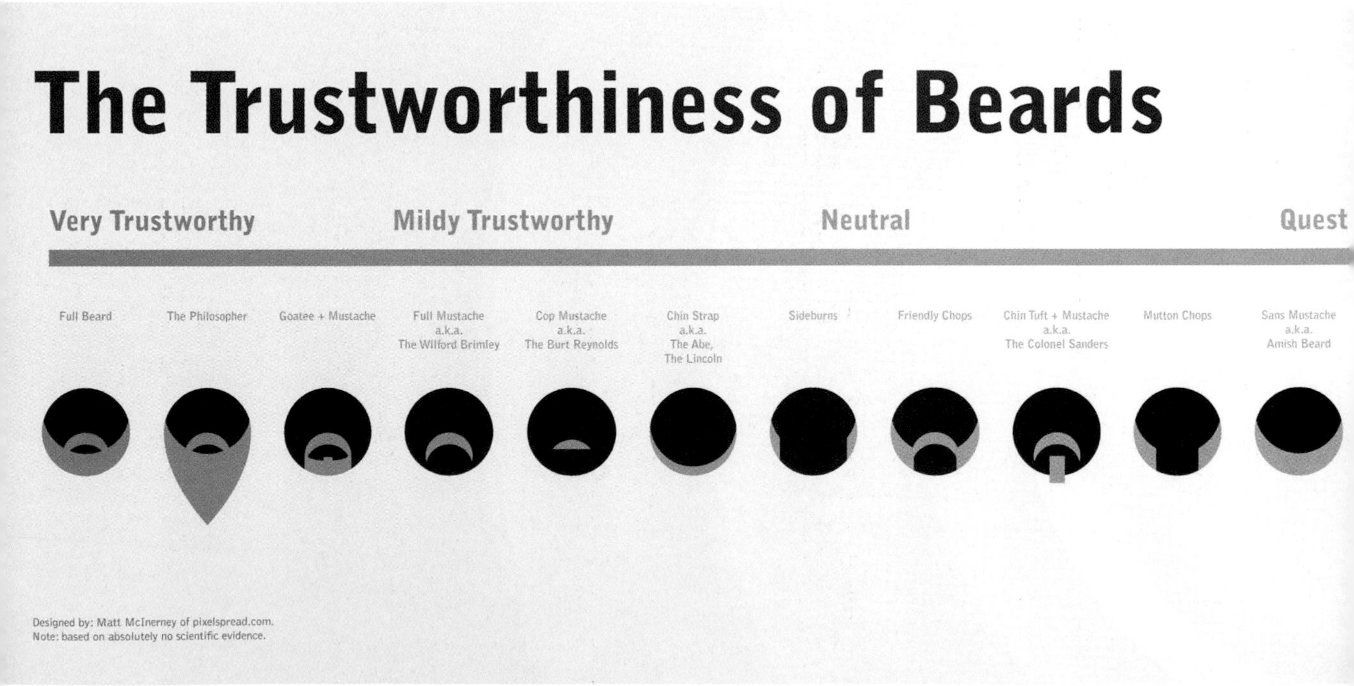

The Trustworthiness of Beards
Very Trustworthy
Mildy Trustworthy
Neutral
Quest
Full Beard
The Philosopher
Goatee + Mustache
Full Mustache a.k.a. The Wilford Brimley
Cop Mustache a.k.a. The Burt Reynolds
Chin Strap a.k.a. The Abe, The Lincoln
Sideburns
Friendly Chops
Chin Tuft + Mustache a.k.a. The Colonel Sanders
Mutton Chops
Sans Mustache a.k.a. Amish Beard
Designed by: Matt McInerney of pixelspread.com.
Note: based on absolutely no scientific evidence.

FIGURE 2-23 The Trustworthiness of Beards *(2010) by Matt McInerney*

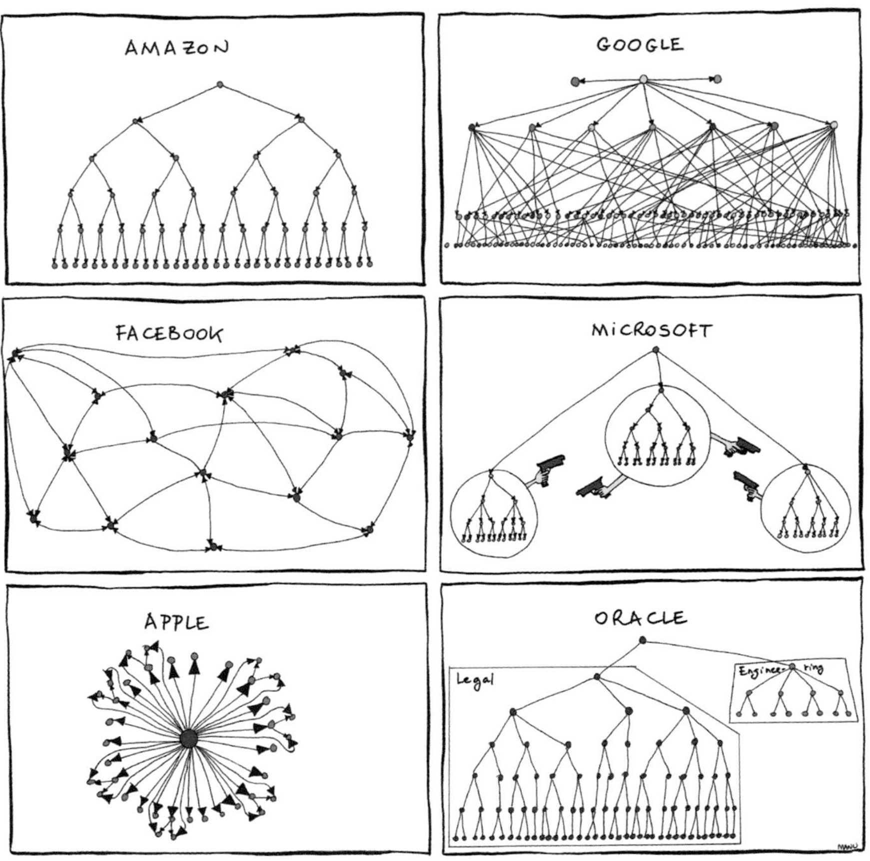

FIGURE 2-22 Organizational Charts *(2011) by Manu Cornet, http://www.bonkersworld.net/organizational-charts/*

Unsavory
Threatening
Dangerous
Disastrous

Goatee
The Horseshoe a.k.a The Hogan
Chin Tuft
Unkempt Beard a.k.a. The Homeless Beard
Neck Beard
Soul Patch a.k.a. The Landing Strip
Patchy Beard a.k.a. The Sidney Crosby
Pencil Thin Chin Strap
Werewolf
Pencil Thin Mustache a.k.a. The John Waters, The Joe Jackson
The Hitler

Whether you call these graphics visualization or put them in a category of their own, there are a lot of things to learn here that apply to traditional visualization. Why do so many people like these graphics? What is it about these graphics that resonates with readers and makes them want to share with others? Does color and readability play a role? How about layout? Some researchers have tried to answer these questions, but so far, they've only scratched the surface.

DATA ART

Now enter the right side of the spectrum, where the imagination runs wild, data and emotion drive together, and creators make for human connection. It's hard to say what data art is exactly, but the work is often less about decision making and more about a relationship to the numbers—or rather what they represent—to experience data, which can feel cold and foreign. Data art is made of the stuff that analysis and information graphics could often use more of.

In 2012, with the Olympics in London starting in a few months, artists Quayola and Memo Akten translated athletic movement, which in itself can be beautiful, to generative animation in *Forms*, as shown in Figure 2-24. On a small screen a video plays of an athlete, such as a gymnast or a diver, floating and spinning in the air. On a large screen, generated particles, sticks, and poles move along accordingly. Sound accompanies the movement to make computer-generated shapes seem more real.

There are no axes, labels, or grids. Instead, it is like real-life activities taking on different forms. As shown in Figure 2-25, Jason Salavon, a Chicago-based artist, used MTV's compilation list for the "greatest music videos of all time," compressing each video to its average colors. You lose the music, but you get a sense of theme and flow from the colors in chronological order.

Graphic designer Frederic Brodbeck did similar work in *Cinemetrics*, which derived movie data—color, motion, and time—to create a "visual fingerprint" for each, as shown in Figure 2-26. Each segment represents the color and time in a part of the movie, and in the animated version, segments move back and forth according to the amount of movement in the segment.

FIGURE 2-24 Forms *(2012) by Memo Akten (http://memoakten.com/) and Quayola (http://www.quayola.com/), https://vimeo.com/37954818*

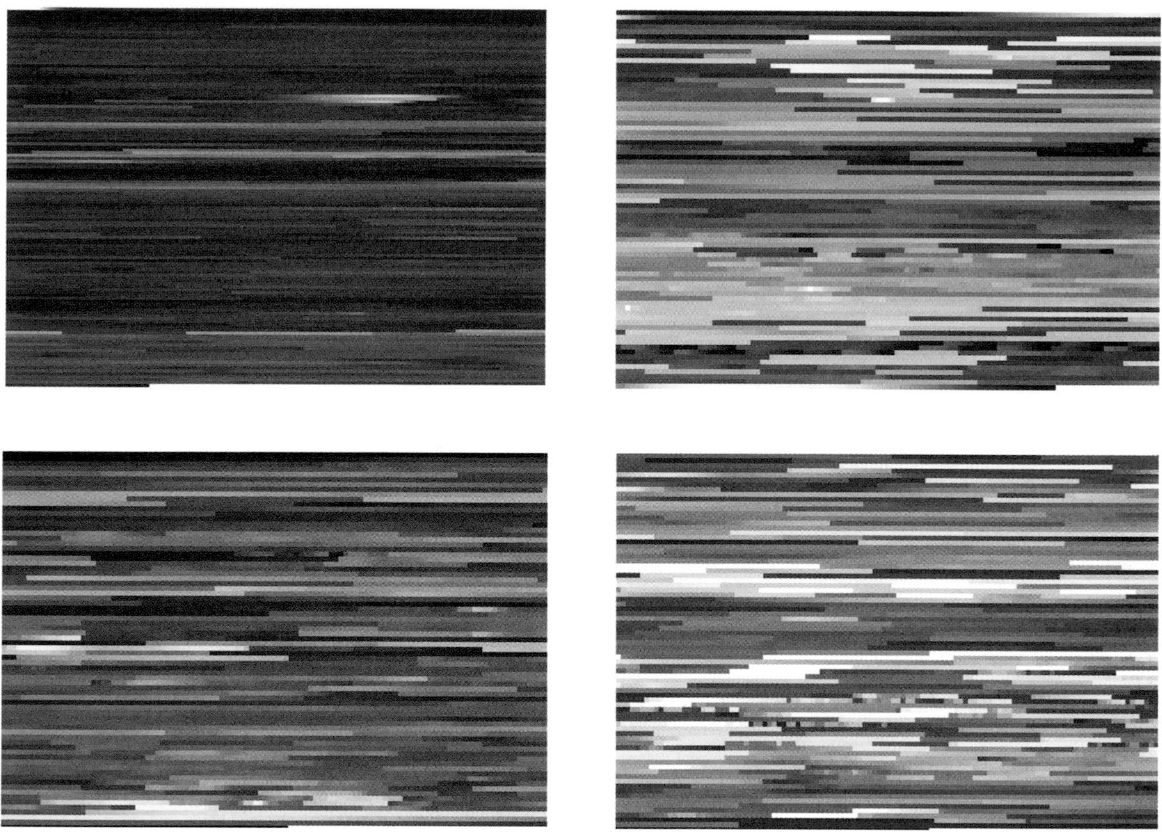

FIGURE 2-25 *Thriller, Vogue, Smells Like Teen Spirit, and Express Yourself from* MTV's 10 Greatest Music Videos of All Time *(2001) by Jason Salavon, http://salavon.com/work/MtvsTop10/*

A History of the Sky, shown in Figure 2-27, by artist Ken Murphy also reconfigured our traditional view of time and space; however, instead of movies, the sky was used as the source of inspiration.

Murphy installed a camera on the San Francisco Exploratorium's roof and programmed the camera to take a picture every 10 seconds for a year. Instead of stringing all the pictures together on a single continuous timeline, Murphy played all the days at once on a 24-hour timeline. In a few minutes, you can see how the length of days change and how the weather fluctuates for a year.

What if the data source is invisible? How are you supposed to visualize it? As shown in Figure 2-28, Timo Arnall, Jørn Knutsen, and Einar Sneve Martinussen fashioned a measuring rod with Wi-Fi sensors and small lights to visualize the networks that we use every day in *Immaterials: Light painting WiFi*. In any given

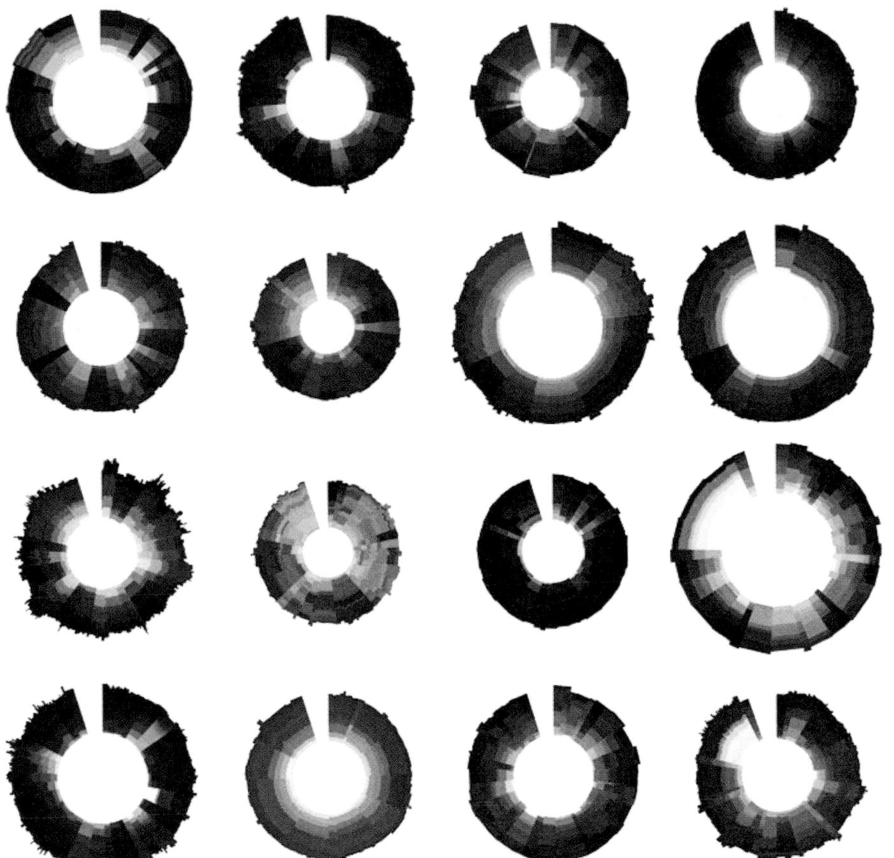

FIGURE 2-26 Cinemetrics (2012) by Frederic Brodbeck, http://cinemetrics.fredericbrodbeck.de

position, the rod displays signal strength as a bar, so the better the signal, the more lights that flash. This, combined with long exposure photography, *Immaterials* paints a picture of Wi-Fi signals in a physical space.

Although these works were made for an art exhibit or to decorate people's walls, it's easy to see how they could be useful to some. For example, an athlete or coach might be interested in perfecting movement, and visual traces can make it easier to see patterns. *Forms* might not be as straightforward as, say, motion-capture software that replays exact movements, but the mechanics are similar. Likewise, filmmakers could use *Cinemetrics* to study use of color and movie dynamics, and engineers might find signal strength in a physical space helpful in research to improve Wi-Fi technology.

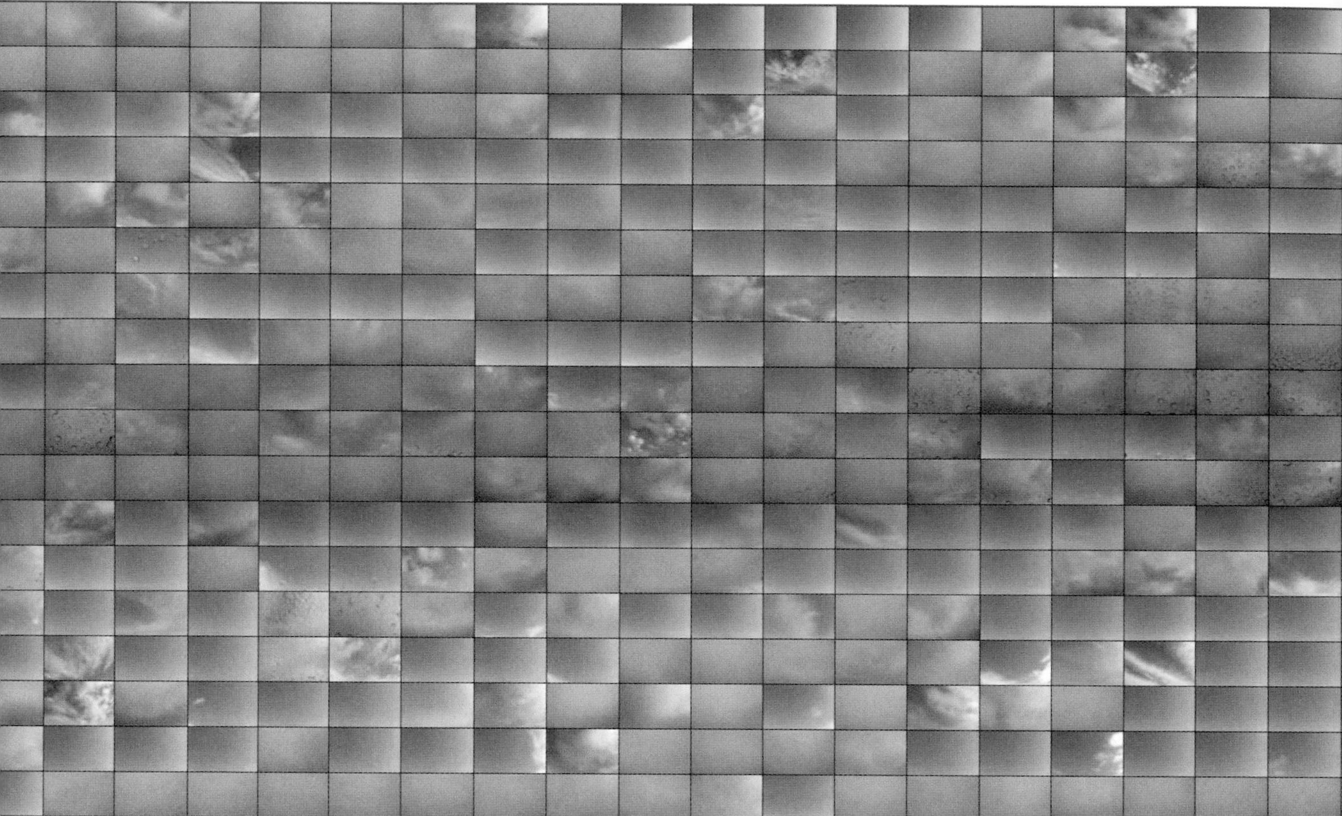

FIGURE 2-27 A History of the Sky *(2011) by Ken Murphy, http://datafl.ws/25s*

Again, this brings you back to, "What is data art?" Or more important, "What is visualization?" It's a medium with a wide range of applications. There are genres of visualization within a spectrum, but there are no clear-cut boundaries (which there doesn't need to be). Like movies, which can be both a documentary and a drama or even a comedy and a horror, a visualization work can be art and factual.

For example, Stamen Design, known for both practical and beautiful interactive maps, put an artistic spin on geography with its experimental *prettymaps*, shown in Figure 2-29. They mapped freely available, community-generated data from Flickr, Natural Earth, and OpenStreetMap, which provided six layers of data. Mapped separately, the geography looks standard, but when you map them all at once with semi-transparency, you get something else.

FIGURE 2-28 Immaterials: Light painting WiFI *(2011) by Timo Arnall, Jørn Knutsen and Einar Sneve Martinussen, https://vimeo.com/20412632; Photo by Timo Arnall*

Real data sources are used and they are mapped geographically, but by combining all the sources and using less traditional aesthetics, the appearance of the geography seems to change.

In *Wind Map*, another work by Fernanda Viégas and Martin Wattenberg, who use visualization as both a tool and expression, wind patterns flow across the United States, as shown in Figure 2-30. The forecasts are updated once per hour from the National Digital Forecast Database, which you can explore by zooming and panning. You can also mouse over flows for wind speed and direction. The more concentrated and faster the streams on the map, the greater the forecasted speed.

FIGURE 2-29 *(following page)* prettymaps *(2010) by Stamen Design, http://prettymaps.stamen.com/*

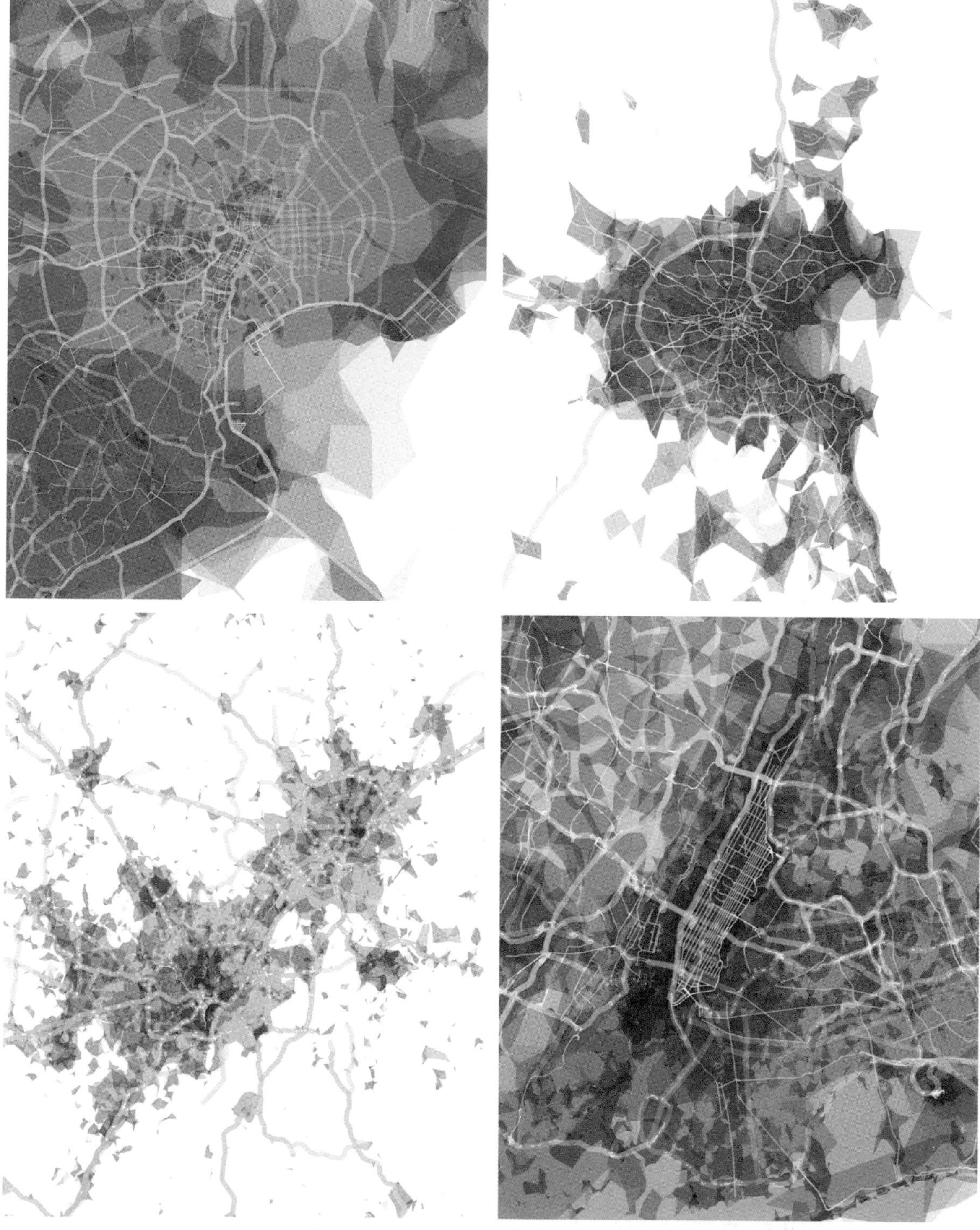

The map could be useful to meteorologists who study wind patterns or to educators who teach weather concepts, but Viégas and Wattenberg consider it an art project. To see the environment as a living, breathing thing is certainly something beautiful to see.

FIGURE 2-30 *(following page)* Wind Map *(2012) by Fernanda Viégas and Martin Wattenberg, http://hint.fm/wind/*

I Want You To Want Me, by Jonathan Harris and Sep Kamvar, shown in Figure 2-31, is an installation that was commissioned by New York's Museum of Modern Art. Like DuBois' map (shown in Chapter 1, "Understanding Data"), Harris and Kamvar's piece uses data collected from online dating sites, which captures how people identify themselves and who they want to be with. *I Want You To Want Me* parses profiles and extracts sentences that start with "I am" or "I am looking for" and represents each sentence with a balloon floating in an interactive sky. Each balloon carries the silhouette of an animated person, almost as if each represents an individual's floating hope to find an ideal partner. (The piece, by the way, was installed on Valentine's Day.)

Although there are statistical breakdowns for aspects like top first dates, desires, and turn-ons, *I Want You To Want Me*, installed on a large, high-resolution touchscreen, is like a story that lets you peek in and explore people's search for relationships. You can easily immerse yourself in the data, which is both personal and easy to relate to. It's harder to do that with a traditional graph. That said, the key to high-quality data art, like any visualization, is still to let the data guide design.

THE EVERYDAY

Visualization has also found its way into the everyday, especially because almost all online content is stored in a database. And as people grow more comfortable with interacting on their computers, developers can create interfaces that display more data at once. From the side of those who build applications, this is great because the growing amounts of data require new views that the old ones can't accommodate, and from the consumer side, the experience improves as the data is easier to digest.

In 2004, Marcos Weskamp created *newsmap*, which is a treemap view into Google News, as shown in Figure 2-32. If you go straight to Google News, you get a standard list of headlines, complemented with a thumbnail. Some of the top stories are listed at the top, and recent ones in the right sidebar.

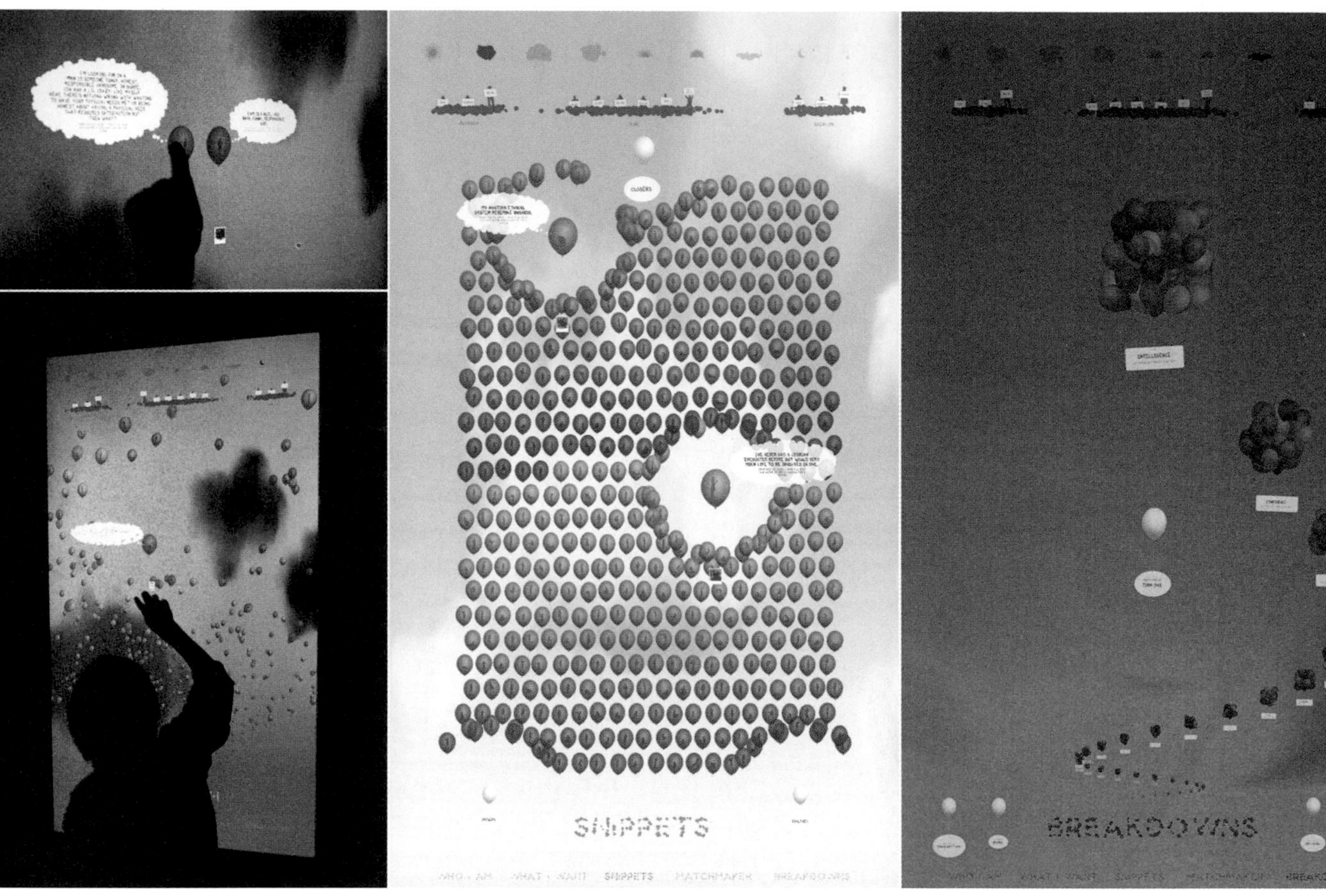

FIGURE 2-31 I Want You To Want Me *(2008) by Jonathan Harris and Sep Kamvar, http://www.iwantyoutowantme.org/*

However, Weskamp's *newsmap* shows headlines sized by popularity, based on a number of related articles. Each rectangle represents a clickable story and is colored by topic, such as world, national, or business. So you get a sense of what's going on in the world at a glance, and there are a variety of options (such as country of interest and time frame) and you can include and exclude topics.

Geographic maps are heavily used as a search tool. Their main use on the web was to look up directions from point A to point B, but as developers add layers of information and integrate deeper functionality, complete applications are developed to provide information and context of areas.

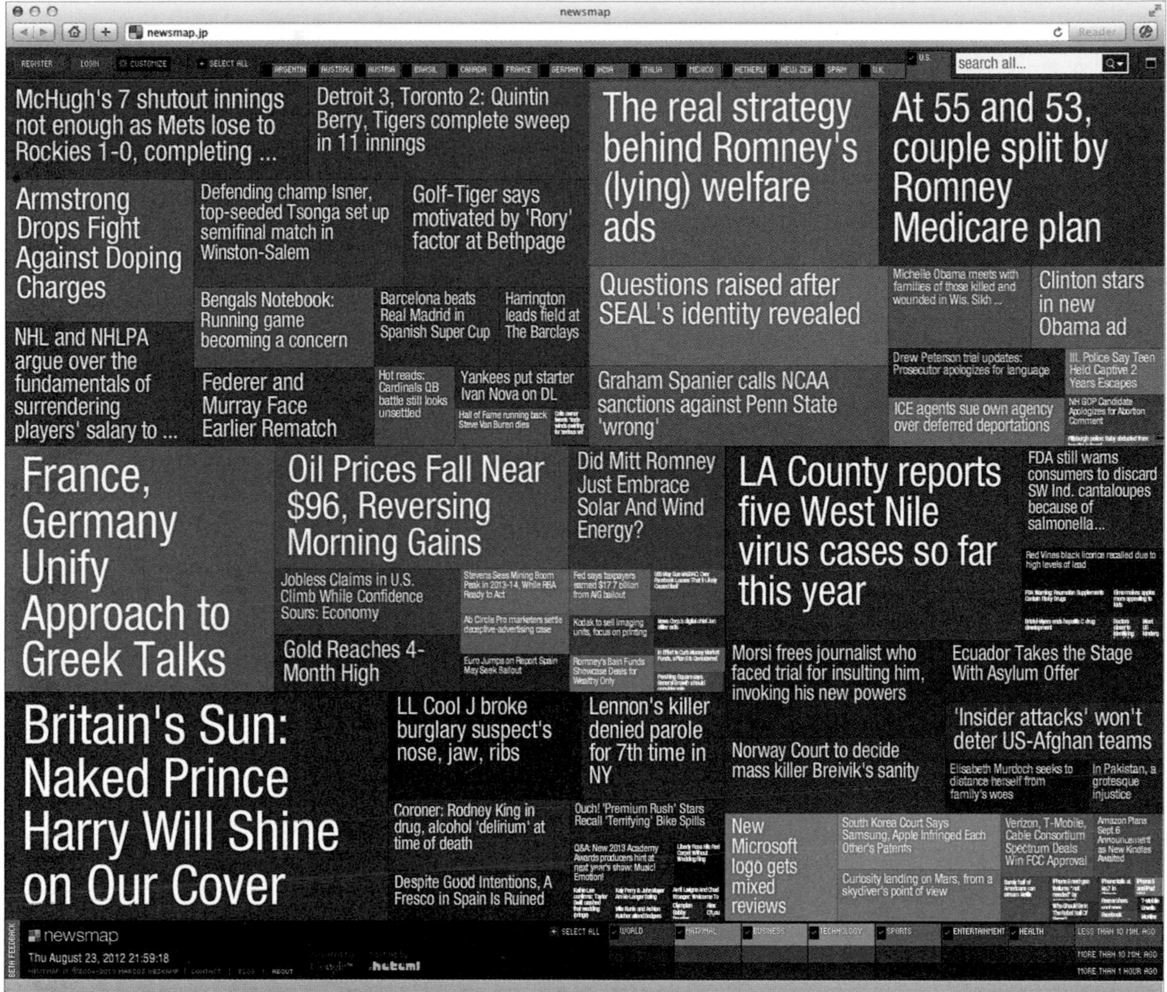

FIGURE 2-32 newsmap *by Marcos Weskamp, http://newsmap.jp*

Google Maps is, of course, the most heavily used because you can look up nearby stores, restaurants, and other businesses, but the application mostly shows only pointers and markers on specific locations. Sometimes, you want to know trends and patterns or just get a general sense of an area, like when you look for a place to live. Trulia, a site that helps you search for real estate, provides useful layers of information other than the properties for sale, as shown in Figure 2-33.

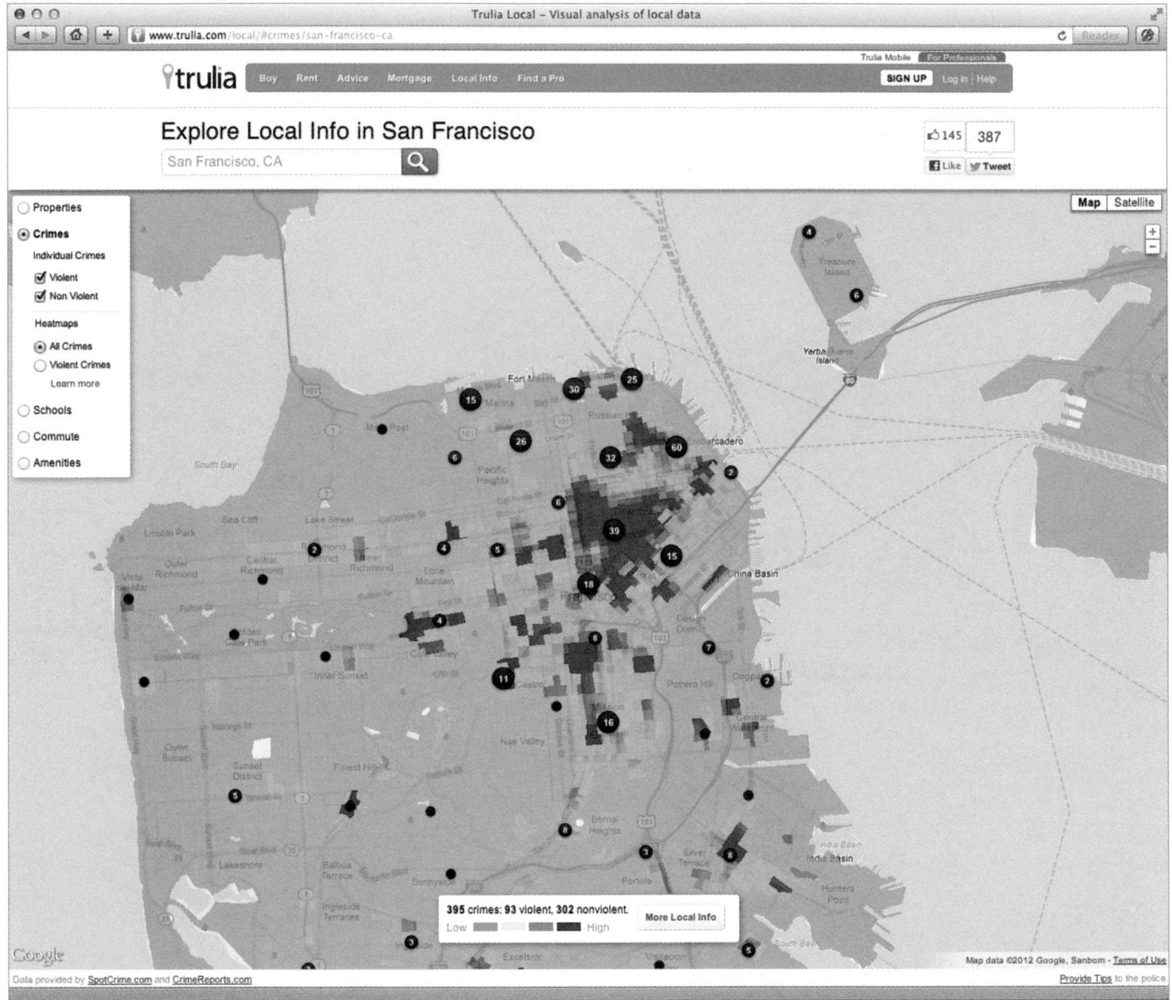

FIGURE 2-33 Trulia Local, *http://www.trulia.com/local*

You can look at crime, violent and nonviolent; filter schools by rating; and see approximate commute times from a given location. The applications helps you make a more informed decision, based on more than just square feet and price, when you buy a house.

Some interfaces completely change how you interact and relate to the data. *Planetary*, by Bloom, is an iPad app that places your iTunes music library into the context of a solar system.

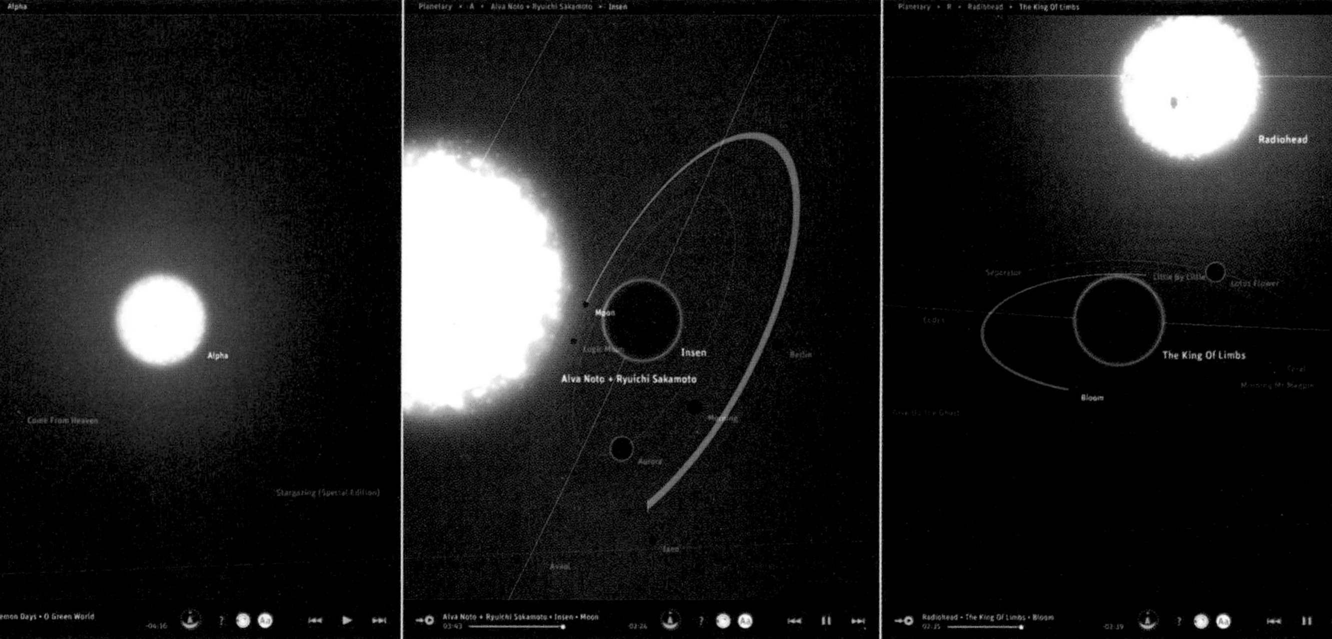

FIGURE 2-34 Planetary *by Bloom, http://planetary.bloom.io/*

Artists are stars, albums are planets that orbit the stars, and tracks are moons that orbit the planets. Instead of a music library to jump to a specific song, your music is transformed into a landscape you can explore and rediscover. And because you use *Planetary* via the iPad's touch interface, the data almost feels tangible.

What happens when data actually is tangible, embedded in a physical object? In 2010, the Really Interesting Group made Christmas ornaments shaped by scrobbles on music site Last.fm, miles traveled according to Dopplr, and apertures used on Flickr, as shown in Figure 2-35. Tada. Instant personalized Christmas gift.

Rachel Binx and Sha Hwang streamlined the process with meshu, a service that enables you to make your own jewelry based on geographic locations. Pick points on a map, and they fabricate your necklace, earrings, or cufflinks out of wood, acrylic, nylon, or silver.

Location data represents where you are, where you've been, and where you're going, so each meshu is like a snapshot of life, that you can wear.

As mobile technology advances, and the gap between digital and physical gets smaller, visualization will play a larger role in connecting the two worlds.

FIGURE 2-35 datadecs *by Really Interesting Group, http://datafl.ws/25v*

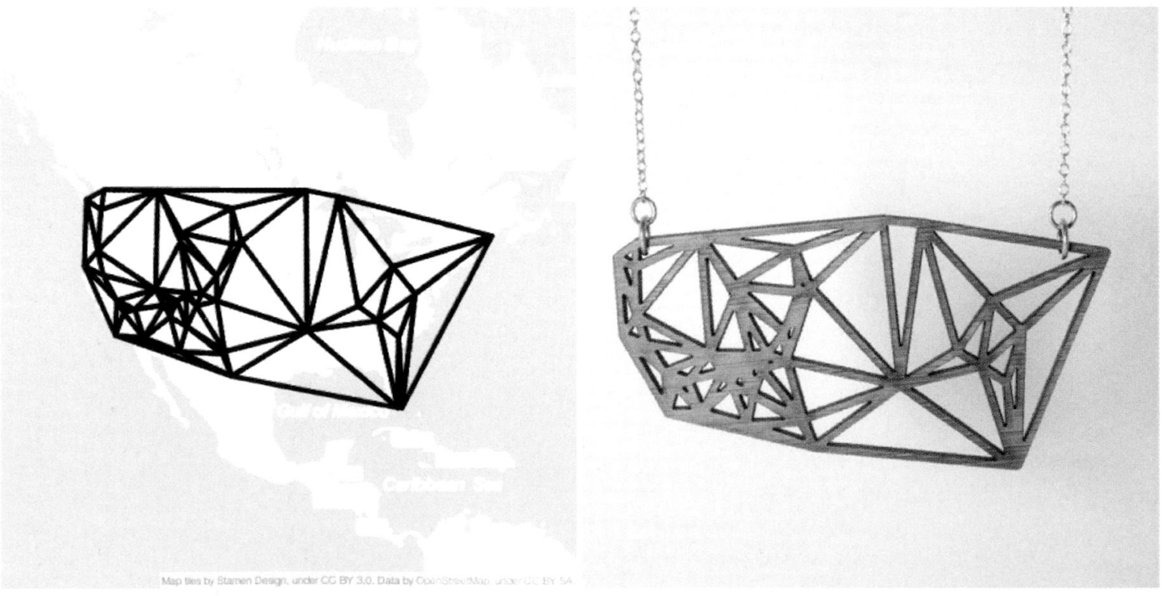

FIGURE 2-36 meshu *by Rachel Binx and Sha Hwang, http://meshu.io/*

WRAPPING UP

The definition of visualization changes by who you ask, and as a whole, the breadth of visualization changes every day. As you come across rules and design suggestions for how to present data, be sure to know their context.

In writing, there is grammar and syntax that is good to know, but where you can bend the rules is also important. Certain movie formulas can work, but often the ones that stray are the biggest success stories.

With visualization, draw from previous work and keep guidelines in mind, but don't let it keep you from what works best to achieve your goal and to communicate to your audience. As you've seen in this chapter, your goals and your audience can change dramatically across applications. It's good that visualization is a medium that you can use for a lot of things.

Representing Data

When you visualize data, you represent it with a combination of visual cues that are scaled, colored, and positioned according to values. Dark-colored shapes mean something different from light-colored shapes, or dots in the top right of a two-dimensional space mean something different than dots in the bottom left.

Visualization is what happens when you make the jump from raw data to bar graphs, line charts, and dot plots. It's the process that takes you from the grid of photos in Chapter 1, "Understanding Data," to a bar graph over time, as shown in Figure 3-1.

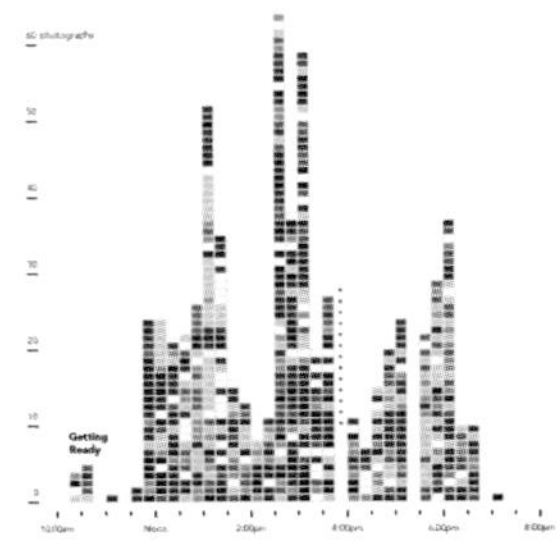

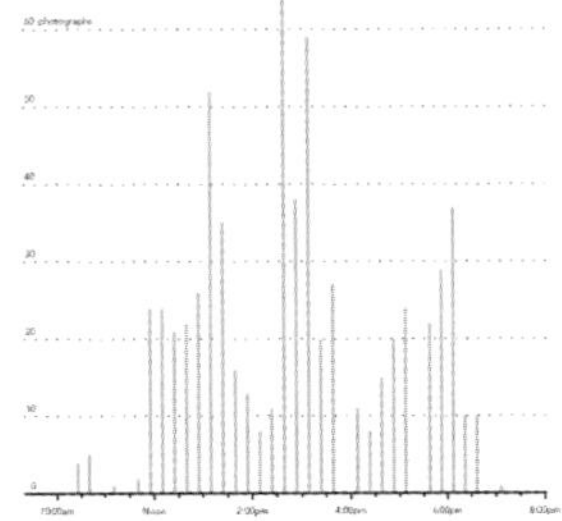

← More literal   More abstract →

FIGURE 3-1 *Abstraction process*

It's easy to think that this process is instant because software enables you to plug data in, and you get something back instantly, but there are steps and choices in between. What shape should you choose to encode your data? What color is most appropriate for the purpose and message? You can let the computer choose everything for you (it can save time), but there are advantages when you choose. At the least, if you know the elements of visualization and how they can be combined and modified, you know what to tell the computer to do rather than let the computer dictate everything you make.

In many ways, visualization is like cooking. You are the chef, and datasets, geometry, and color are your ingredients. A skilled chef, who knows the process of how to prepare and combine ingredients and plate the cooked food, is likely to prepare a delicious meal. A less skilled cook, who heads to the local freezer section to see what microwave dinners look good, might nuke a less savory meal. Of course, some microwave dinners taste good, but there are a lot that taste bad.

Whereas the person who is only familiar with entering the time and power level on a microwave must either endure poor-tasting meals or stick only to the handful of good ones, people who understand the ingredients and actually know how to cook have fewer limitations. The skilled chef might even transform an average frozen dinner into a gourmet meal.

Likewise, with visualization, when you know how to interpret data and how graphical elements fit and work together, the results often come out better than software defaults.

VISUALIZATION COMPONENTS

What are the ingredients of visualization? Figure 3-2 shows a breakdown into four components, with data as the driving force behind them: visual cues, coordinate system, scale, and context. Each visualization, regardless of where it is on the spectrum, is built on data and these four components. Sometimes, they are explicitly displayed, and other times they form an invisible framework. The components work together, and your choice with one affects the others.

Note: Cartographer Jacques Bertin described a similar breakdown in *Semiology of Graphics*, and statistician Leland Wilkinson later provided a variation in *The Grammar of Graphics*.

VISUAL CUES

In its most basic form, visualization is simply mapping data to geometry and color. It works because your brain is wired to find patterns, and you can switch back and forth between the visual and the numbers it represents. This is the important bit. You must make sure that the essence of the data isn't lost in that back and forth between visual and the value it represents because if you can't map back to the data, the visualization is just a bunch of shapes.

You must choose the right visual cue, which changes by purpose, and you must use it correctly, which depends on how you perceive the varied shapes, sizes, and shades. Figure 3-3 shows what's available.

Position

When you use position as a visual cue, you compare values based on where others are placed in a given space or coordinate system. For example, when you look at a scatterplot, as shown in Figure 3-4, you judge a data point based on its x- and y-coordinate and where it is relative to others.

FIGURE 3-2 *(following page)* *The components of visualization*

Working parts

Several pieces work together to make a graph. Sometimes these are explicitly shown in the visualization and other times they form a visual in the background. They all depend on the data.

Title of this Graph

A description of the data or something worth highlighting to set the stage.

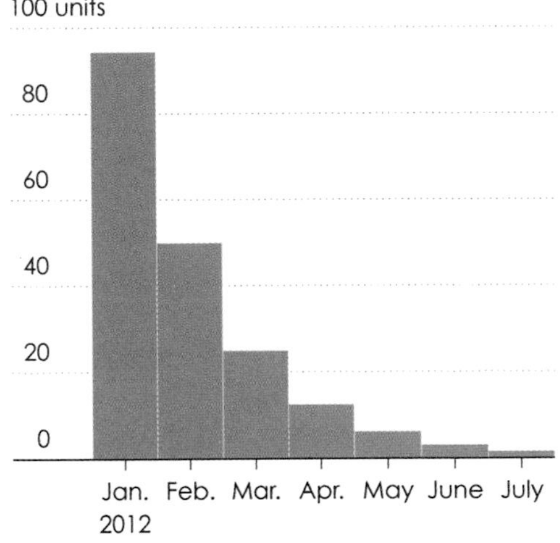

Source: Somewhere reputable

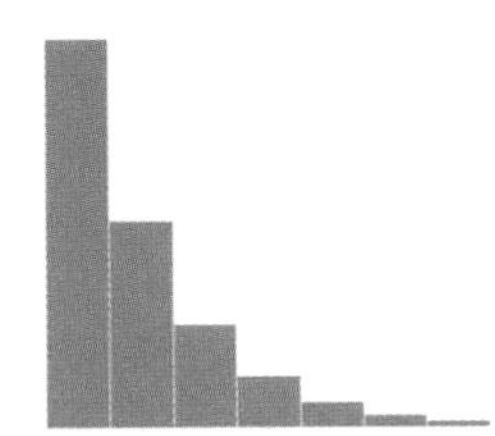

Visual Cues

Visualization involves encoding data with shapes, colors, and sizes. Which cues you choose depends on your data and your goals.

Coordinate System

You map data differently with a scatterplot than you do with a pie chart. It's x- and y-coordinates in one and angles with the other; it's cartesian versus polar.

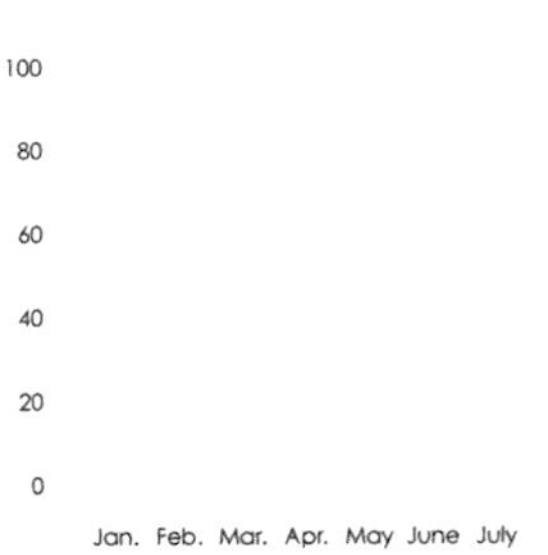

Scale

Increments that make sense can increase readability, as well as shift focus.

Title of this Graph

A description of the data or something worth highlighting to set the stage.

units

2012

Source: Somewhere reputable

Context

If your audience is unfamiliar with the data, it's your job to clarify what values represent and explain how people should read your visualization.

Visual cues

When you visualize data, you encode values to shapes, sizes, and colors.

Position
Where in space the data is

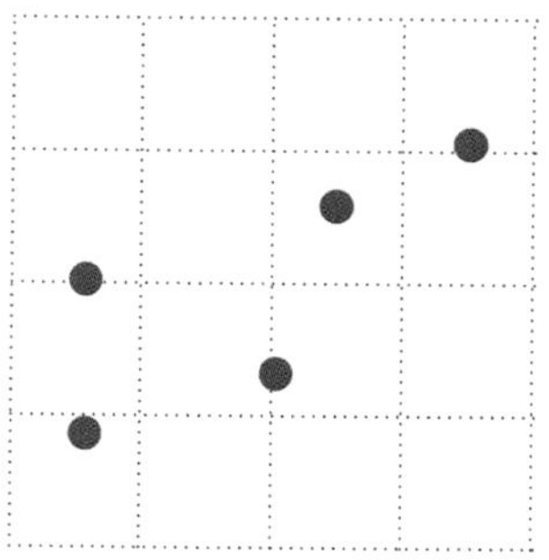

Length
How long the shapes are

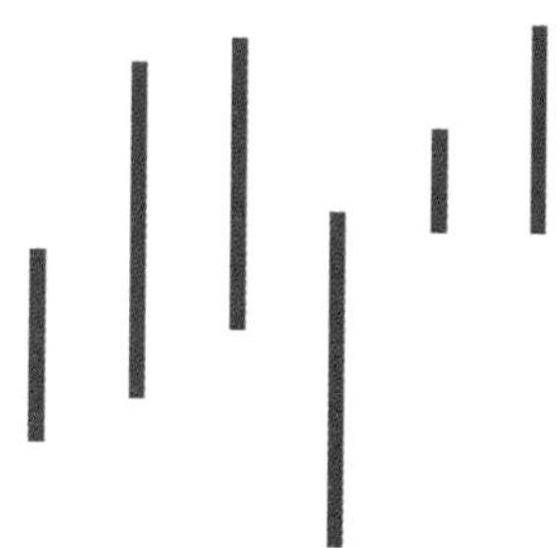

Angle
Rotation between vectors

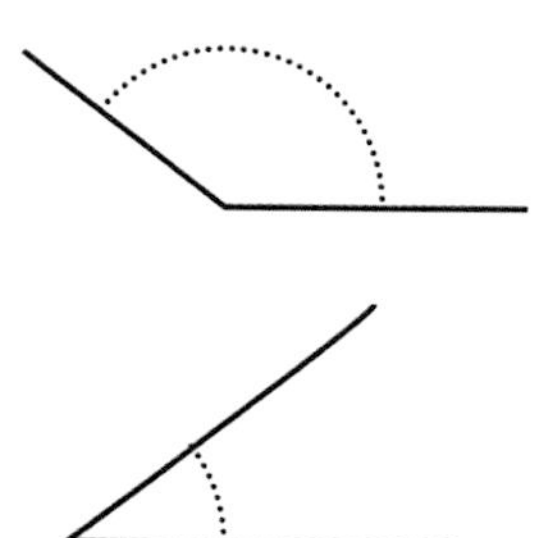

Direction
Slope of a vector in space

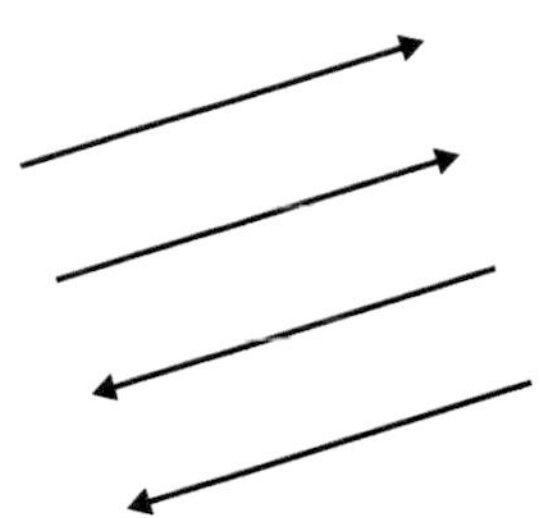

Shapes
Symbols as categories

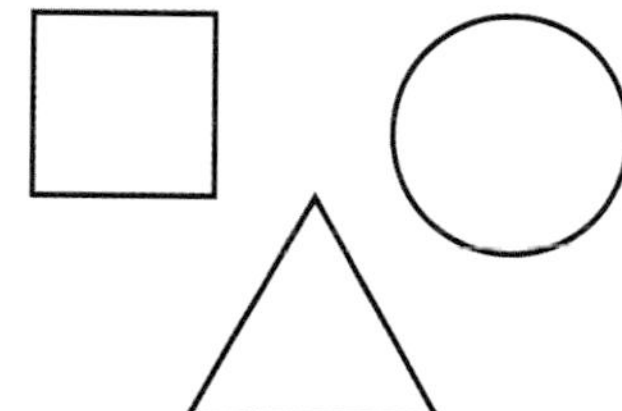

Area
How much 2-D space

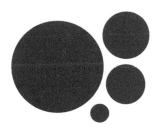

Volume
How much 3-D space

Color saturation
Intensity of a color hue

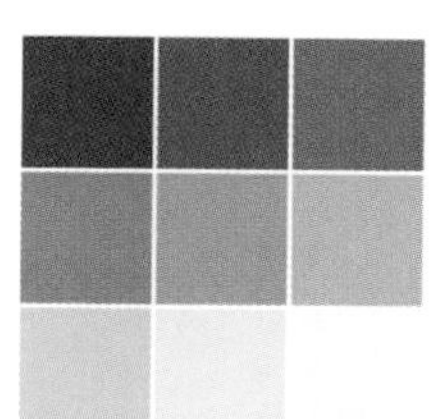

Color hue
Usually referred to as color

FIGURE 3-3 *Visual cues*

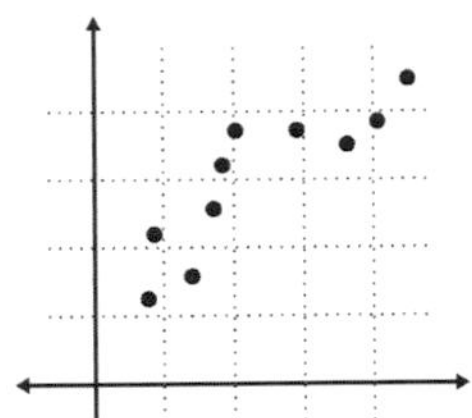

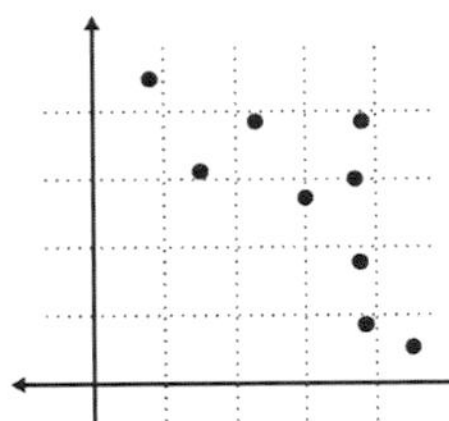

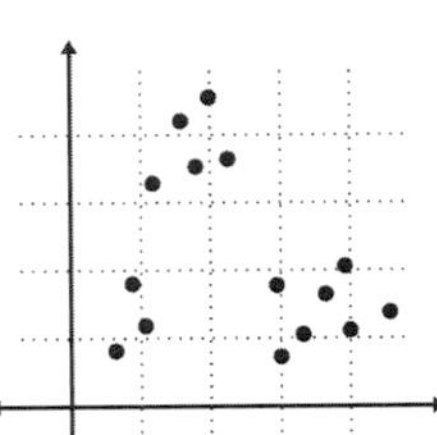

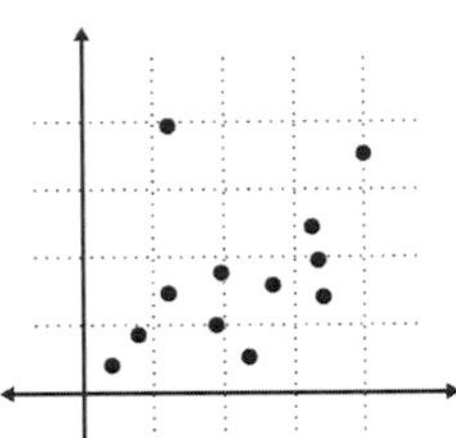

FIGURE 3-4 *Scatterplots*

One of the advantages of using just position is that it tends to take up less space than the other visual cues because you can draw all the data within the x- and y-plane, and you can represent each point with a dot. Unlike other visual cues that use size to compare values, all points in a position-based plot are the same size. In turn, you can spot trends, clusters, and outliers by plotting a lot of data at once.

However, the advantage of using position alone can also be a disadvantage. If you look at a lot of points at once in a scatterplot, it can be a challenge to identify what each point represents. Even in an interactive plot, you still must mouse over or select a point to find out more information, and overlap can cause more problems.

Length

Length is most commonly used in the context of bar charts. The longer a bar is, the greater the absolute value, and it can work in all directions: horizontal, vertical, or even at different angles on a circle.

How do you judge length visually? You figure out the distance from one end of a shape to the other end, so to compare values based on length, you must see both ends of the lines or bars. Otherwise, you end up with a skewed view of maximums, minimums, and everything in between.

As a simple example, as shown in Figure 3-5, a major news outlet displayed a bar graph on television that compared a tax rate before and after a date.

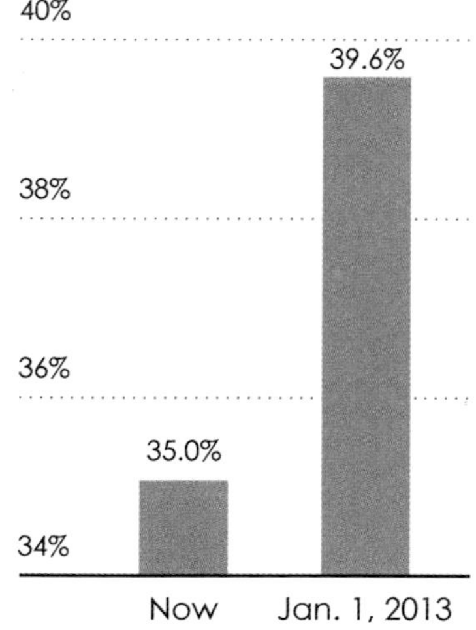

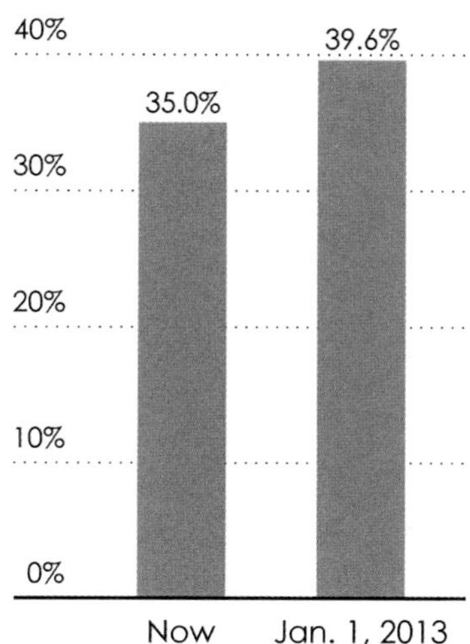

FIGURE 3-5 *Incorrect bar graph on left and correct one on the right*

The difference between the two values looks like a huge increase—the length of the right bar is about five times the length as the other—because the value axis starts at 34 percent. The chart on the right shows the change when the axis starts at zero, which looks less dramatic. Of course, you can always look at the axis to verify what you see (and you always should), but that defeats the purpose of showing the values with length, and if the chart is shown quickly on television, most people won't notice the misstep.

Angle

Angles range from zero to 360 degrees on a circle. There's the 90-degree right angle, the obtuse angle that is greater than 90 degrees, and the acute angle that is less than 90 degrees. A straight line is 180 degrees.

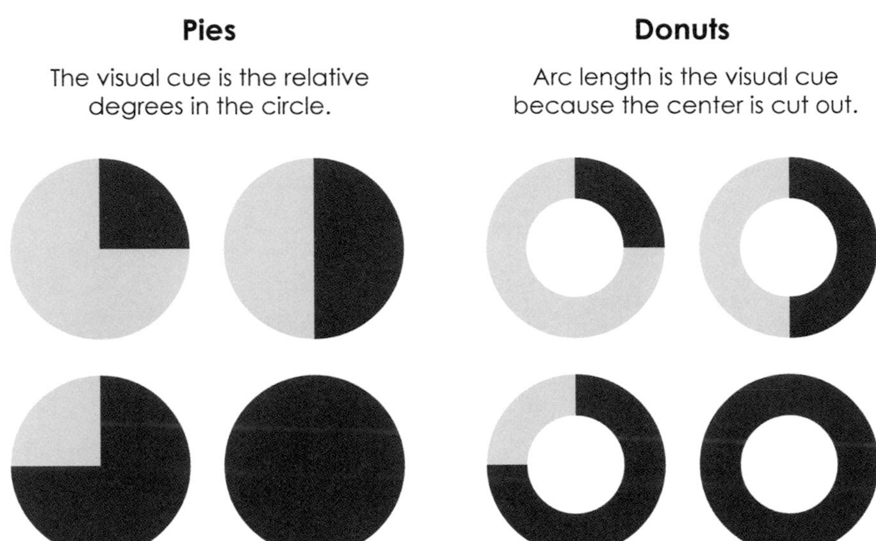

Pies
The visual cue is the relative degrees in the circle.
Donuts
Arc length is the visual cue because the center is cut out.

FIGURE 3-6 *Mmm, pies and donuts*

Note: Although the donut chart is often considered the pie chart's close cousin, arc length is the former's visual cue because the center of the circle, which indicates angles, is removed.

For each angle in between zero and 360 degrees, there is an implied opposite angle that completes the rotation, and together those two angles are considered conjugates. This is why angles are commonly used to represent parts of a whole, using the fan favorite, but often maligned, pie chart shown in Figure 3-6. The sum of the wedges makes a complete circle.

Direction

Direction is similar to angle, but instead of relying on two vectors joined at a point, direction relies on a single vector's orientation in a coordinate system. You can see which way is up, down, left, and right and everything in between.

This helps you determine slope, as shown in Figure 3-7. You can see increases, decreases, and fluctuations.

The amount of perceived change depends a lot on the scale, as shown in Figure 3-8. For example, you can make a small change in percentage look like a lot by stretching out the scale. Likewise, you can make a big change look like a little by compressing the scale.

A rule of thumb is to scale your visualization so that direction fluctuates mostly around 45 degrees, but this is hardly a concrete rule. The best thing to do is to start with this suggestion and then adjust accordingly based on context. If a small change is significant, then it might be appropriate to stretch the scale so that you can see the shift. In contrast, if a small change is not significant, don't stretch out the scale just to make a shift look dramatic.

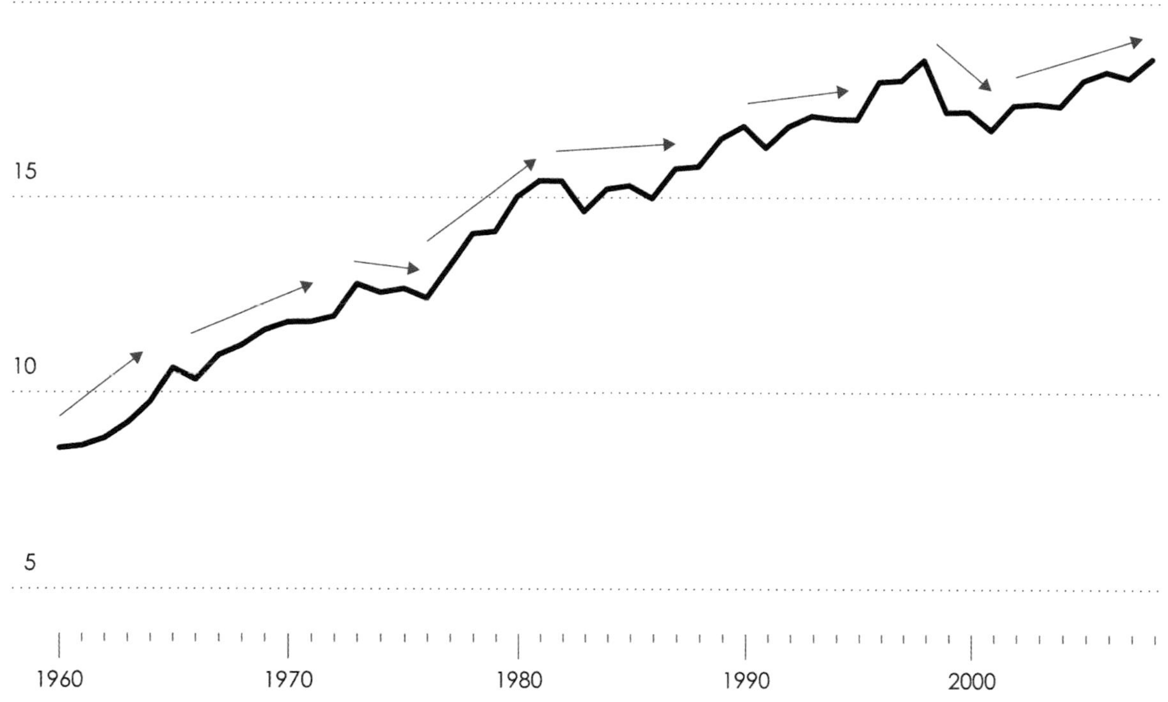

FIGURE 3-7 *Slope and time series*

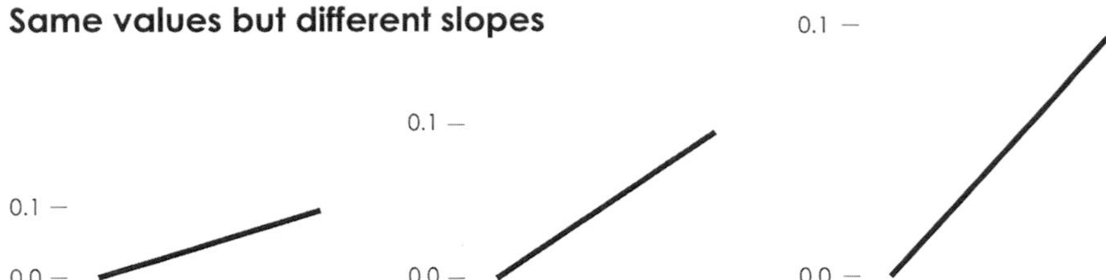

FIGURE 3-8 *Same amount of change shown on varied scales*

Shapes

Shapes and symbols are commonly used with maps to differentiate categories and objects. Location on a map can be directly translated to the real world, so it makes sense to use icons to represent things in the real world. You might represent forests with trees or residential areas with houses.

In a chart context, shapes to show variation are used less frequently than they used to be. When graphs were drawn with paper and a pencil and computers still worked with punch cards, symbols were an easy way to differentiate categories. For example, as shown in Figure 3-9, triangles and squares could be used in a scatterplot, which is quicker to draw than to switch between colored pencils and pens or fill a single shape with a solid or cross-hatched pattern.

Nevertheless, varied shapes can provide context that points alone can't, and it's typically not more difficult to try with your favorite software.

Area and Volume

Bigger objects represent greater values. Like length, area and volume can be used to represent data with size, but with two and three dimensions, respectively. For the former, circles and rectangles are commonly used, and with the latter, cubes and sometimes spheres. You can also size more detailed icons and illustrations.

Be sure to mind how many dimensions you use. The most common mistake is to size a two- or three-dimensional object by only one dimension, such as height, but to maintain the proportions of all dimensions. This results in shapes that are too big and too small, which makes it impossible to fairly compare values.

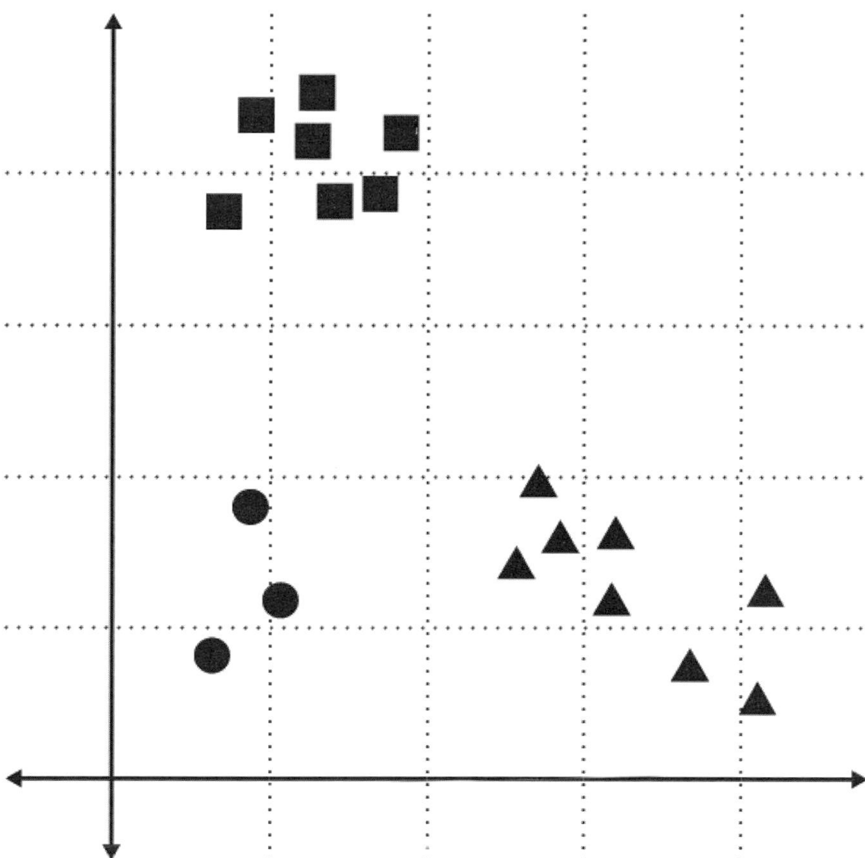

FIGURE 3-9 *Different shapes in scatterplot*

Say you use squares, shapes with two dimensions—width and height—to represent your data. The greater a value, the greater the area of a square, so if one value is 50 percent greater than another, you want the area of the square to be 50 percent greater than the other. However, if you increase the sides of the square by 50 percent instead of the area, which is what some software does by default, the larger square is too big. Instead of an increase in 50 percent, it's an increase of 125 percent. See the jump in difference in Figure 3-10.

You run into the same problem with three-dimensional objects, but the mistake is more pronounced. Increase the width, height, and depth of a cube by 50 percent, and the volume of the cube increases by approximately 238 percent.

Sizing by area

This is one unit.

Four units sized by area

4 times the area as unit square

Four units *incorrectly* sized by side length

16 times the area as unit square

Sizing by volume

This is one unit.

Four units sized by volume

4 times the volume as unit cube

Four units *incorrectly* sized by edge length

32 times the volume as unit square

FIGURE 3-10 *Squares and cubes sized by different dimensions*

Color

Color as a visual cue can be spilt into two categories: hue and saturation. They can be used individually or in combination.

Color hue is what you usually just refer to as color. That's red, green, blue, and so on. Differing colors used together usually indicates categorical data, where each color represents a group. Saturation is the amount of hue in a color, so if your selected color is red, high saturation would be very red, and as you decrease saturation, it looks more faded. Used together, you can have multiple hues that represent categories, but each category can have varying scales.

Careful color selection can lend context to your data, and because there is no dependency on size or position, you can encode a lot of data at once. However, keep color blindness in mind if you want to make sure that everyone can interpret your graphics. Approximately 8 percent of men and 0.5 percent of women are red-green deficient, so when you encode your data only with those colors, this segment of your audience will have trouble decoding your visualization, if they can at all. Figure 3-11 shows how some shades are perceived by those who are color-deficient.

Does this mean you aren't allowed to use red and green in your graphics? No. You can combine visual cues so that everyone can make out differences, which you'll get to in a few sections.

Note: Something to think about: If someone is red-green deficient, how do they follow traffic signals that use red for stop and green for go? They note the order of the lights.

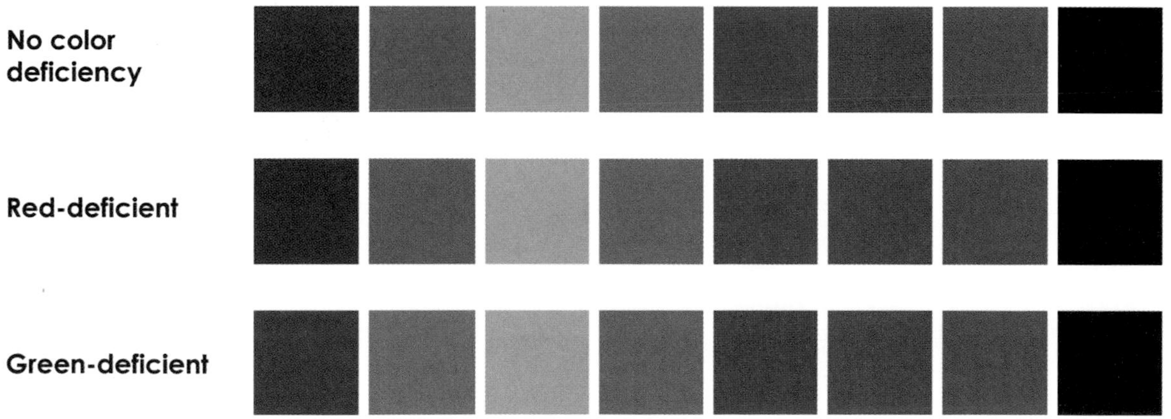

FIGURE 3-11 *Colors as perceived by those who have color vision deficiencies*

Perception of Visual Cues

In 1985, William Cleveland and Robert McGill, then statistical scientists at AT&T Bell Laboratories, published a paper on graphical perception and methods. The focus of the study was to determine how accurately people read the visual cues above (excluding shapes), which resulted in a ranked list from most accurate to least accurate, as shown in Figure 3-12.

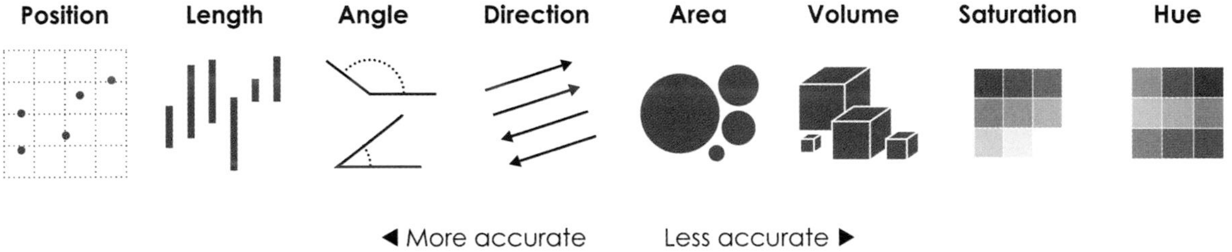

FIGURE 3-12 *Visual cues ranked by Cleveland and McGill*

A lot of visualization suggestions (and current research) stem from this list, which places bar charts above pie charts, heat maps at the bottom, and so on. This is sound advice, and you see more on this in Chapter 5, "Visualizing with Clarity," but remember that this list doesn't mean that dot plots are always better than bubble plots or that pie charts are evil.

Following this list blindly is an oversimplification of what visualization is. As you saw in the previous chapter, efficiency and exactness are not always the goal. That said, regardless of what you want to visualize data for, it's good to know how well people can read your visual cues and what information they can extract. In other words, use these rankings as a guide rather than a rule book.

COORDINATE SYSTEMS

When you encode data, you eventually must place the objects somewhere. There's a structured space and rules that dictate where the shapes and colors go. This is the coordinate system, which gives meaning to an x-y coordinate or a latitude and longitude pair. There are several systems, but as shown in Figure 3-13, there are three that cover most of your bases: Cartesian, polar, and geographic.

Coordinate systems

There are a variety of them, from cylindrical to spherical, but these three will cover most of your bases.

Cartesian

If you've ever made a graph, the x- and y-coordinate system will look familiar to you.

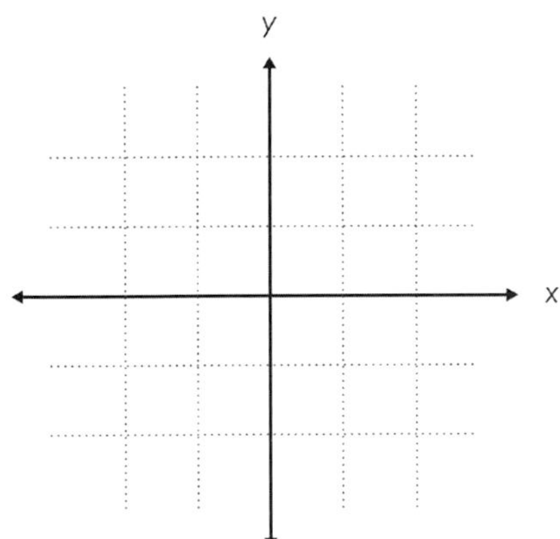

Polar

Pie charts use this system. Coordinates are placed based on radius *r* and angle *θ*.

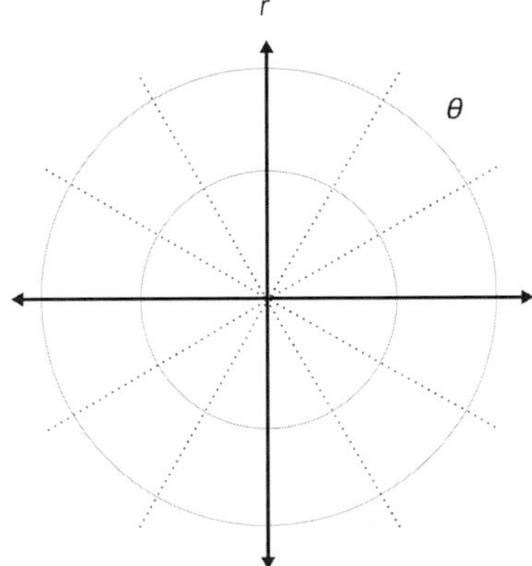

Geographic

Latitude and longitude are used to identify locations in the world. Because the planet is round, there are multiple projections to display geographic data in two dimensions. This one is the Winkel tripel.

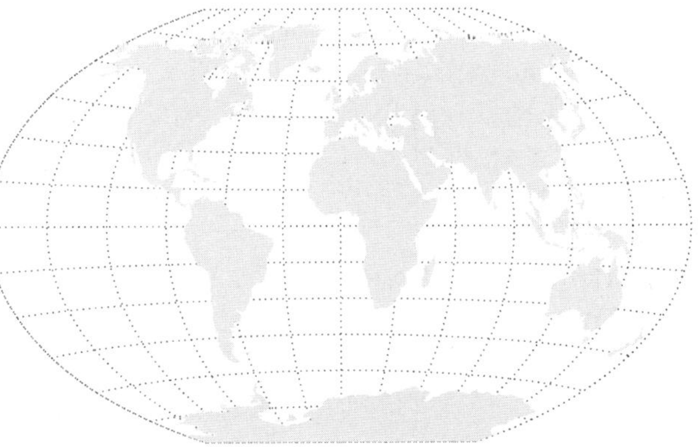

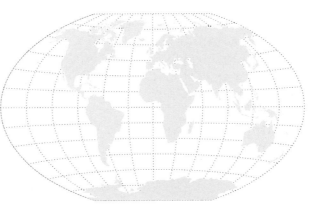

FIGURE 3-13 *Commonly used coordinate systems*

Cartesian

The Cartesian coordinate system is the most commonly used one with charts. If you've made a traditional graph, such as a bar chart or a dot plot, you've used Cartesian coordinates.

You typically think of coordinates in the system as an *x* and *y* pair that is denoted as (x, y). Two lines that are perpendicular to each other, and range from negative to positive, form the axes. The place the lines intersect is the origin, and the coordinate values indicate the distance from that origin. For example, the (x, y) pair at (0, 0) is at the intersection of the lines, and the (1, 2) pair is one unit away from the origin on the horizontal and two units away on the vertical.

To make this high school geometry flashback complete, you can find the distance between any two points, denoted as (x_1, y_1) and (x_2, y_2), with the distance formula.

$$distance = \sqrt{(x_2 - x_1)^2 + (y_2 - y_1)^2}$$

You can also extend the Cartesian space to more than two dimensions. For example, a three-dimensional space would use a (x, y, z) triplet instead of just a (x, y) pair.

The takeaway is that you can describe geometric shapes using Cartesian coordinates, which makes it easier to draw in the space. From an implementation standpoint, the coordinate system enables you to encode values to paper or a computer screen.

Polar

Made a pie chart? You've used the polar coordinate system, too. Although you might have used only the angle part and not the radius. Referring to Figure 3-13, the polar coordinate system consists of a circular grid, where the rightmost point is zero degrees. The greater the angle is, the more you rotate counterclockwise. The farther away from the circle you are, the greater the radius is.

Place yourself on the outer-most circle, and increase the angle. This rotates you counterclockwise toward the vertical line (or the y-axis if this were Cartesian coordinates), which is 90 degrees (that is, a right angle). Rotate one-quarter more, and you get to 180 degrees. Rotate back to where you started, and that's a 360-degree rotation. Your radius would be smaller if you rotated along the smaller circle.

This system is used less than the Cartesian coordinate system, but it can be useful in cases in which the angle or direction is important.

Geographic

Location data has the added benefit of a connection to the physical world, which in turn lends instant context and a relationship to that point, relative to where you are. With a geographic coordinate system, you can map these points. Location data comes in many forms, but it's most commonly described as latitude and longitude, which are angles relative to the Equator and Prime Meridian, respectively. Sometimes elevation is also included.

Latitude lines run east and west, which indicates north and south position on a globe. Longitude lines run north and south and indicate the east and west position. Elevation can be thought of as a third dimension. Compared with Cartesian coordinates, latitude is like the horizontal axis, and longitude is like the vertical axis. That is, if you use a flat projection.

The tricky part about mapping the surface of Earth is that it's wrapped around a spherical mass, but you usually need to display it on a two-dimensional surface, like a computer screen. The variety of ways to do this are called projections, and as shown in Figure 3-14, each has its advantages and disadvantages.

When you project something that is three-dimensional onto a two-dimensional plane, some information is lost, whereas other information is preserved.

The Mercator projection, for example, preserves angles in local regions. It was created in the 16th century by cartographer Geradus Mercator primarily for navigation on the seas and is still the most-used projection for online direction lookup. On the other hand, the Albers projection preserves area but distorts shape. So the projection you choose depends on what you want to focus on.

SCALES

Whereas coordinate systems dictate the dimensions of a visualization, scale dictates where in those dimensions your data maps to. There's a variety of them, and you can even define your own scales based on mathematical functions, but most likely you'll rarely stray from the ones in Figure 3-15. These can be grouped into three categories: numeric, categorical, and time.

FIGURE 3-14 *(following page)* *Map projections*

Map projections

Equirectangular

Typically used for thematic mapping, but doesn't preserve area or angle

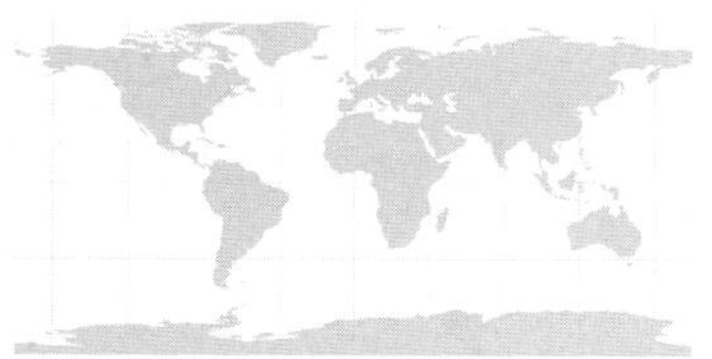

Albers

Scale and shape not preserved; angle distortion is minimal

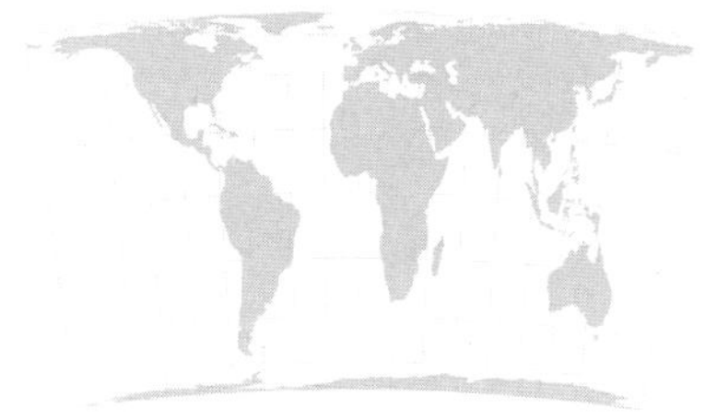

Mercator

Preserves angles and shapes in small areas, making it good for directions

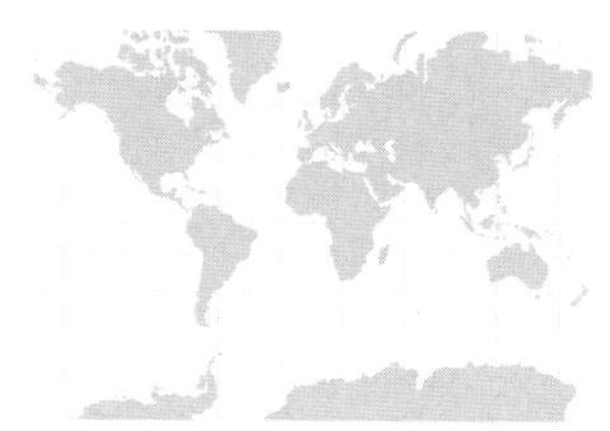

Lambert conformal conic

Better for showing smaller areas and often used for aeronautical maps.

Sinusoidal

Preserves area; useful for areas near the prime meridian

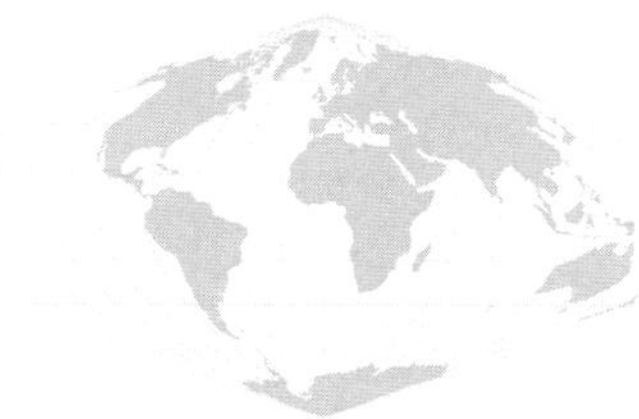

Polyconic

Was often used to show US in the mid-1900s; little distortions in small areas near merdian

Winkel Tripel

Minimized area, angle, and distance distortion; good choice for world map

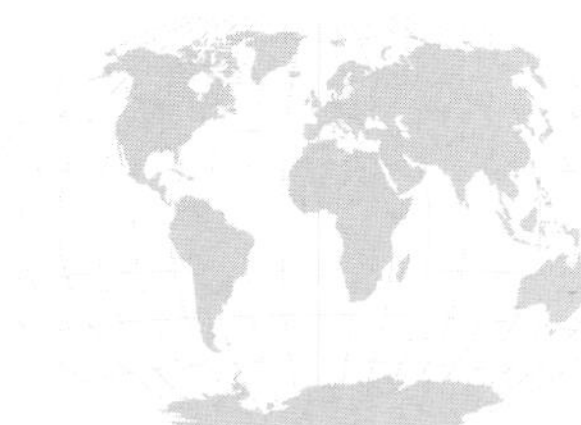

Robinson

A compromise between preserving areas and angles; good to show world map

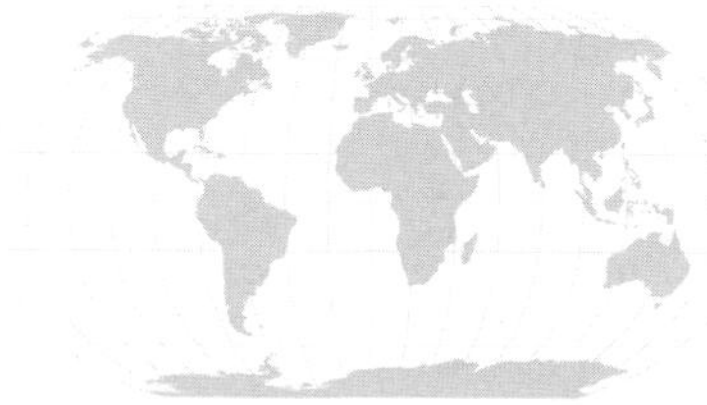

Orthographic

Representing a 3-D object in 2-D, need to rotate to area of interest

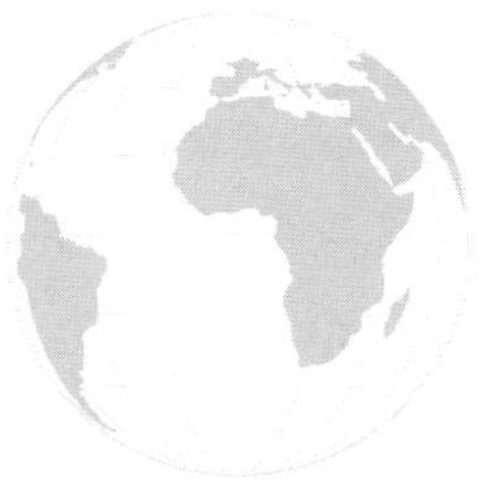

Source: Natural Earth

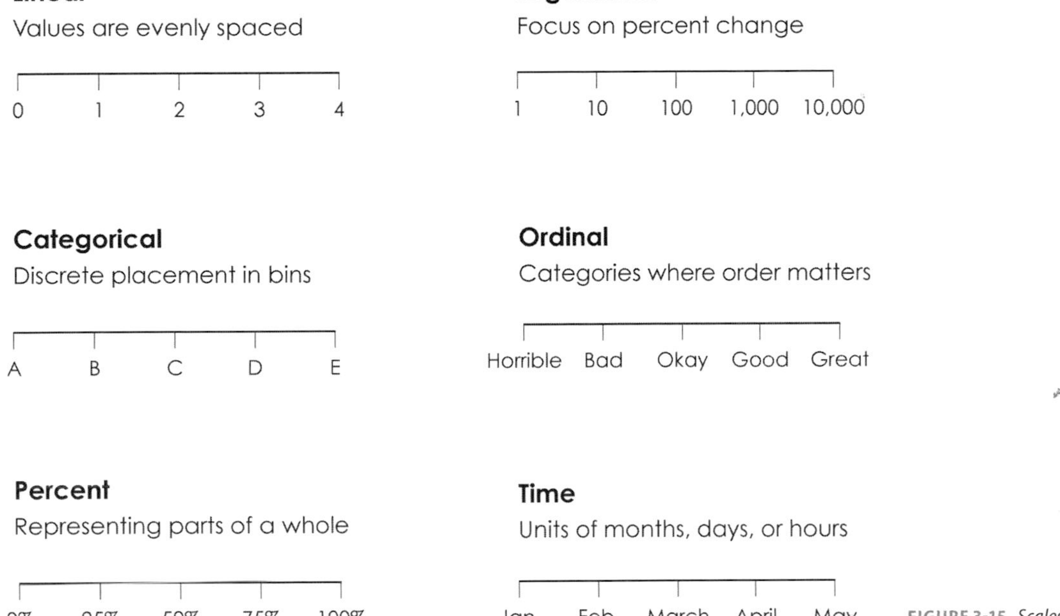

FIGURE 3-15 *Scales*

Numeric

The visual spacing on a linear scale is the same regardless of where you are on the axis. So if you were to measure the distance between two points on the lower end of the scale, it'd be the same if they were at the high end of the scale.

On the other hand, a logarithmic scale condenses as you increase values. This scale is used less than the linear scale and is not as well understood or straightforward for those who don't regularly work with data, but it's useful if you're interested in percent differences more than you are raw counts or your data has a wide range.

For example, when you compare state populations in the United States, you deal with numbers from the hundreds of thousands up to the tens of millions. As of this writing, California has a population of approximately 38 million people, whereas Wyoming has a population of approximately 600,000. As shown in Figure 3-16, with a linear scale, states with smaller populations are clustered on the bottom, and then a few states rest on top. It's easier to see points on the bottom with a logarithmic scale.

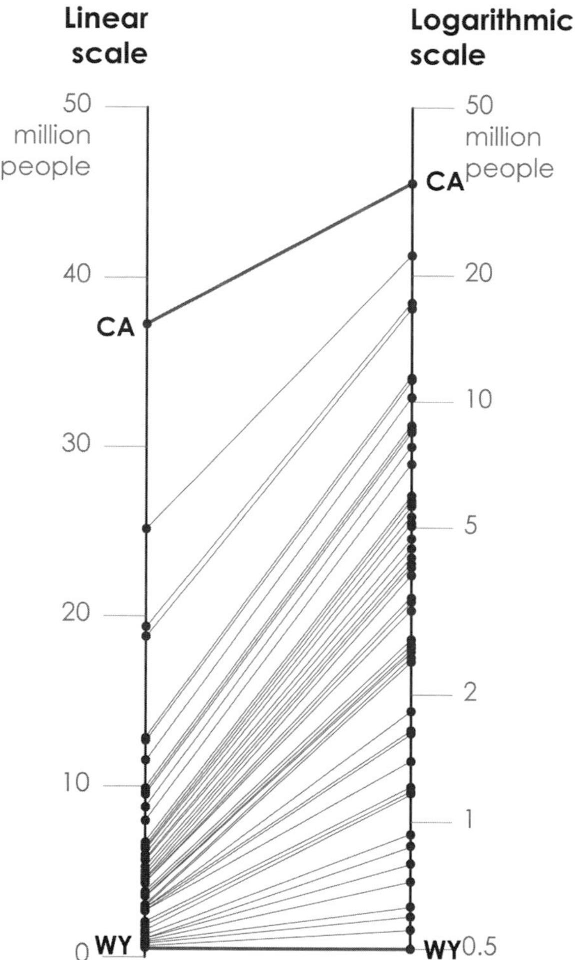

FIGURE 3-16 *Linear versus logarithmic scale*

A percent scale is usually linear, but when it's used to represent parts of a whole, its maximum is 100 percent. As shown in Figure 3-17, the sum of all the parts is 100 percent. This seems obvious—that the sum of percentages in a pie chart, represented with wedges, should not exceed 100 percent—but the mistake seems to come up occasionally. Sometimes it's due to mislabeling, but some people just aren't familiar with the concept.

Correct

The sum of the parts equals 100 percent.

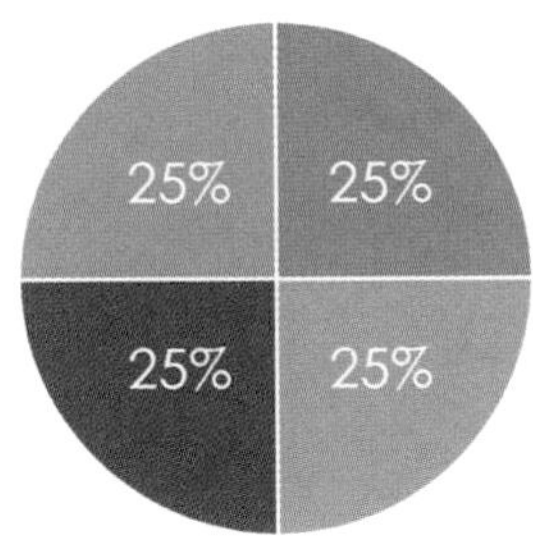

Wrong

The sum of the parts is more than 100 percent.

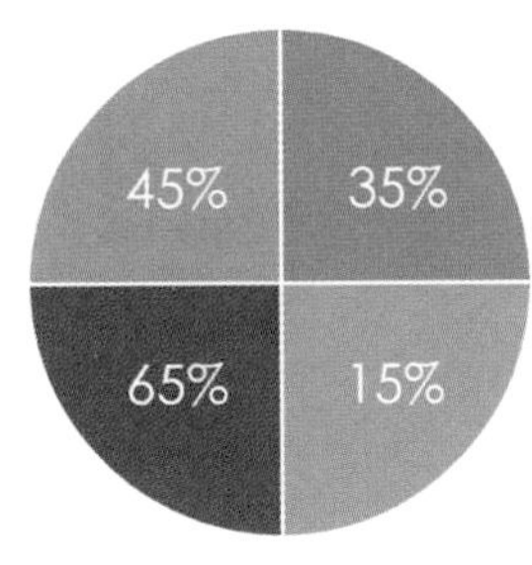

Also, the labels don't match the wedge sizes.

FIGURE 3-17 *Incorrect and correct pie charts*

Categorical

Data doesn't always need to be numeric. It can be categorical, such as people's cities of residence or the political parties of government officials. A categorical scale provides visual separation for these different groups and often works with a numeric scale. A bar plot for example, can use a categorical scale on the horizontal axis and a numeric scale on the vertical to show counts or measurements for different groups, as shown in Figure 3-18.

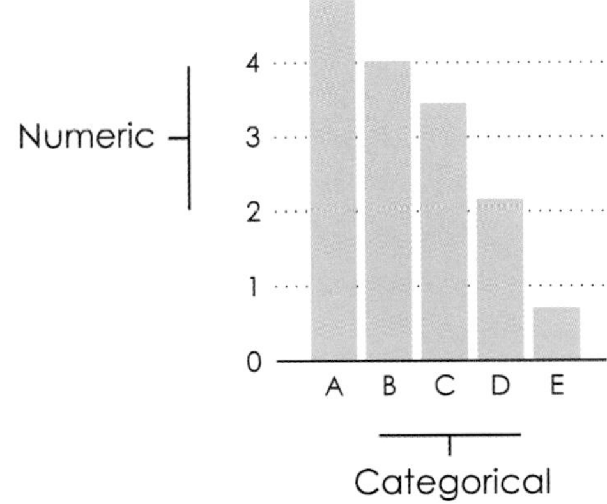

FIGURE 3-18 *Numeric and categorical scales on a bar plot*

Spacing between each category is arbitrary because it does not depend on a numeric value, but it is typically adjusted to increase clarity, which is discussed in Chapter 6, "Visualizing with Clarity."

Ordering should be used in the context of the data. Although this can also be arbitrary, for an ordinal scale that uses categories, order of course matters. If your data is a categorical ranking on movies that ranges from horrible to great, then it makes sense to keep that order visually, which makes it easier to compare and judge quality.

Time

Time is a continuous variable, which lets you plot temporal data on a linear scale, but you can divide it into categories such as months or days of the week, which lets you visualize it as a discrete variable. Also, it cycles, as shown in Figure 3-19. There's always another noon time, Saturday, and January.

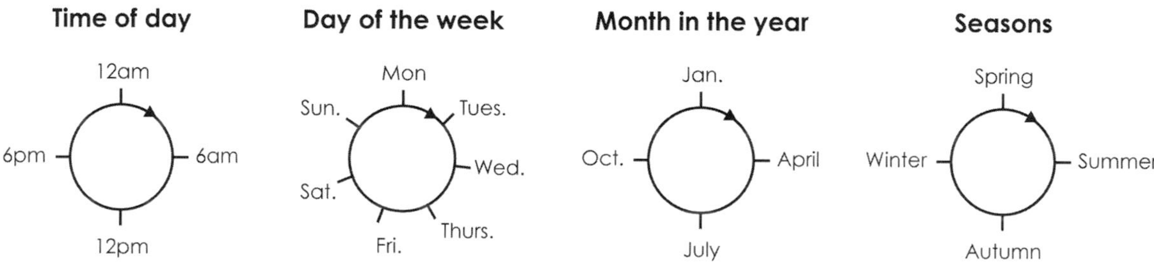

FIGURE 3-19 *Time cycles*

You saw this in Chapter 1, which showed fatal crashes over time, by year, by month, by day, and by hour. Data was plotted continuously in these cases. However, aggregates by time of day, day of the week, and month (over multiple years) showed a different picture.

When communicating data to an audience, the time scale, like geographic maps, gives you an advantage of lending a reader connection because time is a part of everyday life. You feel and experience time internally and through your clocks and calendars, and as the sun rises and sets.

CONTEXT

Context (information that lends to better understanding the who, what, when, where, and why of your data) can make the data clearer for readers and point them in the right direction. At the least, it can remind you what a graph is about when you come back to it a few months later.

Sometimes context is explicitly drawn, and other times it's implied through the medium. For example, as shown in Figure 3-20, designers Matt Robinson and Tom Wrigglesworth drew "sample" on a wall, with ballpoint pens and different typefaces. Because ink usage varies by typeface, each pen had a different amount of ink left, which made for an interesting bar graph. There's no need to label the numeric axis because it's implied by the pens and their ink.

Designer George Kokkinidis approached iPad usage in a similar way; however, as shown in Figure 3-21, instead of comparing remaining ink, he looked at fingerprint traces while he used different apps. For example, in Mail, he typed messages most of the time, so the keyboard pattern is most evident, with some scrolling on the side. In contrast, most interaction is in the bottom-left corner for the game Angry Birds.

Of course, you can't always draw on familiar physical objects for context, so you must provide familiarity and a sense of scale in other ways. The easiest and most straightforward way is to label your axes and specify units of measure, or provide a description that tells others what each visual cue represents. Otherwise, when the data is abstracted, there's no way to decode the shapes, sizes, and colors, and you might as well show an amorphous blob.

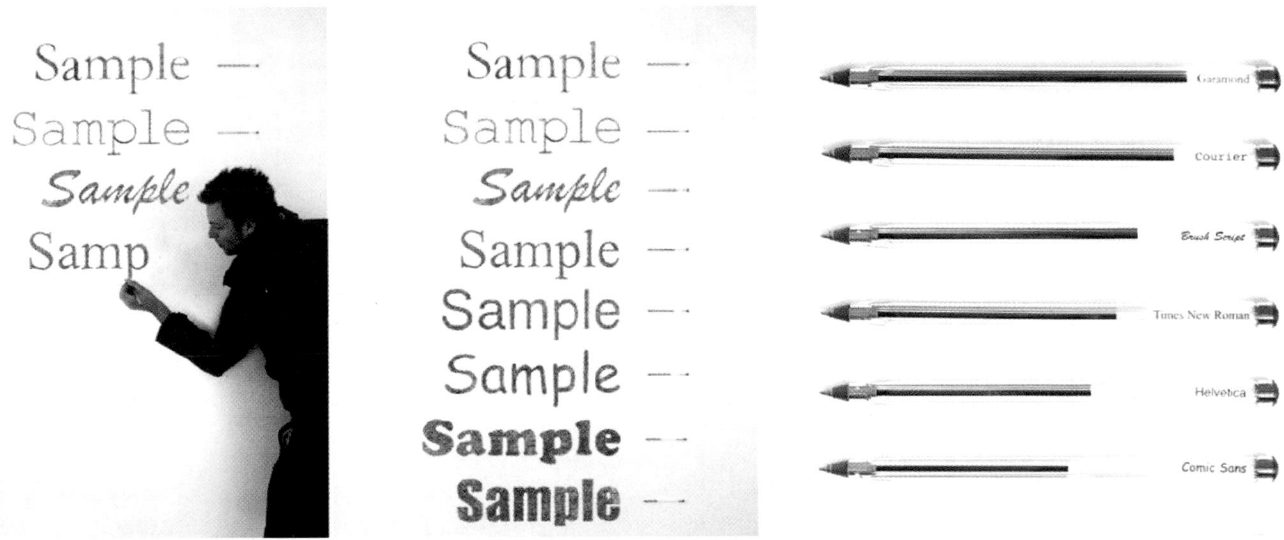

FIGURE 3-20 Measuring Type *(2010) by Matt Robinson and Tom Wrigglesworth, http://datafl.ws/27m*

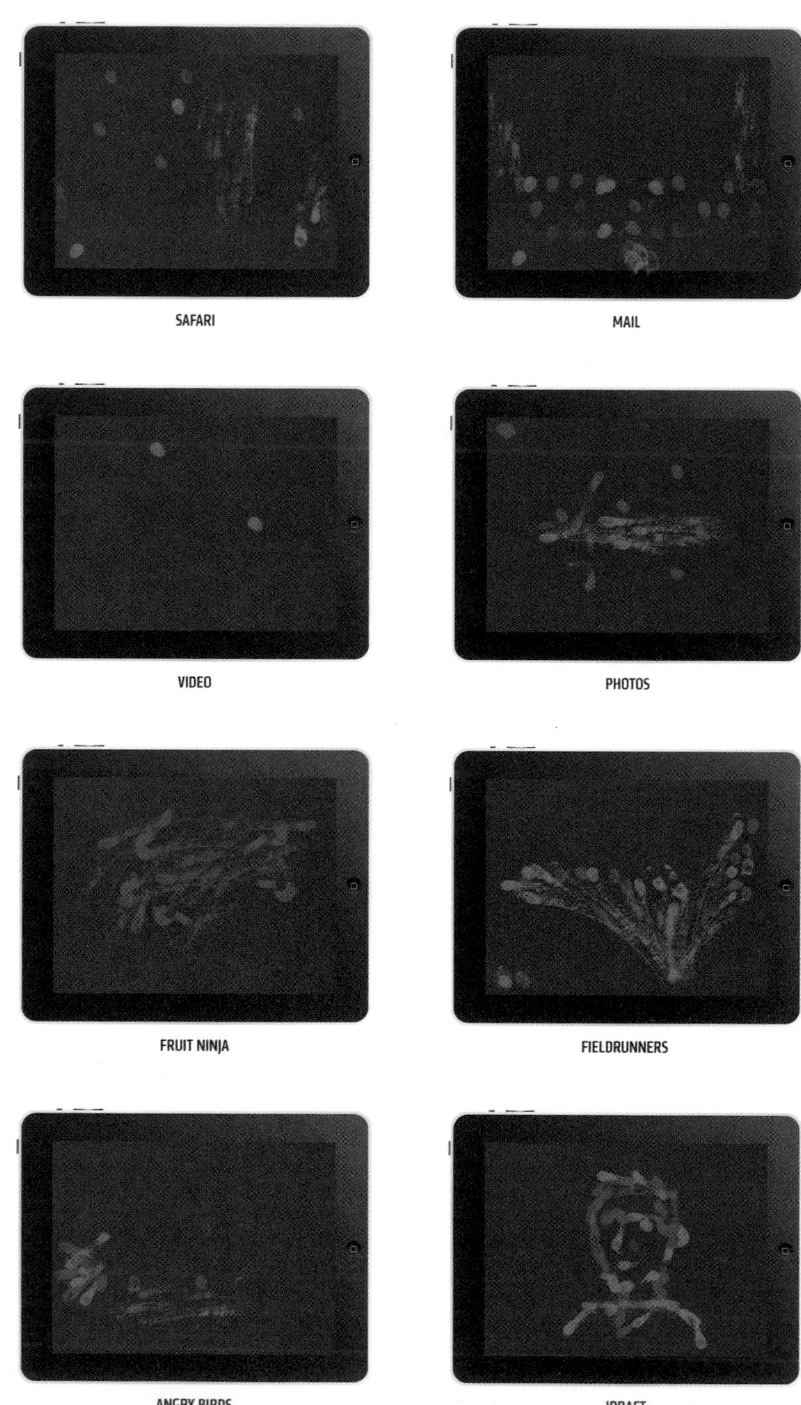

FIGURE 3-21 Remnants of a Disappearing UI *(2010) by George Kokkinidis, http://datafl.ws/27n*

A descriptive title is a small but easy thing you can create to set up readers for what they're about to look at. Imagine you produce a time series plot for gas prices that shows an upward trend. You could just title it "Gas Prices" and that would be a fair title. That's what it is, but you could also title it "Rising Gas Prices," which says what data is used and what is shown. You could also include lead-in text underneath the title that describes fluctuations or by how much gas prices rose.

Your choice of visual cues, a coordinate system, and scale can implicitly provide context. Bright, cheery, and contrasting colors says something different than dark, neutral, and blending colors. Similarly, a geographic coordinate system places you within the context of physical space, whereas an x-y plot using Cartesian coordinates keeps you within a virtual space. A logarithmic scale could suggest a focus on percentage changes and reduce focus on absolute values.

This is why it's important to pay attention to software defaults.

Programs are designed to be flexible and fast and they work outside the context of the data. This is great to draw a visualization base and explore your data, but it's up to you to make the right decisions along the way and to make the computer output something for humans. This comes partly from knowing how you perceive geometry and colors, but mostly it comes from practice and the experience gained from seeing a lot of data and evaluating how others, who aren't familiar with your data, interpret your work. Common sense also goes a long way.

Note: The people who visualize data best got to that level because they examined and visualized a lot of data. They gained experience with each graph made. Reading books will inform better decisions, but it's not until you put what you learn into practice when you really improve.

PUTTING IT TOGETHER

You know what ingredients are available. Now it's time to cook the meal. Viewed separately, the visualization components aren't that useful because they are just bits of geometry floating in an empty space without context. However, when you put the components together, you get a complete visualization worth looking at.

For example, what do you get when you use length as a visual cue, a Cartesian coordinate system, and a categorical scale on the horizontal axis and a linear scale on the vertical? You get a bar chart. Use position with a geographic coordinate system, and you get points on a map.

What do you get when you use a polar coordinate system with the area as the visual cue, a percentage scale on the radius, and a time scale on the rotation? That's a polar area diagram. The most famous one, as shown in Figure 3-22, is Florence Nightingale's chart that visualizes deaths from treatable diseases over time.

On the Origin of Species: The Preservation of Favoured Traces, by designer and developer Ben Fry, uses color and length, Cartesian coordinates, and a linear scale, as shown in Figure 3-23. The interactive and animated visualization shows how Charles Darwin's theory of evolution changed through six editions. The gray blocks represent the original text, and each subsequent color represents a revision in an edition, so you can see what changed and by how much.

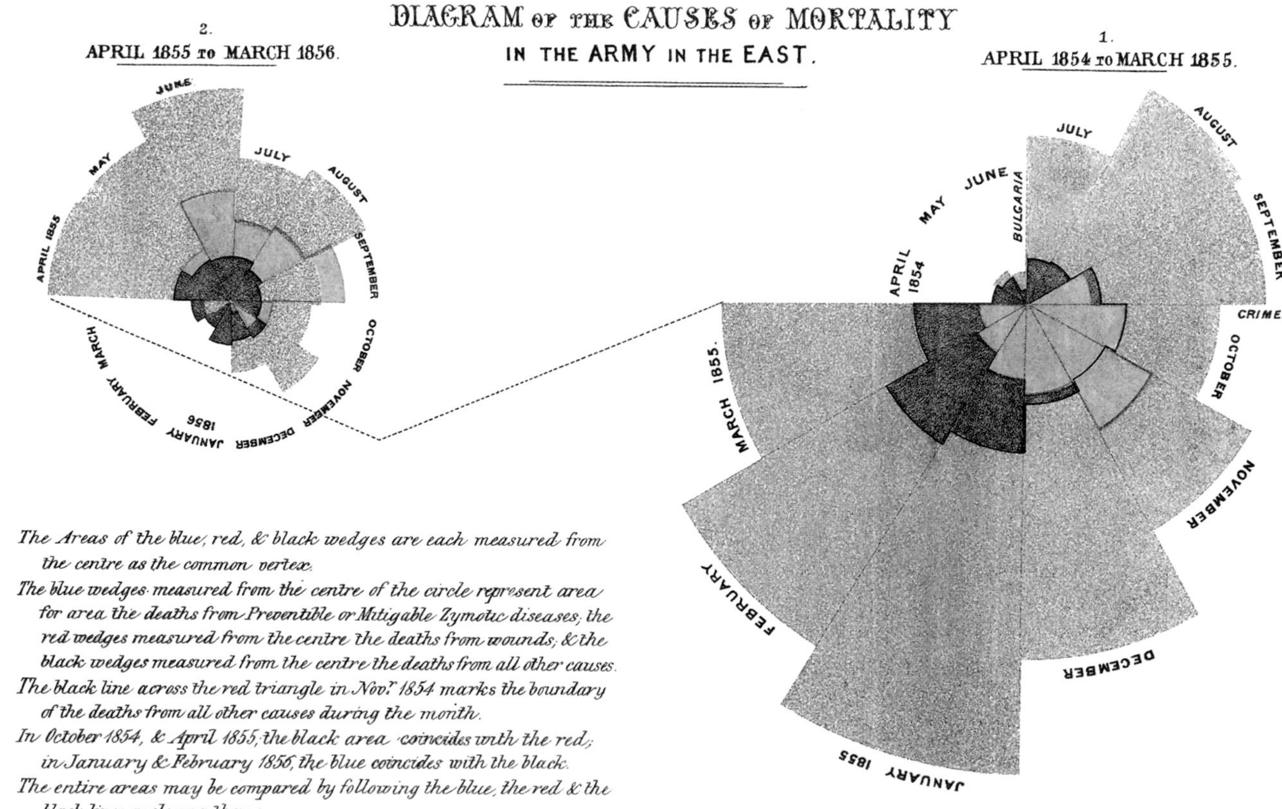

FIGURE 3-22 Diagram of the Causes of Mortality in the Army in the East *(1858) by Florence Nightingale*

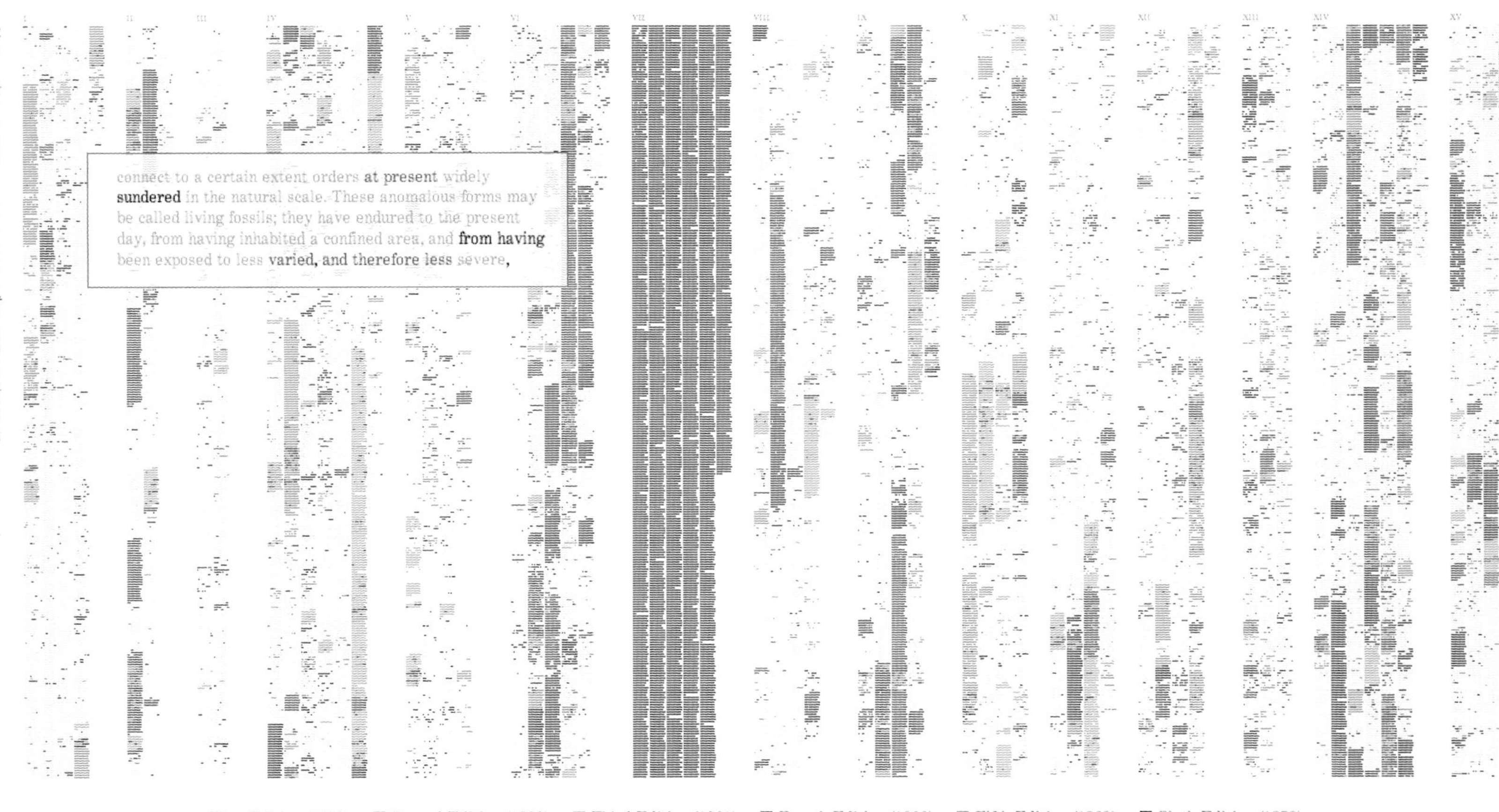

FIGURE 3-23 On the Origin of Species: The Preservation of Favoured Traces *(2009) by Ben Fry, http://benfry.com/traces/*

In the deaths chart shown in Figure 3-24, from the *Statistical Atlas of the United States* published in 1874, length is used to show the distribution of deaths for each state, by age and gender. The horizontal axis on each plot represents the number of deaths on a linear scale, and the vertical axis represents numeric categories that represent age groups.

Figure 3-25 shows generalized combinations, which cover common visualization types such as the line chart, bubble plot, and choropleth map. The key is learning how each component fits together—how each ingredient can complement and enhance others—to make something more useful than the separate parts.

Now try to fit components together, starting with the data and then building on that foundation. Figure 3-26 is a data table from the United States Census Bureau that shows educational attainment (high school graduate or more, bachelor's degree or more, and advanced degree or more) by state, in 1990, 2000, and 2009. The values are percentages for people 25 years old and over. These are important bits about the data that you need to know before actually looking at the data.

The "or more" for each column means you can't just add the values from each column because there's overlap between them. If you want to make a pie chart that shows the values of each column, you must do some math. For example, the United States estimate for people with a high school degree (or equivalent) or more is 75.2 percent. Subtract those with a bachelor's degree or more, 20.3 percent, to get rid of the "or more" part of the high school value, which gives you 54.9 percent of people with only a high school degree.

It's also useful to know the sample population. If it were everyone in America, the percentages would be lower, or if for some odd reason the sample was those under 18, the percentages for an advanced degree or more would represent a tiny group of people who skipped or advanced quickly through elementary and high school.

So you have the most important part of any visualization: the data. There are nine columns, spread out over 3 years and three subcategories, plus one more column for state names, so you can visualize the data on multiple dimensions. You might want to focus on educational attainment in 2009, in which case, a few bar charts, as shown in Figure 3-27, could work.

FIGURE 3-24 *(facing page) Chart showing the distributions of deaths, based on United States census of 1870 by Francis A. Walker*

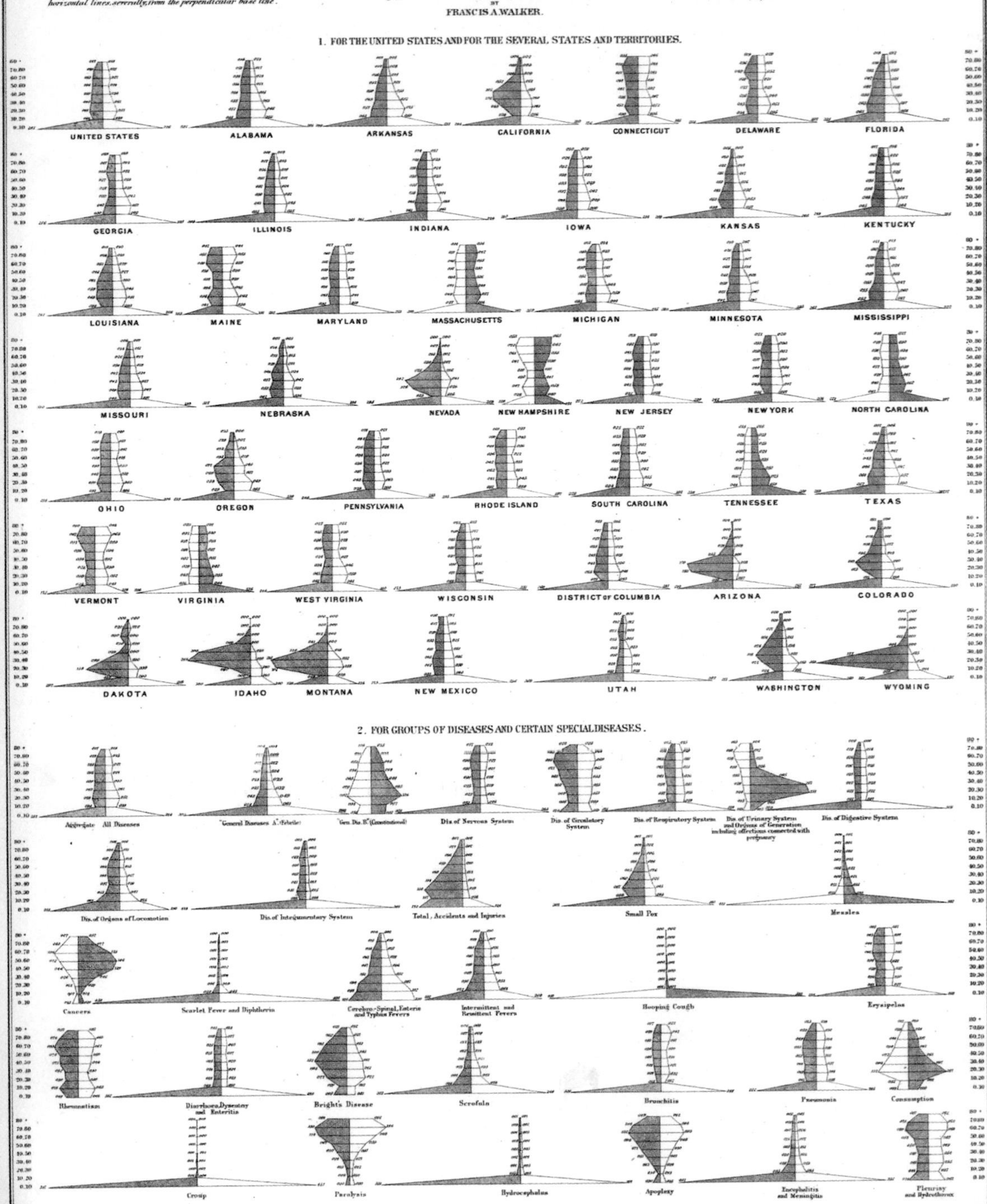

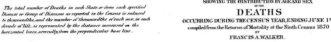

CHART
SHOWING THE DISTRIBUTION BY AGE AND SEX
OF THE
DEATHS
OCCURRING DURING THE CENSUS YEAR, ENDING JUNE 1ST.
compiled from the Returns of Mortality at the Ninth Census 1870
BY
FRANCIS A. WALKER.
The total number of Deaths in each State or from each specified Disease or Group of Diseases as reported in the Census is reduced to thousandths, and the number of thousandths of each sex, in each decade of life, is represented by the distance measured on the horizontal lines, severally, from the perpendicular base line.
The males are on the left of the perpendicular base line and the females on the right.
The lowest horizontal line represents the deaths in the first decade, under ten years of age, and the highest the deaths over eighty years.
The sex which preponderates is shaded.
1. FOR THE UNITED STATES AND FOR THE SEVERAL STATES AND TERRITORIES.
UNITED STATES
ALABAMA
ARKANSAS
CALIFORNIA
CONNECTICUT
DELAWARE
FLORIDA
GEORGIA
ILLINOIS
INDIANA
IOWA
KANSAS
KENTUCKY
LOUISIANA
MAINE
MARYLAND
MASSACHUSETTS
MICHIGAN
MINNESOTA
MISSISSIPPI
MISSOURI
NEBRASKA
NEVADA
NEW HAMPSHIRE
NEW JERSEY
NEW YORK
NORTH CAROLINA
OHIO
OREGON
PENNSYLVANIA
RHODE ISLAND
SOUTH CAROLINA
TENNESSEE
TEXAS
VERMONT
VIRGINIA
WEST VIRGINIA
WISCONSIN
DISTRICT OF COLUMBIA
ARIZONA
COLORADO
DAKOTA
IDAHO
MONTANA
NEW MEXICO
UTAH
WASHINGTON
WYOMING
2. FOR GROUPS OF DISEASES AND CERTAIN SPECIAL DISEASES.
Aggregate All Diseases
"General Diseases A" (Febrile)
"Gen. Dis. B" (Constitutional)
Dis. of Nervous System
Dis. of Circulatory System
Dis. of Respiratory System
Dis. of Urinary System and Organs of Generation including affections connected with pregnancy
Dis. of Digestive System
Dis. of Organs of Locomotion
Dis. of Integumentary System
Total, Accidents and Injuries
Small Pox
Measles
Cancers
Scarlet Fever and Diphtheria
Cerebro-Spinal, Enteric and Typhus Fevers
Intermittent and Remittent Fevers
Hooping Cough
Erysipelas
Rheumatism
Diarrhoea, Dysentery and Enteritis
Bright's Disease
Scrofula
Bronchitis
Pneumonia
Consumption
Croup
Paralysis
Hydrocephalus
Apoplexy
Encephalitis and Meningitis
Pleurisy and Hydrothorax

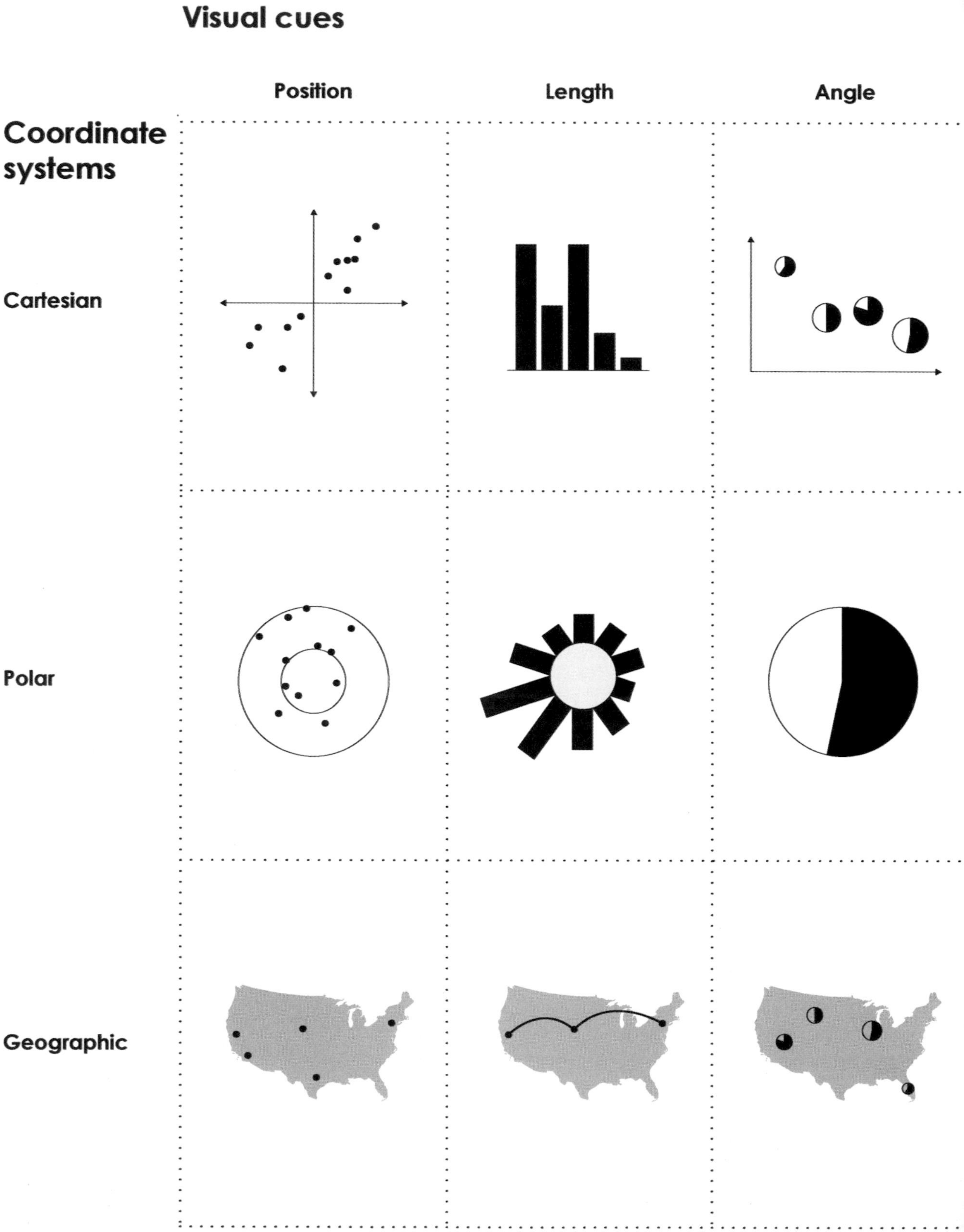

Visual cues

Position

Length

Angle

Coordinate systems

Cartesian

Polar

Geographic

FIGURE 3-25 *Visualization component combinations*

irection	Shapes	Area or Volume	Color

Table 233. Educational Attainment by State: 1990 to 2009

[In percent. 1990 and 2000 as of April. 2009 represents annual averages for calendar year. For persons 25 years old and over. Based on the 1990 and 2000 Census of Population and the 2009 American Community Survey, which includes the household population and the population living in institutions, college dormitories, and other group quarters. See text, Section 1 and Appendix III. For margin of error data, see source]

State	1990			2000			2009		
	High school graduate or more	Bachelor's degree or more	Advanced degree or more	High school graduate or more	Bachelor's degree or more	Advanced degree or more	High school graduate or more	Bachelor's degree or more	Advanced degree or more
United States	**75.2**	**20.3**	**7.2**	**80.4**	**24.4**	**8.9**	**85.3**	**27.9**	**10.3**
Alabama	66.9	15.7	5.5	75.3	19.0	6.9	82.1	22.0	7.7
Alaska	86.6	23.0	8.0	88.3	24.7	8.6	91.4	26.6	9.0
Arizona	78.7	20.3	7.0	81.0	23.5	8.4	84.2	25.6	9.3
Arkansas	66.3	13.3	4.5	75.3	16.7	5.7	82.4	18.9	6.1
California	76.2	23.4	8.1	76.8	26.6	9.5	80.6	29.9	10.7
Colorado	84.4	27.0	9.0	86.9	32.7	11.1	89.3	35.9	12.7
Connecticut	79.2	27.2	11.0	84.0	31.4	13.3	88.6	35.6	15.5
Delaware	77.5	21.4	7.7	82.6	25.0	9.4	87.4	28.7	11.4
District of Columbia	73.1	33.3	17.2	77.8	39.1	21.0	87.1	48.5	28.0
Florida	74.4	18.3	6.3	79.9	22.3	8.1	85.3	25.3	9.0
Georgia	70.9	19.3	6.4	78.6	24.3	8.3	83.9	27.5	9.9
Hawaii	80.1	22.9	7.1	84.6	26.2	8.4	90.4	29.6	9.9
Idaho	79.7	17.7	5.3	84.7	21.7	6.8	88.4	23.9	7.5
Illinois	76.2	21.0	7.5	81.4	26.1	9.5	86.4	30.6	11.7
Indiana	75.6	15.6	6.4	82.1	19.4	7.2	86.6	22.5	8.1
Iowa	80.1	16.9	5.2	86.1	21.2	6.5	90.5	25.1	7.4
Kansas	81.3	21.1	7.0	86.0	25.8	8.7	89.7	29.5	10.2
Kentucky	64.6	13.6	5.5	74.1	17.1	6.9	81.7	21.0	8.5
Louisiana	68.3	16.1	5.6	74.8	18.7	6.5	82.2	21.4	6.9
Maine	78.8	18.8	6.1	85.4	22.9	7.9	90.2	26.9	9.6
Maryland	78.4	26.5	10.9	83.8	31.4	13.4	88.2	35.7	16.0
Massachusetts	80.0	27.2	10.6	84.8	33.2	13.7	89.0	38.2	16.4
Michigan	76.8	17.4	6.4	83.4	21.8	8.1	87.9	24.6	9.4
Minnesota	82.4	21.8	6.3	87.9	27.4	8.3	91.5	31.5	10.3
Mississippi	64.3	14.7	5.1	72.9	16.9	5.8	80.4	19.6	7.1
Missouri	73.9	17.8	6.1	81.3	21.6	7.6	86.8	25.2	9.5
Montana	81.0	19.8	5.7	87.2	24.4	7.2	90.8	27.4	8.3
Nebraska	81.8	18.9	5.9	86.6	23.7	7.3	89.8	27.4	8.8
Nevada	78.8	15.3	5.2	80.7	18.2	6.1	83.9	21.8	7.6
New Hampshire	82.2	24.4	7.9	87.4	28.7	10.0	91.3	32.0	11.2
New Jersey	76.7	24.9	8.8	82.1	29.8	11.0	87.4	34.5	12.9
New Mexico	75.1	20.4	8.3	78.9	23.5	9.8	82.8	25.3	10.4
New York	74.8	23.1	9.9	79.1	27.4	11.8	84.7	32.4	14.0
North Carolina	70.0	17.4	5.4	78.1	22.5	7.2	84.3	26.5	8.8
North Dakota	76.7	18.1	4.5	83.9	22.0	5.5	90.1	25.8	6.7
Ohio	75.7	17.0	5.9	83.0	21.1	7.4	87.6	24.1	8.8
Oklahoma	74.6	17.8	6.0	80.6	20.3	6.8	85.6	22.7	7.4
Oregon	81.5	20.6	7.0	85.1	25.1	8.7	89.1	29.2	10.4
Pennsylvania	74.7	17.9	6.6	81.9	22.4	8.4	87.9	26.4	10.2
Rhode Island	72.0	21.3	7.8	78.0	25.6	9.7	84.7	30.5	11.7
South Carolina	68.3	16.6	5.4	76.3	20.4	6.9	83.6	24.3	8.4
South Dakota	77.1	17.2	4.9	84.6	21.5	6.0	89.9	25.1	7.3
Tennessee	67.1	16.0	5.4	75.9	19.6	6.8	83.1	23.0	7.9
Texas	72.1	20.3	6.5	75.7	23.2	7.6	79.9	25.5	8.5
Utah	85.1	22.3	6.8	87.7	26.1	8.3	90.4	28.5	9.1
Vermont	80.8	24.3	8.9	86.4	29.4	11.1	91.0	33.1	13.3
Virginia	75.2	24.5	9.1	81.5	29.5	11.6	86.6	34.0	14.1
Washington	83.8	22.9	7.0	87.1	27.7	9.3	89.7	31.0	11.1
West Virginia	66.0	12.3	4.8	75.2	14.8	5.9	82.8	17.3	6.7
Wisconsin	78.6	17.7	5.6	85.1	22.4	7.2	89.8	25.7	8.4
Wyoming	83.0	18.8	5.7	87.9	21.9	7.0	91.8	23.8	7.9

Source: U.S. Census Bureau, 1990 Census of Population, CPH-L-96; 2000 Census of Population, P37. "Sex by Educational Attainment for the Population 25 Years and Over"; 2009 American Community Survey, R1501, "Percent of Persons 25 Years and Over Who Have Completed High School (Includes Equivalency)," R1502, "Percent of Persons 25 Years and Over Who Have Completed a Bachelor's Degree," and R1503, "Percent of Persons 25 Years and Over Who Have Completed an Advanced Degree," <http://factfinder.census.gov/>, accessed February 2011.

FIGURE 3-26 *Data table on educational attainment in the United States*

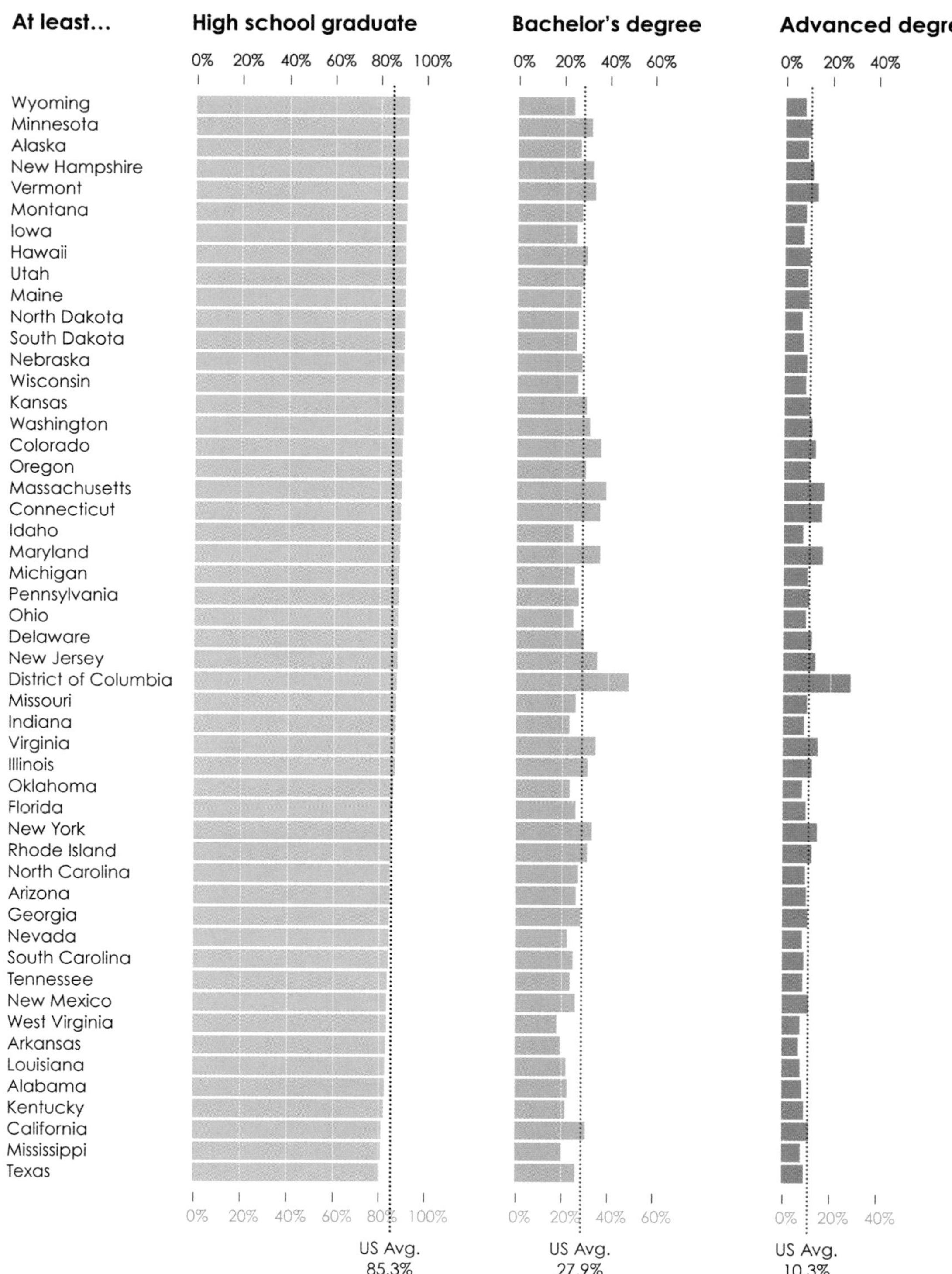

FIGURE 3-27 *Bar charts on educational attainment*

This is practically a direct translation of the last three columns in the table. Each row represents the values for a state, and each column is a level of attainment. Each bar chart has its own linear scale, but the increments are spaced equally and start at zero percent. States are sorted by estimated percent of people with a high school diploma or equivalent, in descending order, rather than alphabetically, like in the table. Instead of giving the national average its own row, it's presented as a vertical dotted line to provide a sense of low and high. Color hue—gray, light blue, and blue—is used to indicate three separate estimates.

Break it down. That's length (bars), color (each bar chart), and position (lines for national averages) as visual cues, a Cartesian coordinate system, linear scales for each of the bar charts, and a categorical scale for the sorted states. The title and subtitles provide context for what the data is about.

If you are more interested in the changes between 2000 and 2009 than you are just the 2009 percentages, Figure 3-28 shows a few options that shift focus. Length and position are still used, as well as a linear scale on the horizontal axis and a categorical scale on the vertical. However, the context and layout are different than the bar charts. Some other visual cues are also incorporated.

An open circle represents the high school attainment in 2000 for each state, and the solid circles represent the same for 2009. The dots are placed in the same position vertically, and a line is used to connect the two dots. The longer the line is, the greater the change, by percentage points, was from 2000 to 2009.

The shift from open circle to closed circle provides a sense of direction. In this example, high school attainment in all states improved, so your eyes always shift from left to right, but if attainment decreased in one of the states, you could use the same visual cue. For example, if there were a decrease from 80 percent to 70 percent, the solid dot would be on the left of the open one. You can also use arrows if you want to highlight direction more prominently. All states showed increases in this example, though, so a focus on the magnitude of the changes and the values of the endpoints was more appropriate.

You can see how a change in sorting can shift focus. States are sorted alphabetically in the first chart, and the lack of visual order makes it more challenging to make comparisons. You can see the increases and it's easy to find a state of interest, but as an overall picture, you don't get much.

In contrast, the second chart shows the same data ordered instead by the highest percentage of attainment in 2009. It starts with Wyoming and goes down to Texas. This focuses on the more recent estimates, whereas still making it easy to pick out the values for 2000 because generally speaking, states with higher percentages in 2009 were higher in the rankings in 2000, too. That said, you can also sort by the 2000 estimates and move the labels to the left to shift focus in this direction.

Finally, the chart on the far right introduces color as a visual cue. This is the same as the second chart that sorts by 2009 estimates, but color is used to highlight states that increase the most by percentage. The District of Columbia, which albeit isn't a state, had the greatest percentage increase, so it is shown in black. The lower the increase, the lighter the states are shown. States in between are shown with varying shades of green. So if you look at the individual components of this chart, you get length, position, direction, and color used as visual cues; it uses a Cartesian coordinate system; and a linear numeric scale is used on the horizontal, with a categorical scale on the vertical.

You don't have to stop here. As shown in Figure 3-29, position and direction can be used differently to show the increases from 2000 to 2009. Unlike the previous charts, states are plotted on a linear scale that represents high school attainment instead of on a categorical scale. Values are categorized by year on the horizontal. This is essentially a couple of ticks on a time series plot. If you were to show years in between, there would be more than two categories on the horizontal axis. In any case, like in a time series plot, a greater slope from point to point means a greater rate of change.

The chart on the right uses the same geometry as the one on the left, and uses color to represent regions in the United States. So although you see improvement with all states, you also see a lot of the states in the South toward the bottom of the scale and Midwest and West states more toward the top. Although, as is usually the case with real data, there are exceptions, such as California in the West that is toward the bottom and Maryland that is in the South is higher up.

Generally speaking though, the higher the attainment in 2000, the higher the attainment was in 2009. This is obvious in Figure 3-30, which uses position as a visual cue and linear scales on both axes.

FIGURE 3-28 *(following page) Change in high school educational attainment between 2000 and 2009*

Alphabetical

Values look scattered

2000 2009

Alabama
Alaska
Arizona
Arkansas
California
Colorado
Connecticut
Delaware
District of Columbia
Florida
Georgia
Hawaii
Idaho
Illinois
Indiana
Iowa
Kansas
Kentucky
Louisiana
Maine
Maryland
Massachusetts
Michigan
Minnesota
Mississippi
Missouri
Montana
Nebraska
Nevada
New Hampshire
New Jersey
New Mexico
New York
North Carolina
North Dakota
Ohio
Oklahoma
Oregon
Pennsylvania
Rhode Island
South Carolina
South Dakota
Tennessee
Texas
Utah
Vermont
Virginia
Washington
West Virginia
Wisconsin
Wyoming

70% 80% 90%

High school or more

Greatest to least, endpoint

Focus on most recent numbers

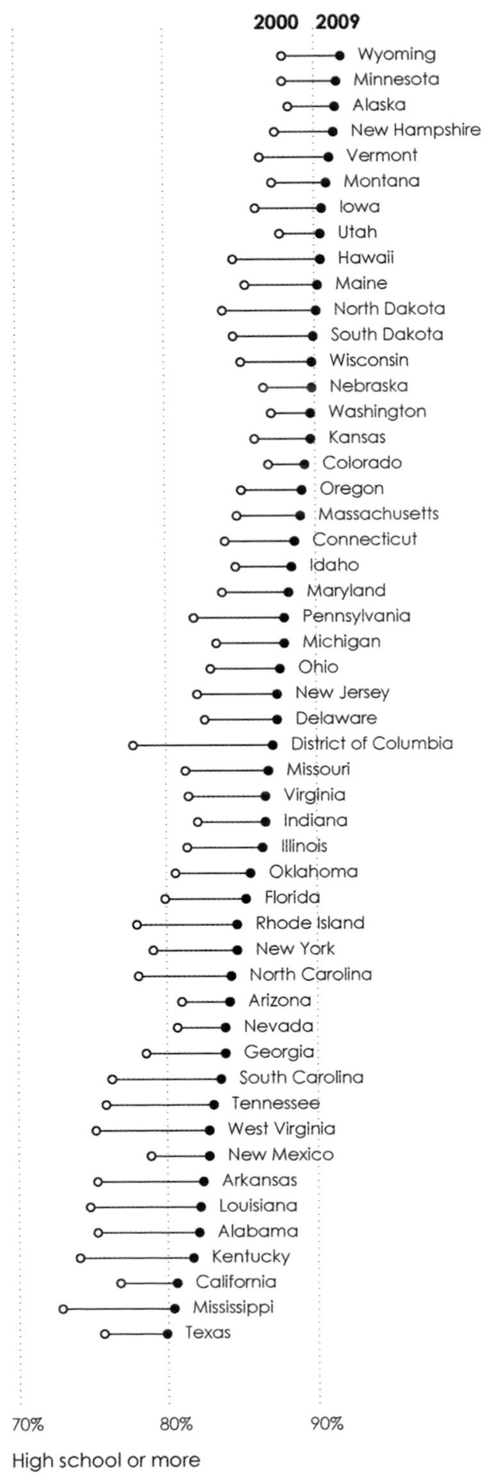

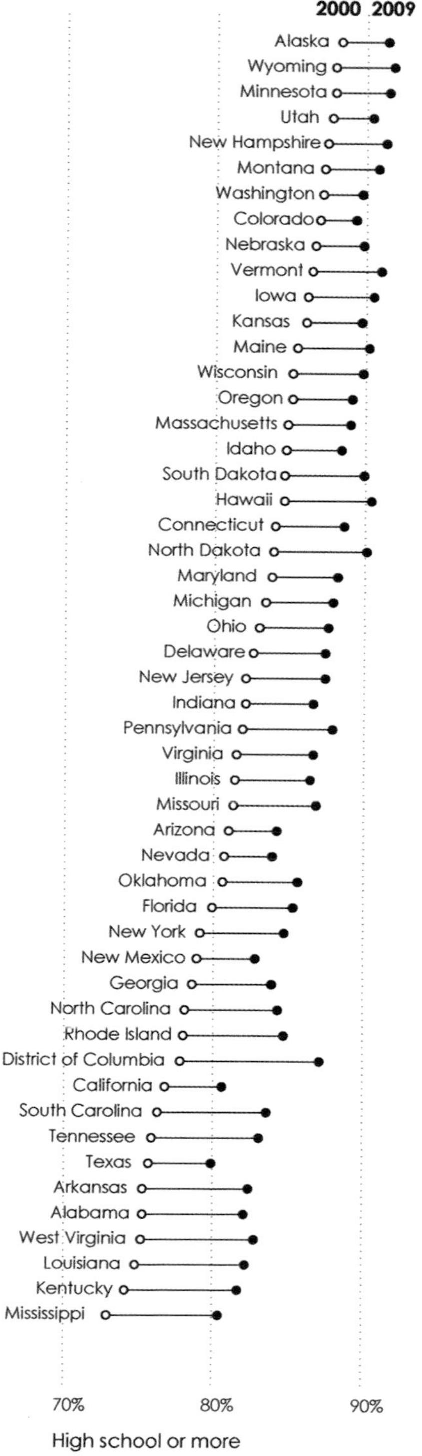
Greatest to least, startpoint
Focus on past data
2000
2009
Alaska
Wyoming
Minnesota
Utah
New Hampshire
Montana
Washington
Colorado
Nebraska
Vermont
Iowa
Kansas
Maine
Wisconsin
Oregon
Massachusetts
Idaho
South Dakota
Hawaii
Connecticut
North Dakota
Maryland
Michigan
Ohio
Delaware
New Jersey
Indiana
Pennsylvania
Virginia
Illinois
Missouri
Arizona
Nevada
Oklahoma
Florida
New York
New Mexico
Georgia
North Carolina
Rhode Island
District of Columbia
California
South Carolina
Tennessee
Texas
Arkansas
Alabama
West Virginia
Louisiana
Kentucky
Mississippi
70%
80%
90%
High school or more

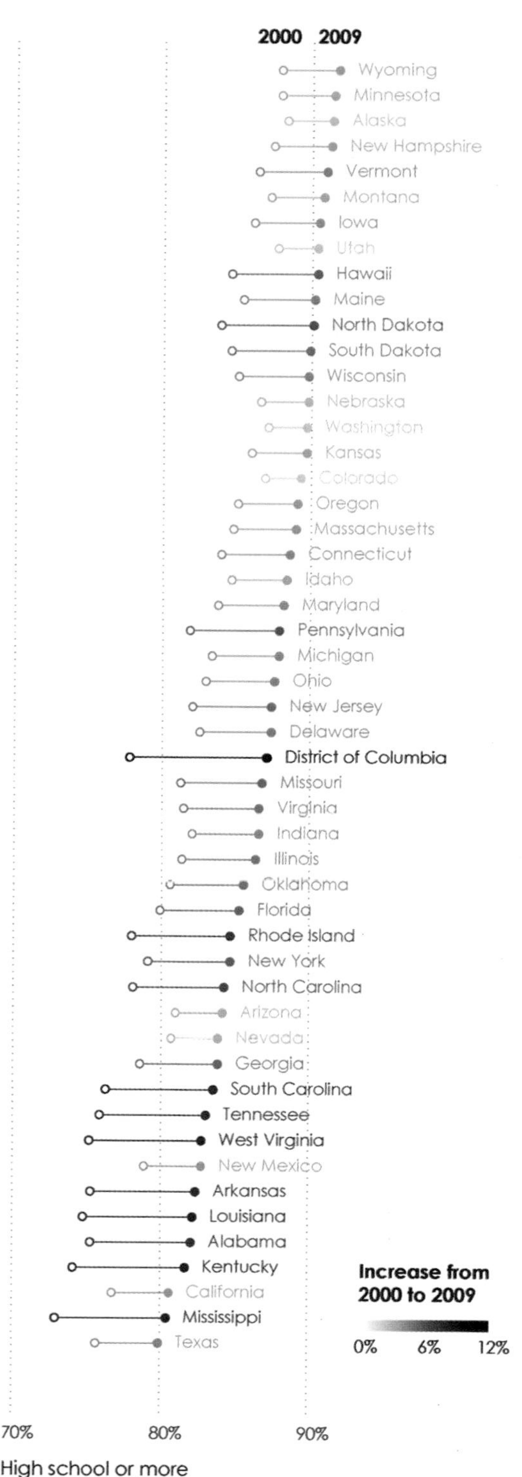
Greatest to least, endpoint + Color
Reference to percent increase
2000
2009
Wyoming
Minnesota
Alaska
New Hampshire
Vermont
Montana
Iowa
Utah
Hawaii
Maine
North Dakota
South Dakota
Wisconsin
Nebraska
Washington
Kansas
Colorado
Oregon
Massachusetts
Connecticut
Idaho
Maryland
Pennsylvania
Michigan
Ohio
New Jersey
Delaware
District of Columbia
Missouri
Virginia
Indiana
Illinois
Oklahoma
Florida
Rhode Island
New York
North Carolina
Arizona
Nevada
Georgia
South Carolina
Tennessee
West Virginia
New Mexico
Arkansas
Louisiana
Alabama
Kentucky
California
Mississippi
Texas
Increase from 2000 to 2009
0%
6%
12%
70%
80%
90%
High school or more

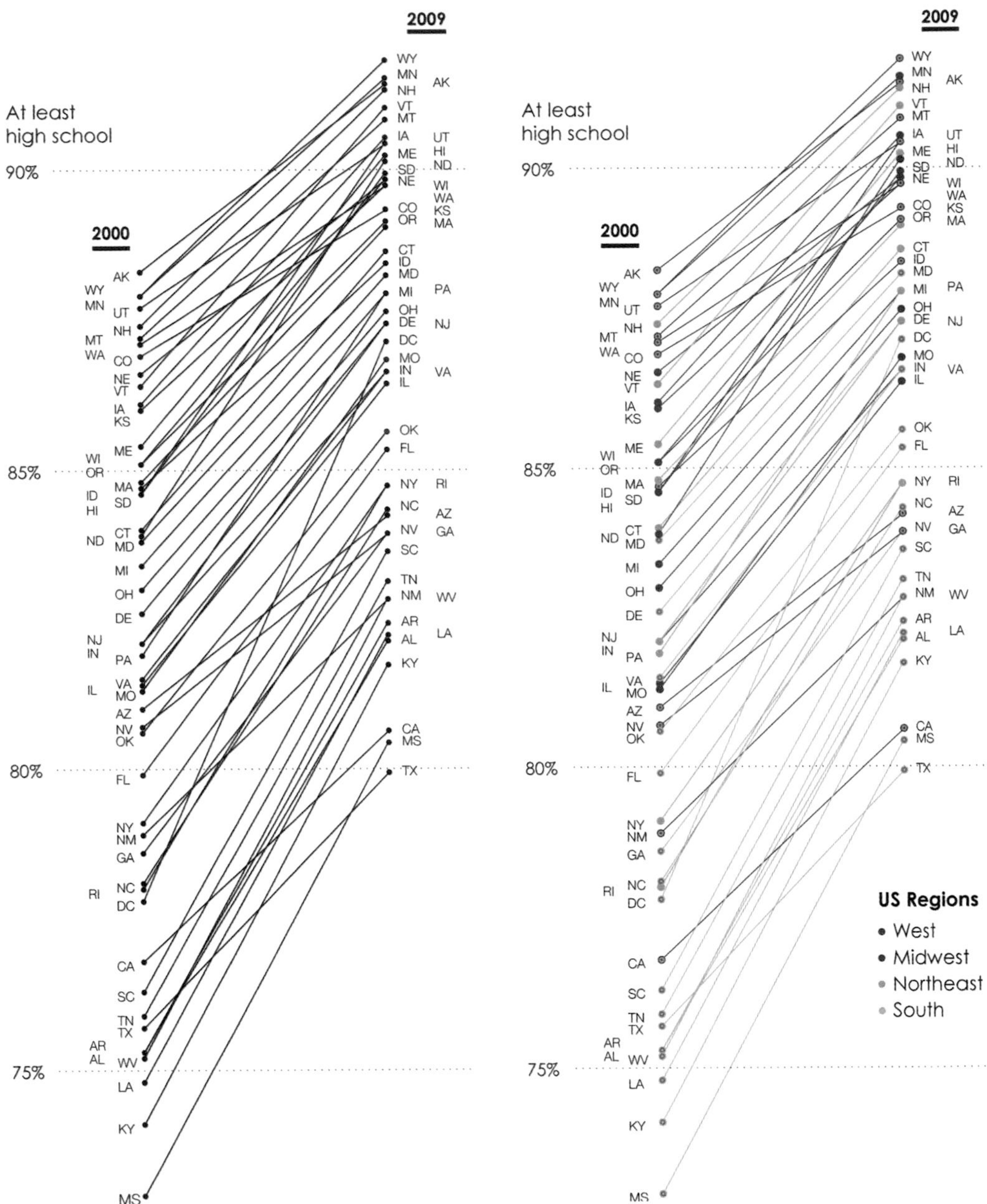

FIGURE 3-29 *Using position and direction*

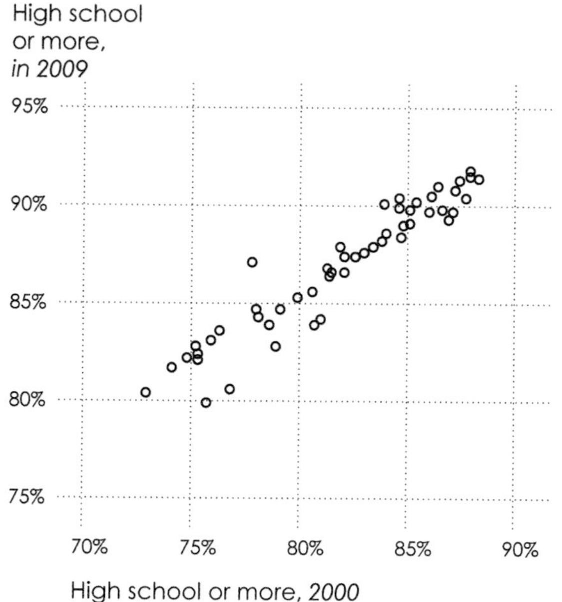

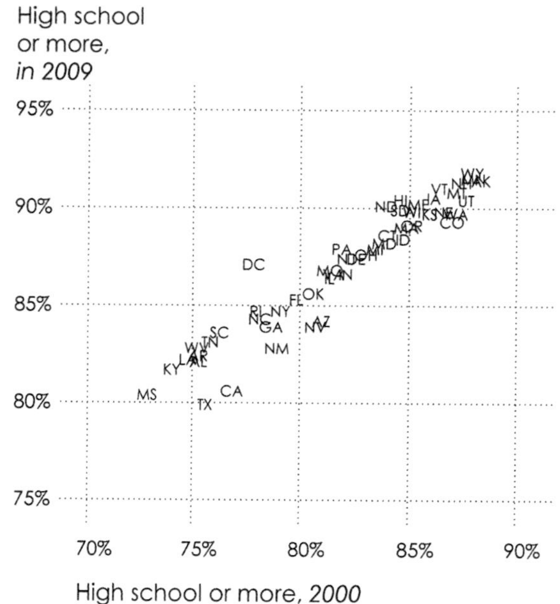

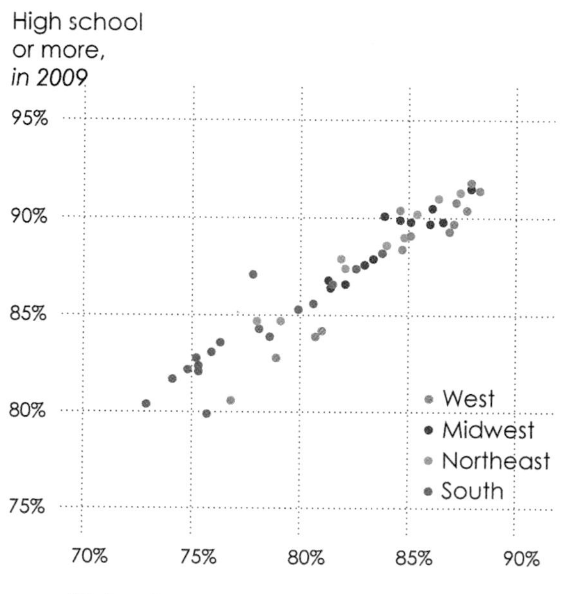

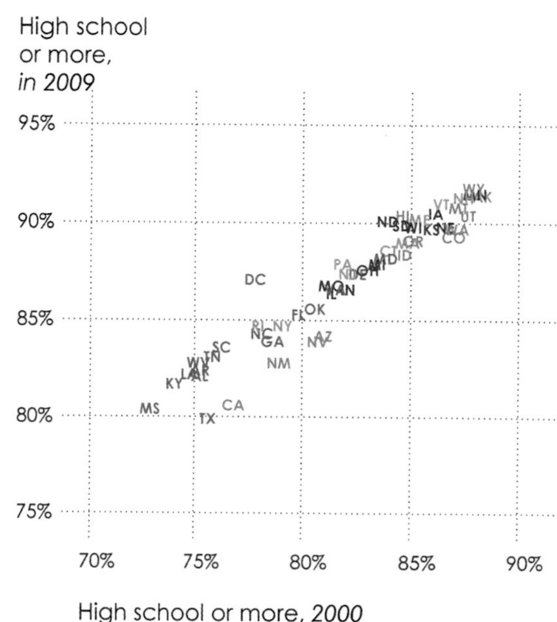

FIGURE 3-30 *Position, symbols, and colors in scatterplots*

High school attainment in 2000 is plotted on the horizontal axis, and attainment in 2009 is on the vertical. There is an obvious upward trend, and you can spot Washington, DC sticking out somewhat, indicating the higher rate of improvement (and probably difference in demographics). You can also see Texas and California lagging around the bottom-left corner. As shown in previous charts—and I'm sure you're getting the hang of it now—you can incorporate other visual cues such as color, symbols, or both to provide additional dimensions of information.

Note: If you want to annoy cartographers, you can also call choropleth maps heat maps, as they are often referred to. The heat map was created to visualize 2-D data, and choropleth maps are the geographic equivalent. I personally keep the terms and methods separate.

Remember this is geographic data, so you must map it, right? (Actually, just because location is attached to your data, which seems like almost always these days, a map is not always the most useful view, which is discussed in the next chapter.) Figure 3-31 shows a handful of maps with states colored using varying scales and metrics, which are called choropleth maps.

Note that although each map uses the same method, the choice of scale can change the map's focus and message. For example, the map on the top left uses a quartile scale, which means the states were split into four even groups based on a metric. In this case, the metric is the percentage of people with a bachelor's degree in 2009. This makes a map with colors that are evenly distributed.

However, the map that shows the same data on a linear scale, with just three shades of green, shows darker shades in the Midwest and Northeast regions. Compare this with the quartile map, and you still get the lighter areas in the South, but the rest of the map tells a different story. Likewise, you can further abstract the data by coloring states by whether they are below or above the average (top right) or whether percentages increased or decreased (bottom right).

As shown in Figure 3-32, you can also show several maps at once to see how something has changed geographically over time. Since you've looked the data from several perspectives already, you know that a high value in 2000 generally means a higher value in 2009, because the states improved at similar rates.

You see about the same thing when you compare 1990 to 2000. In 1990, you see a more lightly colored map, where several states showed 15 percent or less of people 25 years or older with a bachelor's degree. Only Wyoming, which had the highest percentage in 2009, shows a percentage higher than 25 percent. As you move left to right, the map gets darker, like you'd expect.

Varying scales

Choice of scale can shift focus and present a different message. The below maps represent how a single dataset can easily change based on this choice.

Quartiles

Breaks decided by splitting into four equally-sized groups

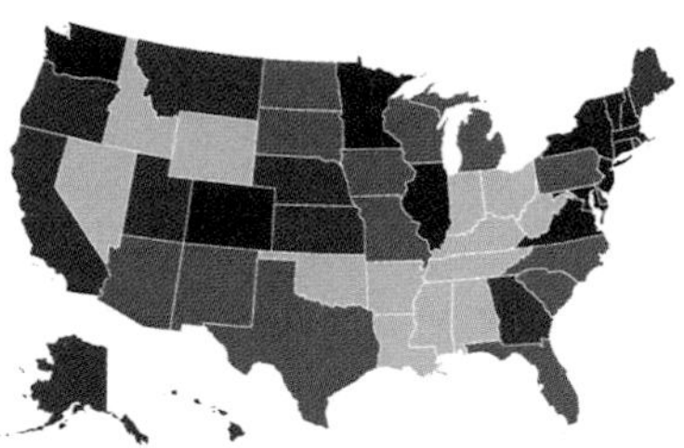

% with at least Bachelor's in 2009

- Greater than 30.6%
- 26.6% – 30.6%
- 24.2% – 26.5%
- Less than 24.2%

Linear

Scale incremented evenly over range

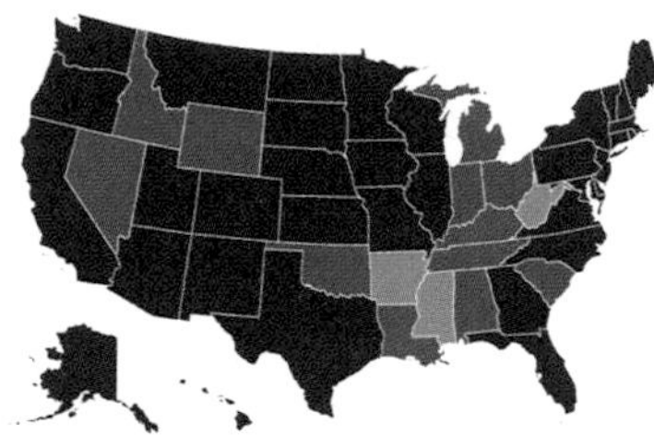

% with at least Bachelor's in 2009

20% >25%

15% 25%

Numeric category

Create category based on a metric in data

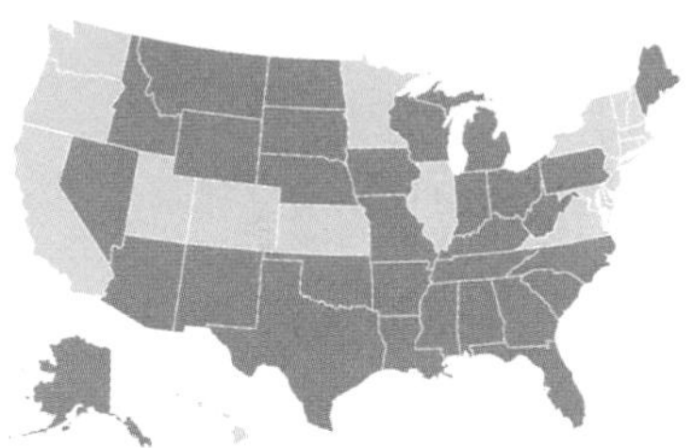

2009 US Avg. of 27.9%

Below avg. Above avg.

Categorical

Groups based on metadata, such as region

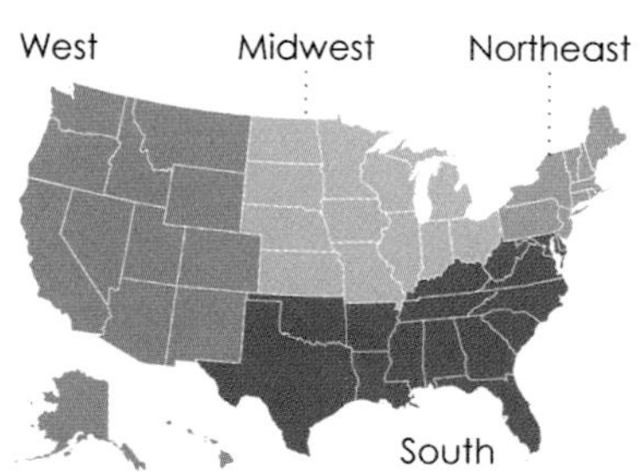

Difference

A linear scale, but based on percent change between years

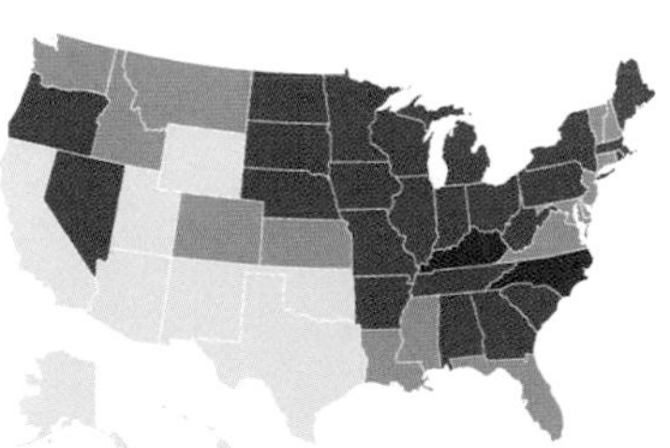

% change from 1990 to 2009

30% 50%

<30% 40% >50%

Categorical difference

Simple split based on increase or decrease (Good news: all increase in this example)

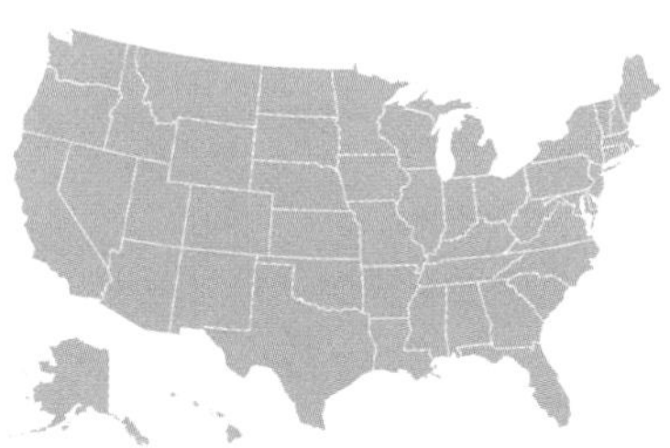

Change from 1990 to 2009

Decrease Increase

FIGURE 3-31 *Choropleth maps*

Geographic coordinates + Time scale + Color

In 1990, about 20 percent of 25+ year olds had at least a bachelor's degree.

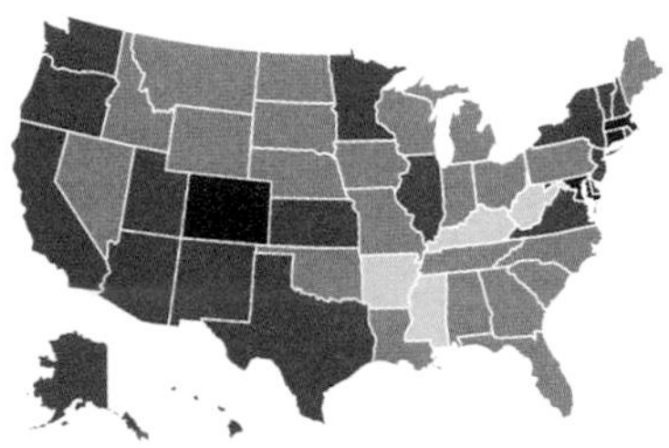

In 2000, the percentage was up to 24.

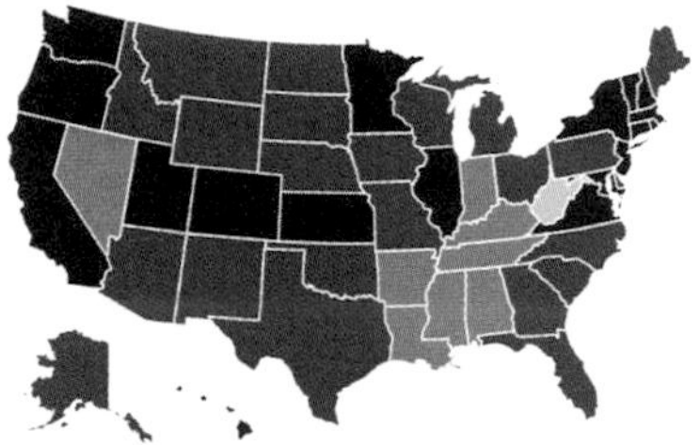

In 2009, the US average was up to 29 percent.

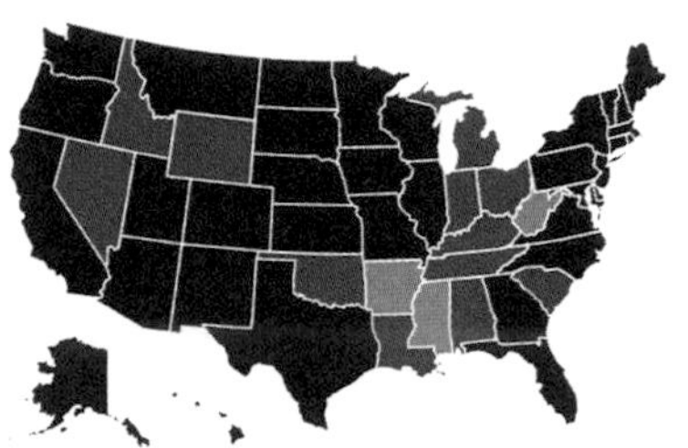

Bachelor's degree or more

15% or less | 20% | More than 25%

FIGURE 3-32 *Maps over time*

WRAPPING UP

At its core, visualization is an abstraction process to map data to color and geometry. This is easy to do from a technical perspective. You can easily draw and color shapes with a pencil and paper. The challenge is to figure out what shapes and colors work best, where to put them, and how to size them.

To make the jump from data to visualization, you must know your ingredients. A skilled chef doesn't just blindly throw ingredients in a pot, turn the stove on high, and hope for the best. Instead, the chef gets to know how each ingredient works together, which ones don't get along, and how long and at what temperature to cook these ingredients.

With visualization, visual cues, coordinate systems, scales, and context are your ingredients. Visual cues are the main thing that people see, and the coordinate system and scale provide structure and a sense of space. Context breathes life into the data and makes it understandable, relatable, and worth looking at.

Get to know how the components work, play with them, and get other people to look at your results and see what information they extract.

Don't forget the main component of every visualization, though. Without data, you have nothing to visualize. Likewise, if you have data with little substance, you get visualization with little substance. However, when you do get data that offers a high number of dimensions or is granular enough to see the interesting details, you still must know what to look for.

The challenge of more data is that you have more visualization options, and many of those options will be poor ones. To filter out the bad and find the worthwhile options—to get to visualization that means something—you must get to know your data. Now on to exploration.

Exploring Data Visually

In 1977, statistician John Tukey published his book Exploratory Data Analysis, which detailed how and encouraged data professionals to analyze data through visualization. This was during a time when most analysis was performed in the context of hypothesis tests and statistical models, one computer filled a room, and graphs were typically drawn by hand. For example, in his book, Tukey provides a tip on how to draw darker symbols with a pen instead of a pencil.

Nevertheless, although the technology was bigger and slower back then, the driving principle is the same. You can see a lot in a picture, and what you see can lead to answers or generate more questions you otherwise never would have thought of.

> *"The greatest value of a picture is when it forces us to notice what we never expected to see."*
>
> *—John W. Tukey, Exploratory Data Analysis (1977)*

The public-facing side of visualization—the polished graphics that you see in the news, on websites, and in books—are fine examples of data graphics at their best, but what is the process to get to that final picture? There is an exploration phase that most people never see, but it can lead to visualization that is a level above the work of those who do not look closely at their data. The better that you understand what your data is about, the better you can communicate your findings.

Note: *The New York Times* and *The Washington Post* discuss the process behind their graphics at http://chartsnthings.tumblr.com/ and http://postgraphics.tumblr.com/, respectively. Work often starts with rough sketches on paper or dry erase board and then moves to exploration and production.

Even if you don't plan to show your results to a wide audience, visualization as an analysis tool enables you to explore data and find stories that you might not find with formal statistical tests. You just need to know what to look for and what questions to ask based on the data that you have available.

The great thing is that tools and access to data are less of a limiting factor than they were in Tukey's time, so you aren't stuck with just pencils, paper, and a ruler to draw thousands of dots and lines.

PROCESS

The specific steps you take in any analysis varies by dataset and project, but generally speaking, you should consider these four things when you explore your data visually.

- What data do you have?
- What do you want to know about your data?

- What visualization methods should you use?
- What do you see and does it makes sense?

The answer to each question depends on the answers that come before, and it's common to jump back and forth between questions. As shown in Figure 4-1, it's an iterative process. For example, if your dataset is only a handful of observations, this limits what you can find in your data and what visualization methods are useful, and you won't see much.

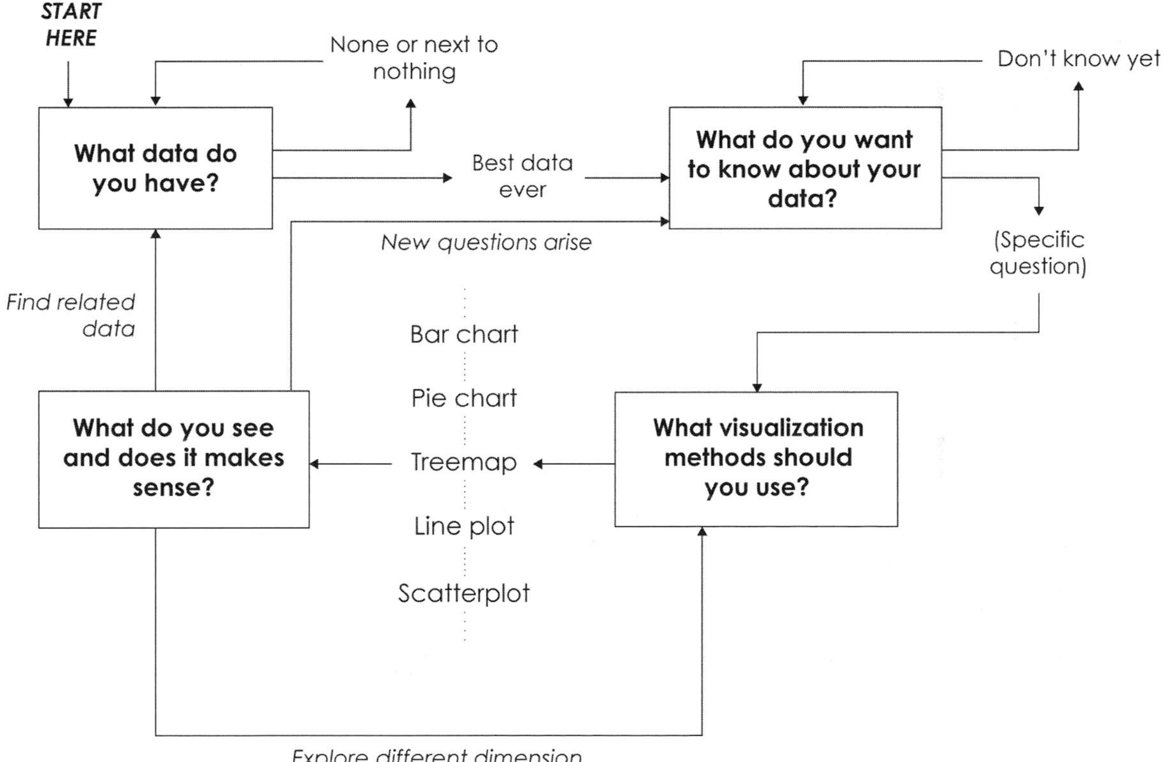

FIGURE 4-1 *The iterative data exploration process*

On the other hand, if you have a lot of data, what you see when you visualize one aspect of it can lead to a curiosity about other dimensions, which in turn leads to different graphics. This is the fun part.

WHAT DATA DO YOU HAVE?

People often form a picture in their head of what a visualization should look like or find an example that they want to mimic. The excitement is great, but when it's time to visualize, they realize they either need more data or their data doesn't work with the chart they want to make.

The common mistake is to form a visual first and get the data later. It should be the other way around—data first and visualization follows.

Often, getting the data that you need is the hardest and most time-consuming part of the visualization process. In school, data is handed to you formatted the way you need so that you can easily load it into the software of choice, but this is hardly ever the case in practice. You might need to scrape data from a website, access an API, or derive values from existing data.

Note: I sometimes spend most of the time gathering data and little time visualizing it. Don't be surprised if you have to do the same. Totally normal.

For example, you might have a list of addresses, but to map them, you need latitude and longitude coordinates. Or you have observations for individuals of a population, but you might be more interested in subpopulations.

Programming can be helpful in this case to automate parts of the process, but there are a growing number of click-and-play applications to manage data, too.

Note: See Chapter 7, "Where to Go from Here," for tools and resources to work with and visualize data.

When you have data you want to explore, pause for a second to consider what values represent, where the data is from, and how variables were measured. Essentially, apply everything you learned in Chapter 1, "Understanding Data."

WHAT DO YOU WANT TO KNOW ABOUT YOUR DATA?

Imagine you have some data to explore. Where do you begin? The answer is easy if you have only one data point. You can just read the value, and most of your findings, if any, will come from outside information and additional data. On the other hand, when you have a dataset with thousands or millions of observations—think spreadsheet with a lot of rows—it can be challenging and often intimidating to figure out what to look at first.

This is where the phrase "drowning in data" comes from. You stare at a bunch of numbers on your computer screen, and values start to blur together the longer you stare. Soon all you see is a blob of data that feels suffocating, but wait; there's hope. Take a step back. Breathe.

To avoid drowning in data, you learn to swim. When you learn to swim, you start at the shallow end and work your way toward the deep end. If you're more adventurous, you snorkel or go deep-sea diving. Even then you don't swim the entire ocean. You explore a little bit at a time, and what you learn during one dive carries over to the next. People drown in the ocean, but when you drown in data, you still get more chances to learn and try again.

Note: Early exploration of a dataset can be overwhelming, because you don't know where to start. Ask questions about the data and let your curiosities guide you.

To start, ask yourself what you want to know about the data. Your answer doesn't need to be complex or profound. Just make it less vague than, "I want to know what the data looks like." The more specific you are the more direction you get. Maybe you want to know the best or worst thing (such as a country, sports team, or school) in your data, so you explore rankings, and if you have multiple variables, you decide what makes something good and something bad. If you have time series data, you might want to know if something has improved or gotten worse over the past decade.

For example, journalist Tim De Chant explores world population densities, as shown in Figure 4-2, and was curious how large a city might be if everyone who lived in the world had the same amount of space. A straightforward method could be to directly map population density around the world, but De Chant put it into a more relatable perspective.

When you ask questions about your data, you give yourself a place to start, and if you're lucky, as you investigate, you'll come up with more questions, and then you dig into those. Coming up with and answering potential questions a user might have while you explore also provides focus and purpose, and helps farther along in the design process when you make graphics for a wider audience.

WHAT VISUALIZATION METHODS SHOULD YOU USE?

As you saw in the previous chapter, there are many chart options and combinations of visual cues to choose from. It's easy to obsess over picking the right chart for your data, but during the early stages of exploration, it's more important to see your data from different angles and to drill down to what matters for your project.

Make multiple charts, compare all your variables, and see if there are interesting bits that are worth a closer look. Look at your data as a whole and then zoom in on categories and individual data points.

THE WORLD'S POPULATION, CONCENTRATED

If the world's 6.9 billion people lived in one city, how large would that city be if it were as dense as...

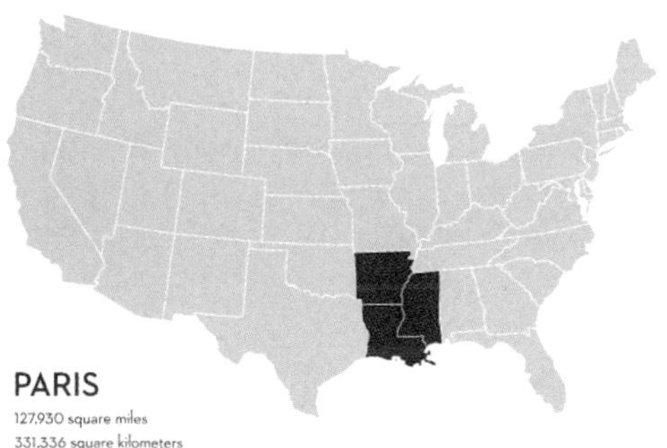

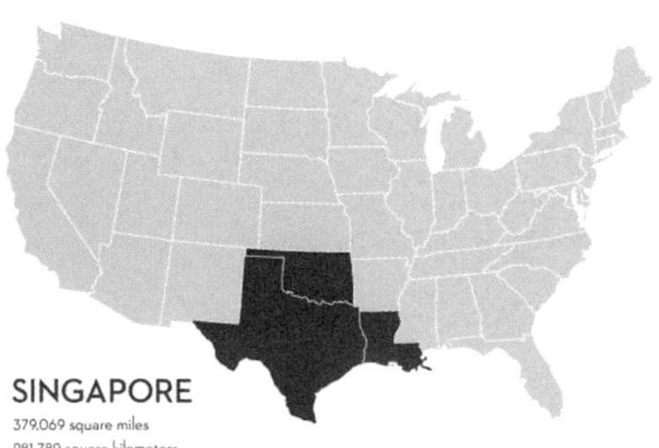

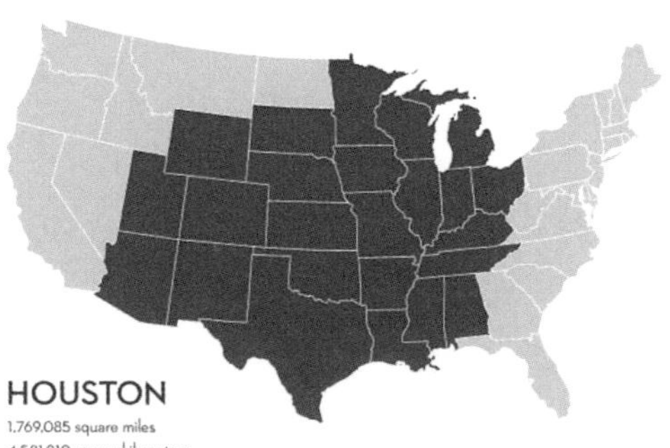

FIGURE 4-2 The World's Population, *Concentrated (2011) by Tim De Chant, http://persquaremile.com/*

This is also a great time to experiment with visual forms. Try different scales, colors, shapes, sizes, and geometries, and you might stumble upon a graphic worth pursuing further. You don't always need to stick to the visual cues that are the "best" at showing data most accurately and are easiest to read. When exploration is your goal, don't let a list of best practices stop you from trying something different because complex data often requires complex visualization.

Note: Ben Shneiderman, a professor of computer science and inventor of the treemap, is often quoted for, "Overview first, zoom and filter, then details-on-demand" in his paper "The Eyes Have It."

For example, Figure 4-3 shows an interactive exploration of article deletions on Wikipedia by Mortiz Stefaner, Dario Taraborelli, and Giovanni Luca Ciampaglia. Wikipedia is a large resource of data with small and large data tables within articles, article edits over time, and user interaction with articles and between each other. The data can be explored on many dimensions, but the topic focus of Notabilia shows a clearer picture.

Note: A common misconception is that you must understand a graphic in under 10 seconds. Relationships and patterns aren't always straightforward, so just because a visualization takes a few minutes to understand doesn't make it a failed attempt.

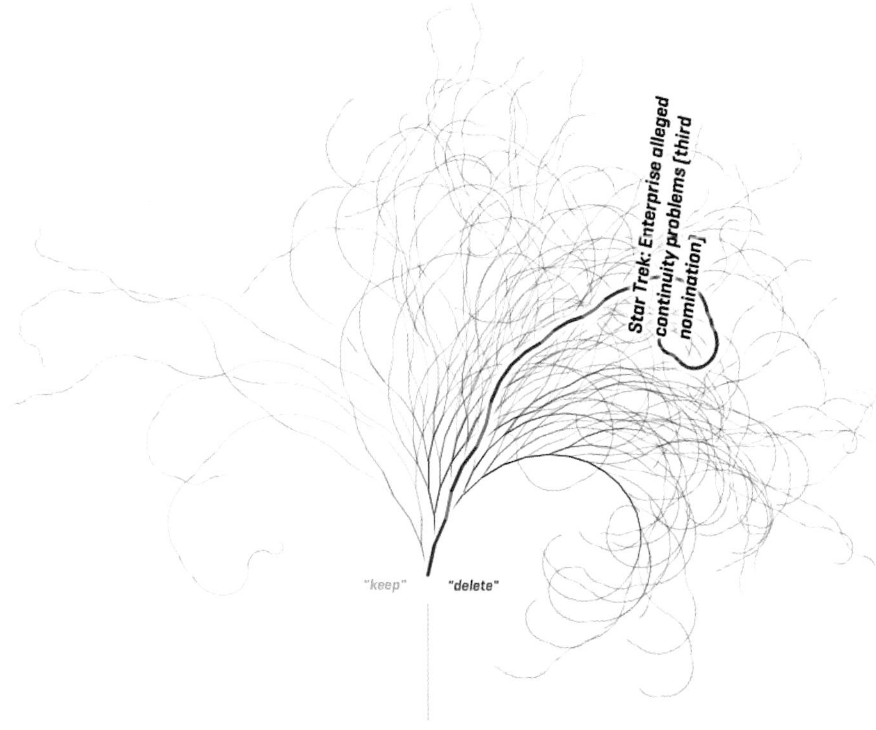

FIGURE 4-3 Notabilia *(2011) by Mortiz Stefaner, Dario Taraborelli, and Giovanni Luca Ciampaglia, http://notabilia.net/*

Each branch represents a user discussion about whether an article should be deleted, and those that curl to the right are discussions that lean strongly for deletion. A curl to the left is a discussion leaning toward keeping an article. The more prominent the curl is, the stronger the agreement between users. Although the visualization isn't traditional, you get still get something out of it.

That said, traditional visualization, such as bar graphs and line charts, can be made easily and read quickly, which makes them great tools to explore data.

As your goals shift, so do your choices of visualization. If you were to design a dashboard that provides the status of a system at a glance, you must visualize the data in a way that is straightforward to digest. On the other hand, if the goal is to encourage reflection or to evoke emotions, efficiency might not be your main concern.

WHAT DO YOU SEE AND DOES IT MAKE SENSE?

After you visualize your data, there are certain things to look for, as shown in Figure 4-4: increasing, decreasing, outliers, or some mix, and of course, be sure you're not mixing up noise for patterns.

Also note how much of a change there is and how prominent the patterns are. How does the difference compare to the randomness in the data? Observations can stand out because of human or mechanical error, because of the uncertainty of estimated values, or because there was a person or thing that stood out from the rest. You should know which it is.

When you find something interesting, ask yourself: Does it make sense? Why does it make sense? This is massively important.

The tendency is to automatically think of data as fact because numbers can't possibly waver. But again, there's uncertainty attached to the data because each data point is a snapshot of what happened during a moment in time. You infer everything else.

Note: Inference and uncertainty: This is what statistics is all about. If you can, take a statistics course. Although you can learn a lot from visual exploration alone, traditional statistics can help you examine data in greater detail.

In the rest of this chapter, you look at specific data types more closely. Keep the process in mind as you make your way through.

VISUALIZING CATEGORICAL DATA

You might like to group people, places, and things. The classifications and categorizations lend structure to what otherwise would be an amorphous blob of stuff. Figure 4-5 shows some of your options to visualize such categories.

The bar graph, of course, is one of the most common ways to show categorical data. Each rectangle represents a category, and the longer the rectangle is, the greater the value that it represents. Whether a higher value means better or worse can, of course, vary by dataset and point of view.

For example, in February 2012, the Pew Internet and American Life Project surveyed approximately 2,200 people about how they use the Internet, social networking sites such as Facebook and Twitter, and whether politics was a regular occurrence on those sites. Figure 4-6 shows the results for four of the fifty questions.

As you might expect, Google was the most common chosen search engine; Facebook was far ahead of Twitter and career-based social network LinkedIn. The responses to the other questions probably aren't that surprising to you either.

Note: The Pew Internet & American Life Project makes its survey data freely available at http://www.pewinternet.org/.

In this example, the bar graph is the visual equivalent of a list. Each bar represents a value, and you use separate bars and charts for separation. Length is your visual cue, with rectangles placed on a linear scale. You could however use a different scale and shapes to represent the same data. Figure 4-7 shows the same poll results with squares sized by area.

Notice that the differences among categories doesn't look as dramatic in the symbols plots as they do in the bar graphs. For example, the bar for Google looks a lot longer than the rest in the search engine bar graph, but when you compare the square for Google, it looks bigger, but not quite by the same magnitude relative to the other squares.

This might be considered a drawback, but it can also be an advantage when you have hundreds of values that vary by orders of magnitude. With symbol plots, you can organize squares and circles in any way you want in two-dimensional space, as shown in Figure 4-8. On the other hand, bar graphs are restricted in that each bar must start at the zero-axis and must extend straight across or upward to the corresponding value.

Note: Because there aren't many categories for each question, the bar chart seems like a better choice in this example, but you don't need to rule out area as a visual cue automatically.

FIGURE 4-4 *(following page)* *Patterns and visual cues*

Visual Cues

Patterns	Position	Length	Angl
Increase			
Decrease			
Combination			
Outlier			
Noise			

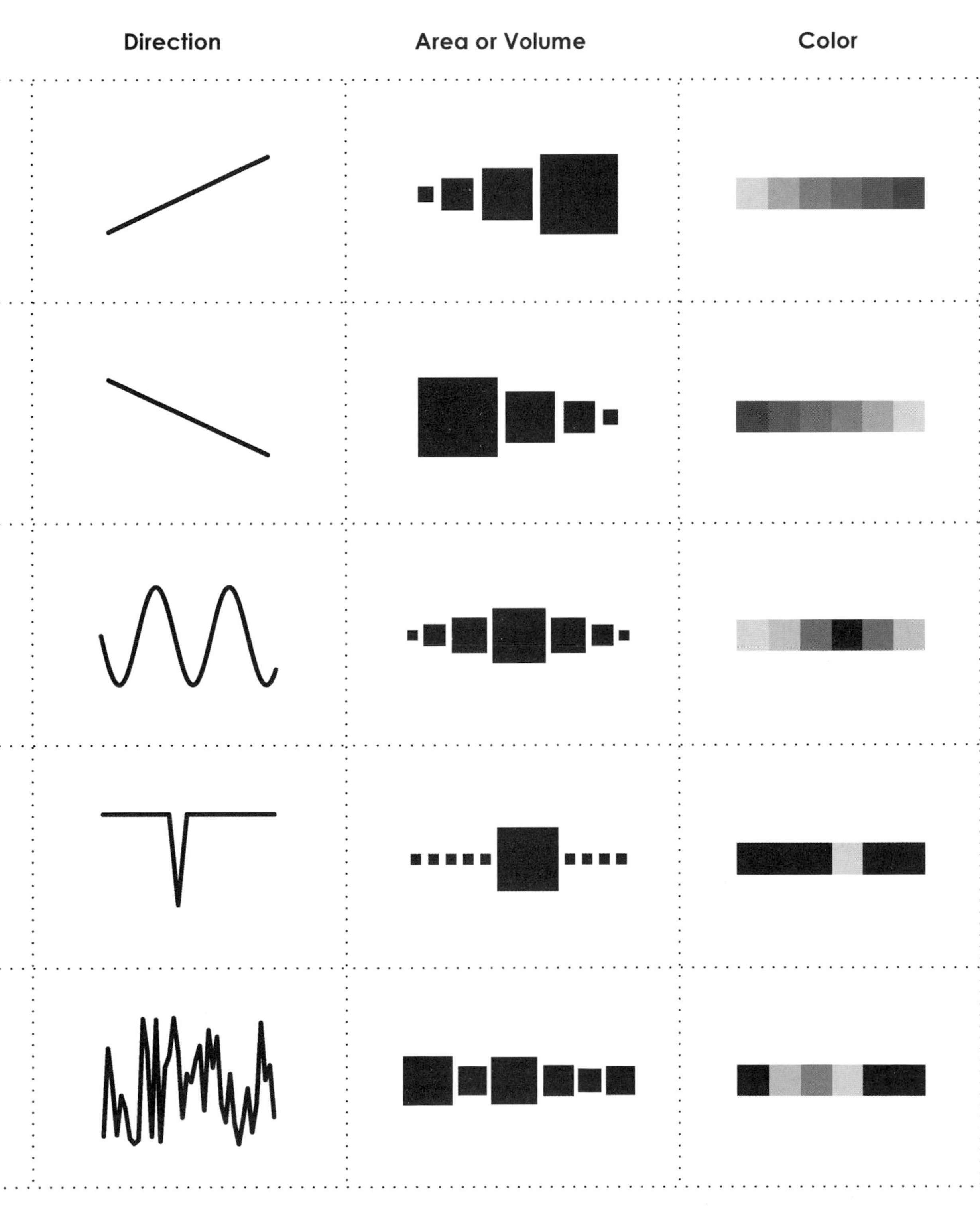
Direction
Area or Volume
Color

Categories

When your data is straightforward, with a value for each category, these are easy to read and create.

Bar graph

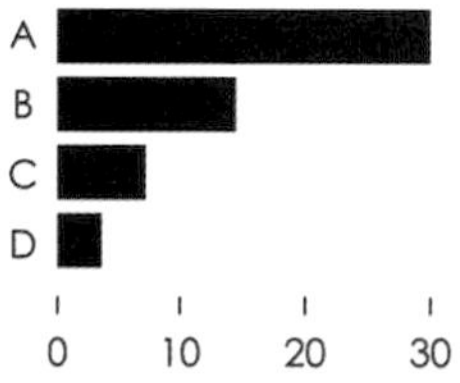

With length as visual cue, useful for straightforward comparisons

Symbol plot

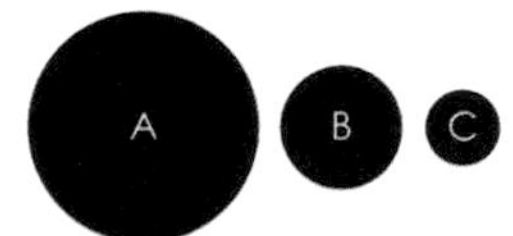

Can be used in place of bars, but can be hard to see small differences

Parts of a whole

The categorical breakdown within a population can be interesting, and you might want to keep the groups together, although often not essential.

Pie chart

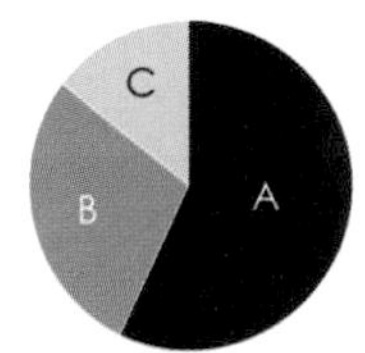

Parts add to 100 percent, typically sorted clockwise for readability

Stacked bar chart

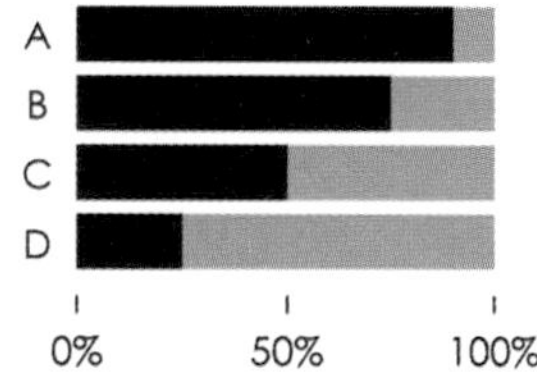

Often used to show poll results and can also be used for raw counts

Subcategories

Data can have a hierarchical structure, which can be important in data interpretation and it often allows for different points of view.

Treemap

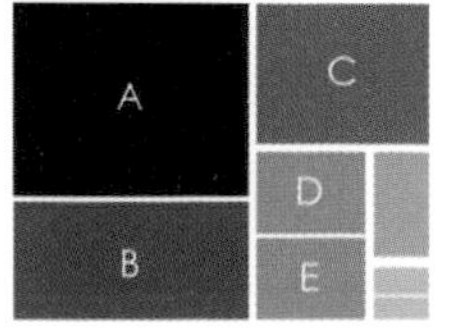

Shows hierarchical structure in a compact space, area often combined with color

Mosaic plot

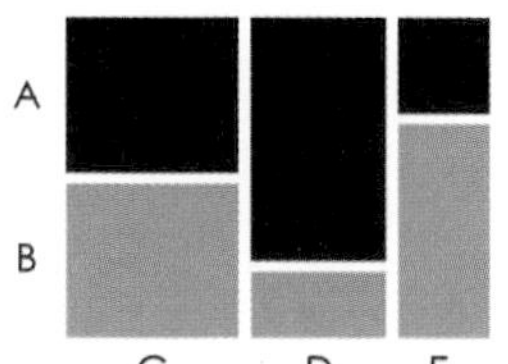

Allows comparisor across multiple categories in one view

FIGURE 4-5 *Visualizing categorical data*

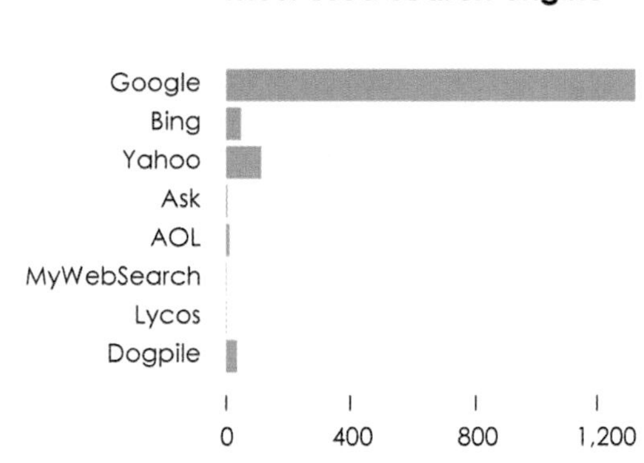

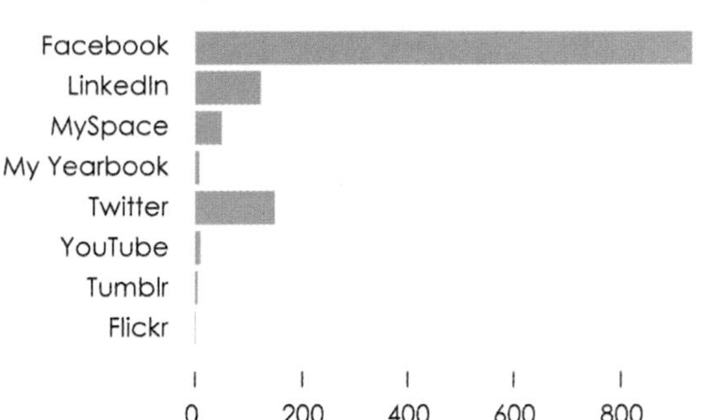

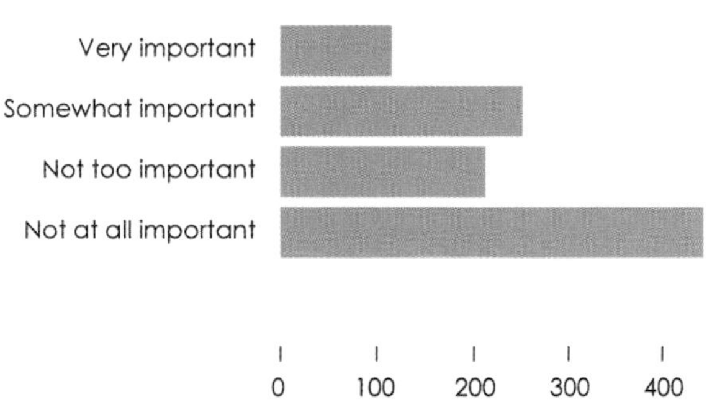

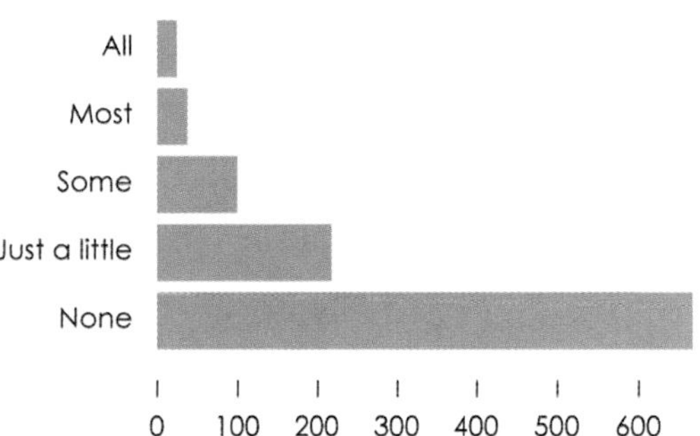

FIGURE 4-6 *Bar graphs for survey results*

PARTS OF A WHOLE

When you put categories together, the sum of the parts can equal a whole. Count everyone in all the states and you have a national aggregate; combine sports divisions and you have a league. Seeing categories as a single unit can be beneficial if you want to see distributions or the spread across a single population.

This is when the pie chart comes into the picture. A full circle represents 100 percent of a whole, and each wedge is a portion of that 100 percent. The sum of all the wedges equals 100 percent. Angle is the visual cue.

Note: You might not be able to get the exact value from a pie chart, but you can still make comparisons when there aren't a lot of categories.

Discussions on whether the pie chart is useful end up running in circles, so you decide if you want to use them. If you do use pie charts, they tend to clutter quickly when you have a lot of categories, simply because there is only so much space in a circle, and small values end up as slivers.

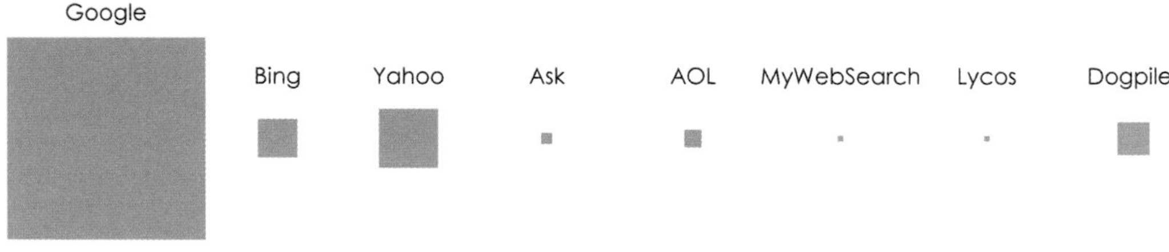

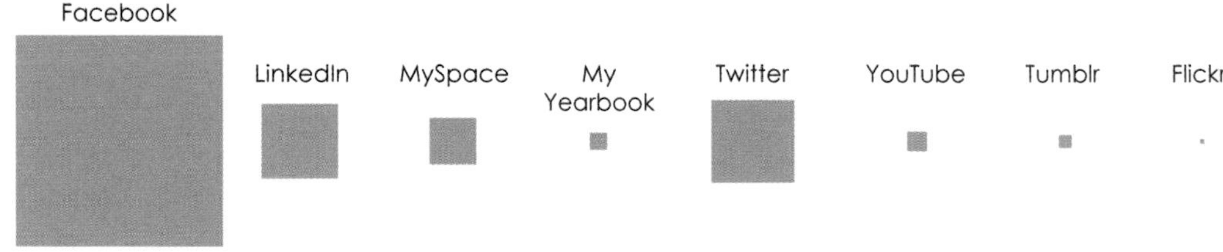

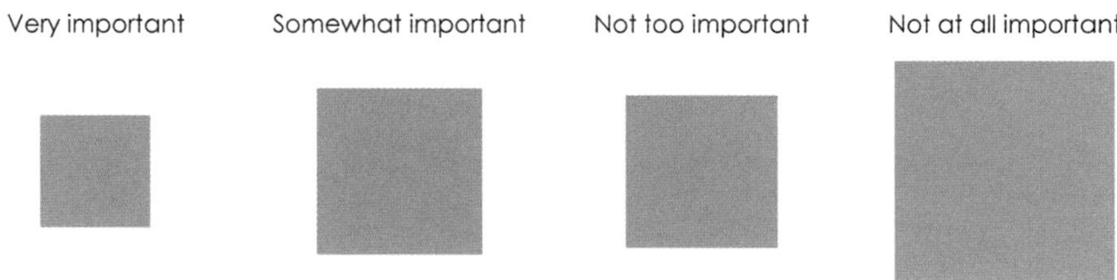

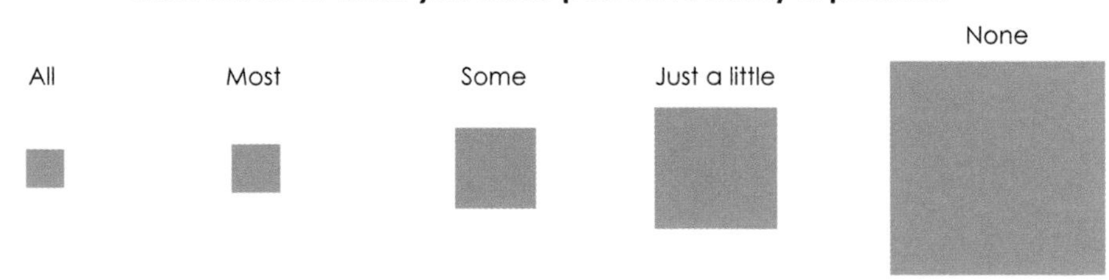

FIGURE 4-7 *Symbols plot for survey results*

Returning to the Pew Research poll on Internet usage, Figure 4-9 shows breakdowns of awareness of targeted advertising online. The first three pie charts show the percentage of respondents who were aware of targeted advertising, those who were okay with it, and those that knew there was a way to limit it, respectively. The next three pie charts show actions that people took, given that they knew they were aware of online tracking.

If you're not fond of pie charts, you can also use stacked bar charts, as shown in Figure 4-10. The full length of the bars represents 100 percent, and each small bar is the equivalent of a wedge in a pie chart.

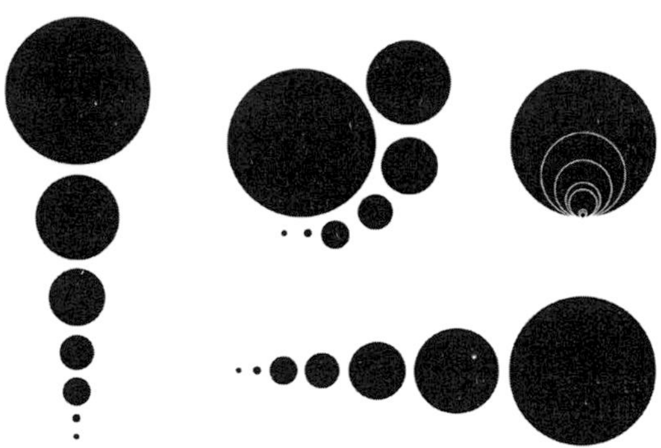

FIGURE 4-8 *Bubble plots organized differently*

Noticed advertising related to recently searched for or visited sites

No answer
No 40%
Yes 58%

Feeling toward targeted advertising

No answer
Yes 26%
No 70%

Aware of ways to limit personal data collected by advertisers

No answer
Yes 37%
No 62%

If aware, have done the following...

Changed browser settings

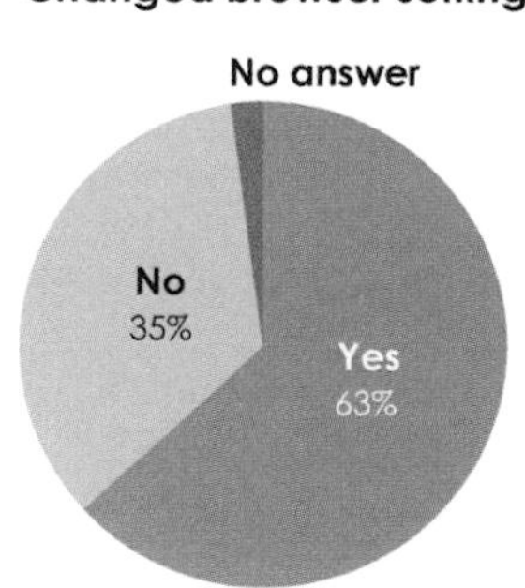

Deleted web history

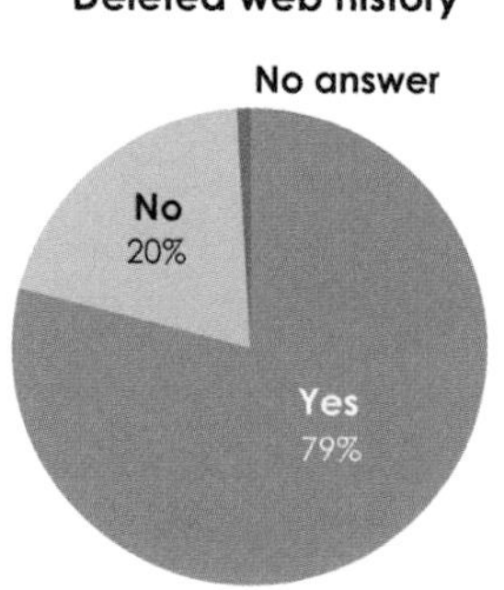

Used privacy settings of website

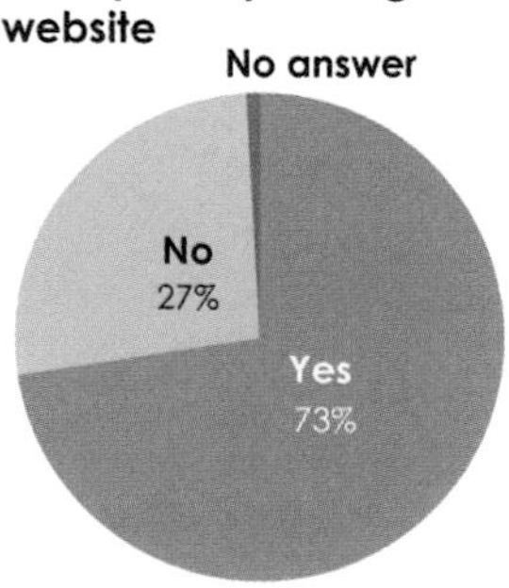

*Sums greater than 100 percent are due to rounding.

FIGURE 4-9 *Pie charts to show categories*

Noticed advertising related to recently searched for or visited sites

Feeling toward targetedd advertising

Aware of ways to limit personal data collected by advertisers

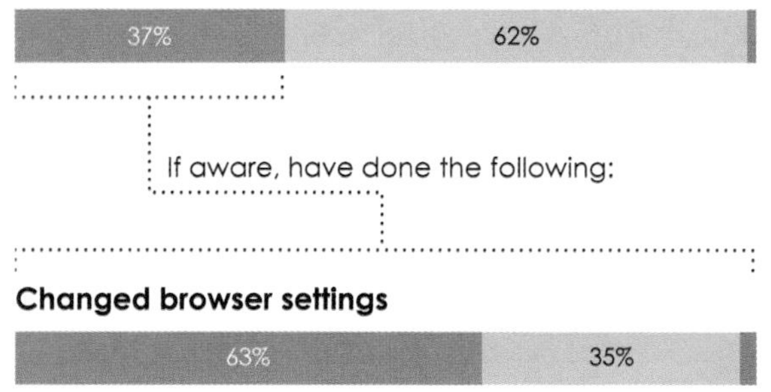

Deleted web history

Used privacy settings of website

*Sums greater than 100 percent are due to rounding.

FIGURE 4-10 *Stacked bars to show categories*

Note: "Map of the Market" by SmartMoney is another popular treemap. It shows the status of the United States stock market in real-time. Check it out at: http://www.smartmoney.com/map-of-the-market/

SUBCATEGORIES

Subcategories, the categories within categories (within categories), are often more revealing than the main categories. As you drill down, there can be higher variability and more interesting things to see.

At the least, showing subcategories can make it easier to browse your data, because you can visually jump to the areas that you care most about. For example, you saw categorical hierachy of the news in Marcos Wescamp's newsmap in Chapter 2, "Visualization: The Medium."

As shown in Figure 4-11, you can use a treemap with the Pew Research survey data. It shows those who use the Internet regularly and those who don't. Within the group of people who use the Internet regularly is a grouping of those who used the Internet the day before the survey and those who did not. However, the survey data doesn't work well with a treemap. Whereas newsmap shows a rectangle for each story sized by current popularity, individuals within a survey are equally weighted.

Instead, a mosaic plot, which shows you proportions within categories and category combinations is more fitting. Like the treemap, you can use a mosaic plot with multiple levels of data, but interpretation can get difficult quickly, so start with the minimum and work your way to the more complex.

Figure 4-12 shows the proportion of people in the survey who said they were the parent or guardian of a child younger than 18 living in the household.

The plot looks like one column from a stacked bar graph. The bigger a section, the more people who gave that answer, so from this view, you see most people said no, some said yes, and there were a few who declined to answer.

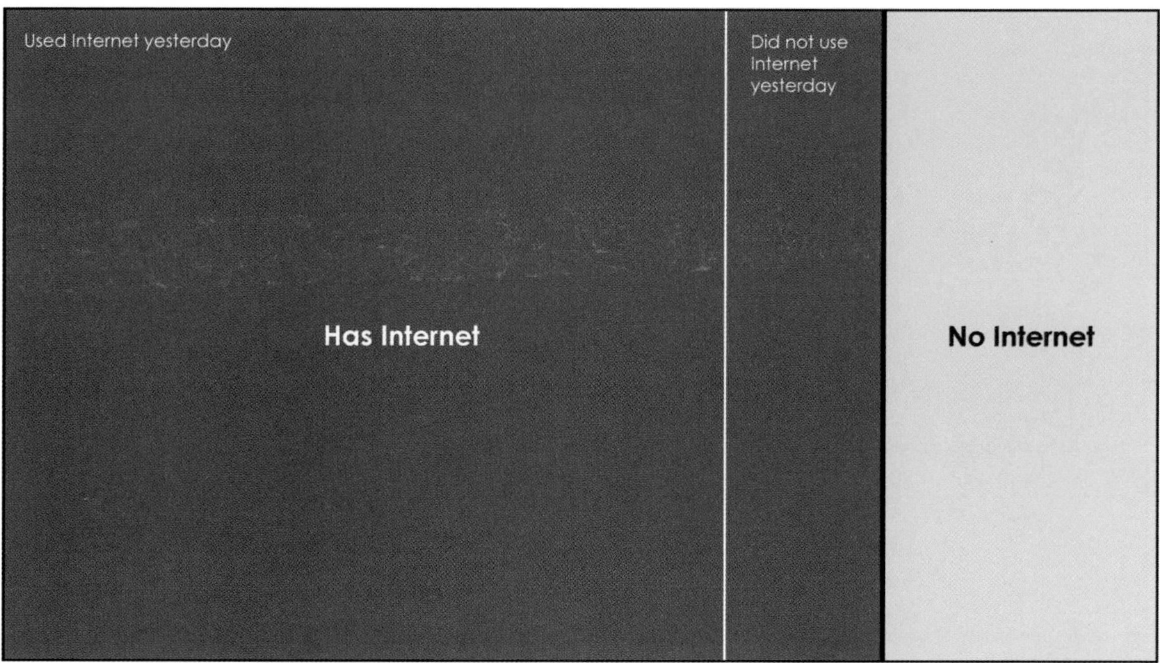

FIGURE 4-11 *Treemap on survey data*

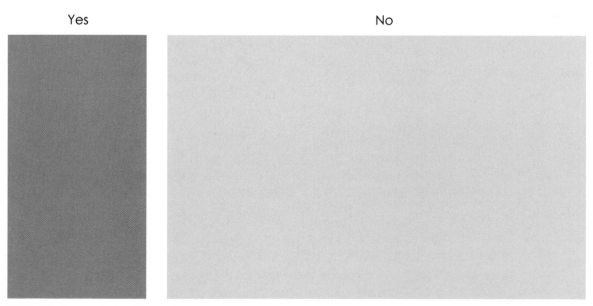

FIGURE 4-12 *Basic mosaic plot with one variable*

What if you want to know the education level of those who are parents or guardians versus those who are not? As shown in Figure 4-13, you can introduce another dimension. It's the same geometry, where more area equals a higher percentage. But now for example, you can see that of those who are parents, a slightly lower percentage were college graduates than those who were not.

You can keep going and bring in a third variable. The orientation of education and parenting are the same, but you can also see e-mail usage. Notice the vertical split on the subsection in Figure 4-14.

You could keep on adding variables, but as you can see, the plot grows more challenging to read, so proceed with caution.

WHAT TO LOOK FOR

With categorical data, you often look for the minimum and maximum right away. This gives you a sense of the range of the dataset, and is easily found with a quick

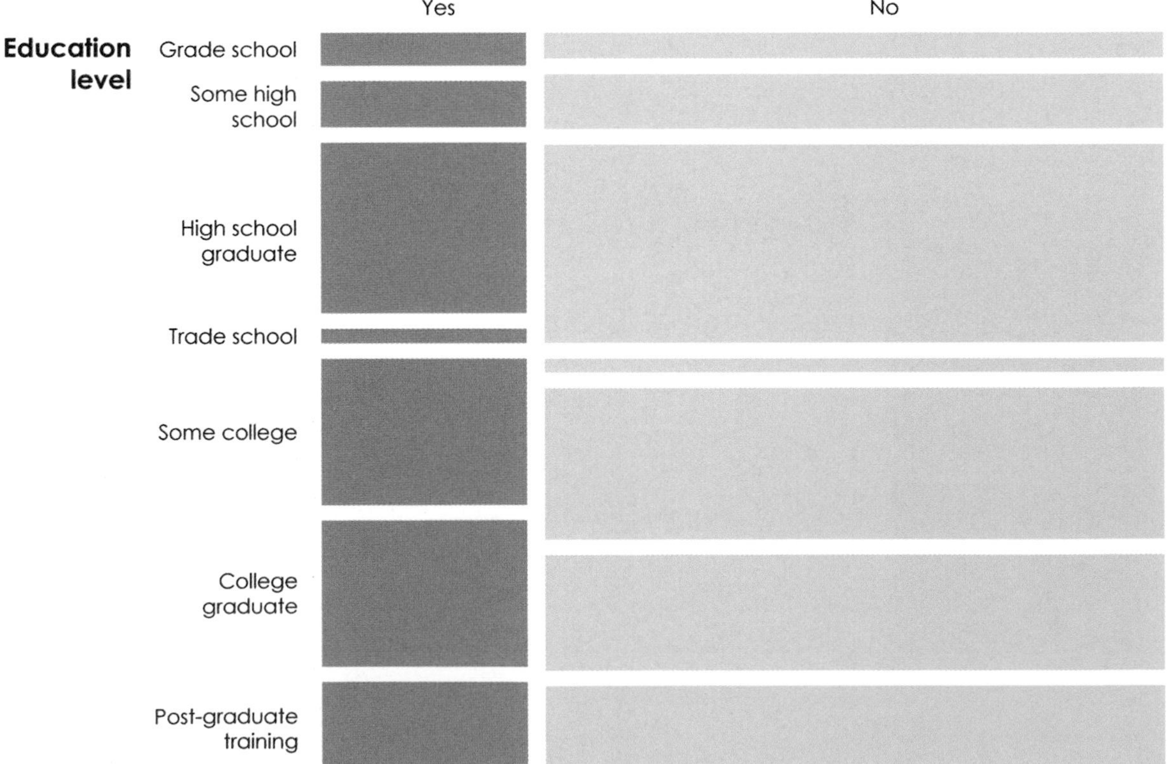

FIGURE 4-13 *Mosaic plot with two variables*

sorting of values. After that, look at the distribution of the parts. Are most values high? Low? Somewhere in between? Finally, look for structure and patterns. If a couple of categories have the same value or high differing ones, it's worth asking why and what makes the categories similar or different, respectively.

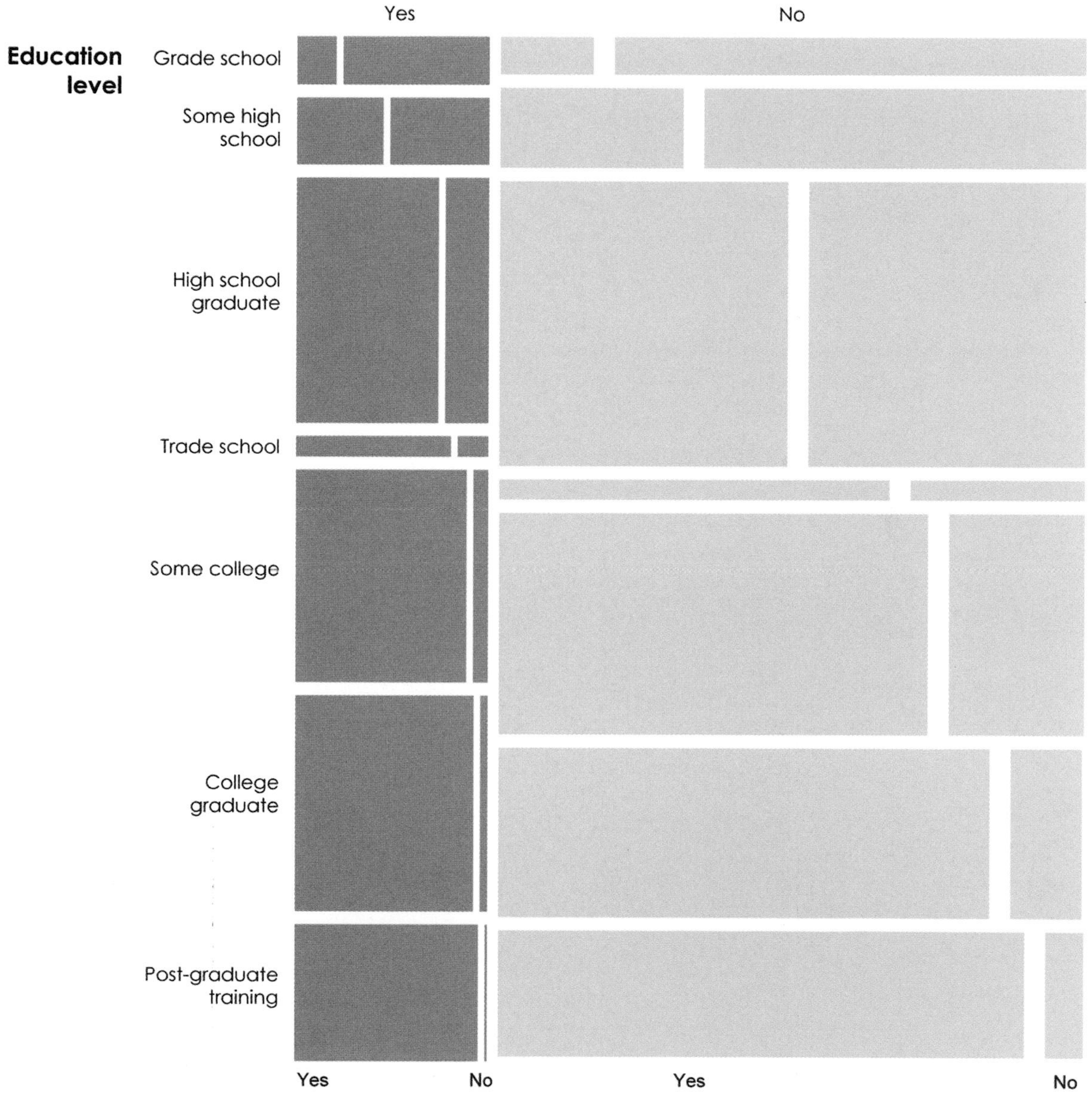

FIGURE 4-14 *Mosaic plot with three variables*

VISUALIZING TIME SERIES DATA

Time passes. Things change, people change, and places change. You can feel time through the sunrise and sunset, your clocks and watches, and the coffee you need to drink when you wake up. When you visualize time series data, as shown in Figure 4-15, your goal is to see what has passed, what is different, and what is the same, and by how much. Compared to last year, is there more or less? What are possible explanations for the increase, decrease, or nonchange? Is there a recurring pattern, and is that good or bad? Expected or unexpected?

As with categorical data, the bar chart is a straightforward way to look at data over time, except instead of categories on one of the axes, you use time.

Figure 4-16 shows the unemployment rate in the United States from 1948 to 2012, according to the Bureau of Labor Statistics. On top is the rate month-to-month, and because there is a high point density, it looks like a continuous area. On the other hand, the graph on the bottom shows only the unemployment rate in January of each year, which allows for space in between bars and makes it easier to distinguish individual points.

In Chapter 1, you saw how car crashes vary over time, and how you can explore time series data at different granularities. The same applies here. You can look at data hourly, daily, annually, by decade, by century, and so on. Sometimes the data format dictates the level of detail because the metric was measured, say, only every 5 years. However, if for example you had measurements by the hour, a high variability might obscure a trend that's more obvious if you take a step back and look at your data by the day.

Usually the magnitude of change between segments of time is more interesting than the value at each point. Although you can interpret trends from a bar graph, you must visually calculate rates. You look at one bar and compare it to the ones before and after.

Time series

There are a variety of ways to see patterns over time, using cues such as length, direction, and position.

Bar graph

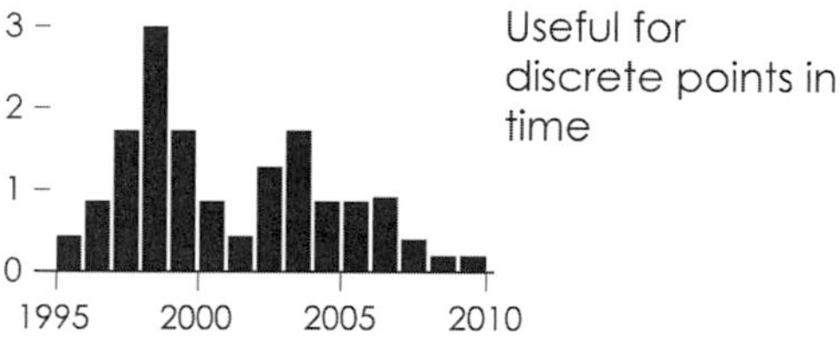

Useful for discrete points in time

Line chart

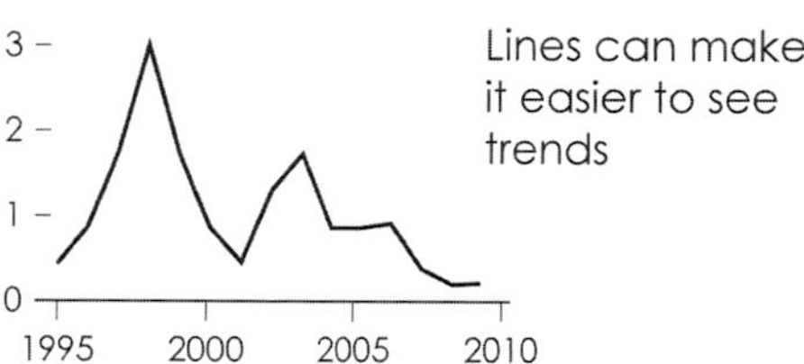

Lines can make it easier to see trends

Dot plot

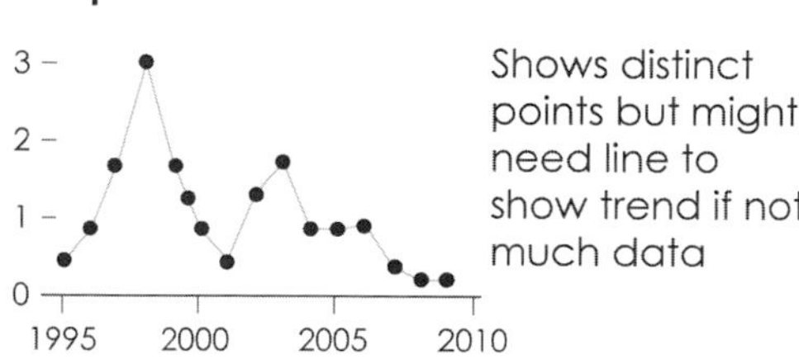

Shows distinct points but might need line to show trend if not much data

Dot-bar graph

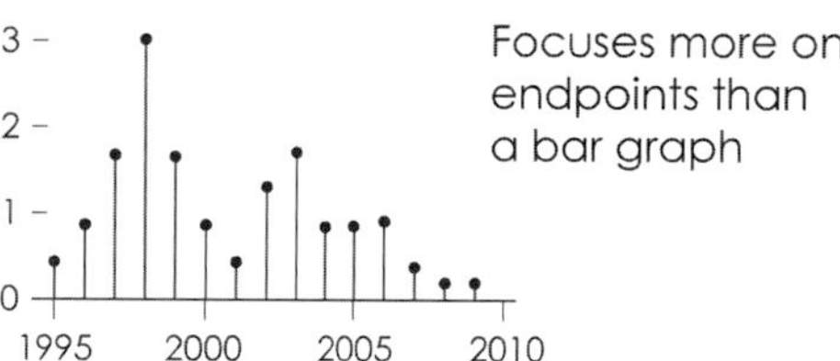

Focuses more on endpoints than a bar graph

Cycles

Time of day, day of the week, and month of the year repeat themselves, so it is often beneficial to align the segments in time.

Radial plot

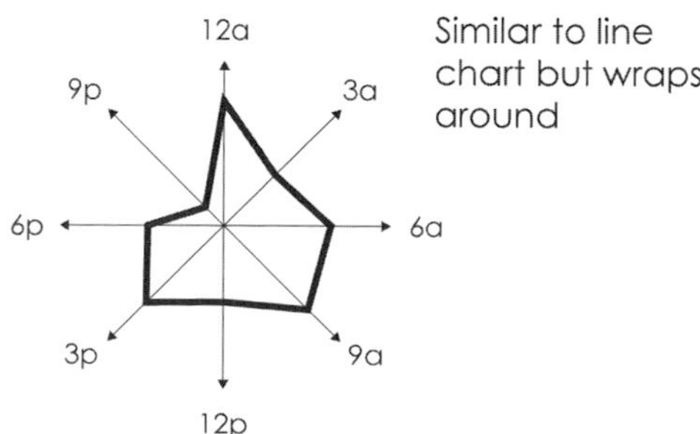

Similar to line chart but wraps around

Calendar

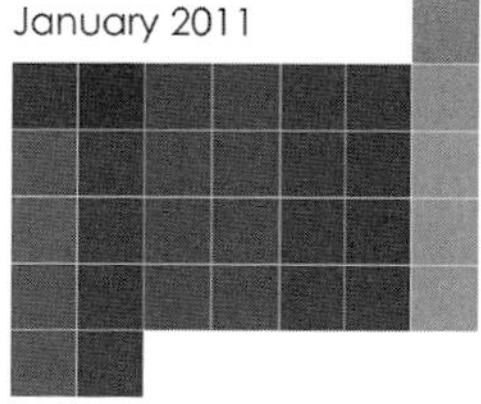

Patterns for days of week seen more easily than views above

FIGURE 4-15 *Visualizing time series data*

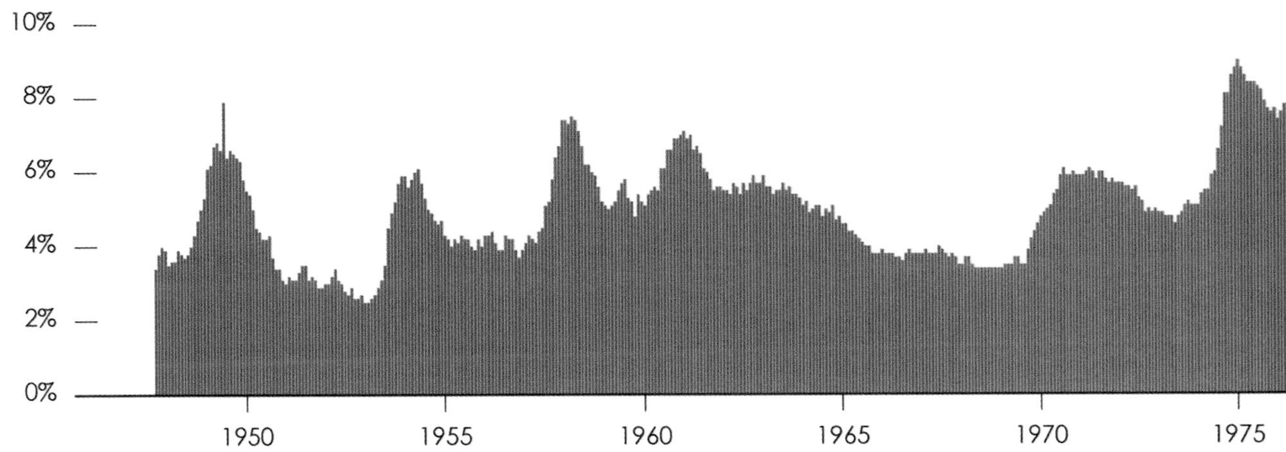

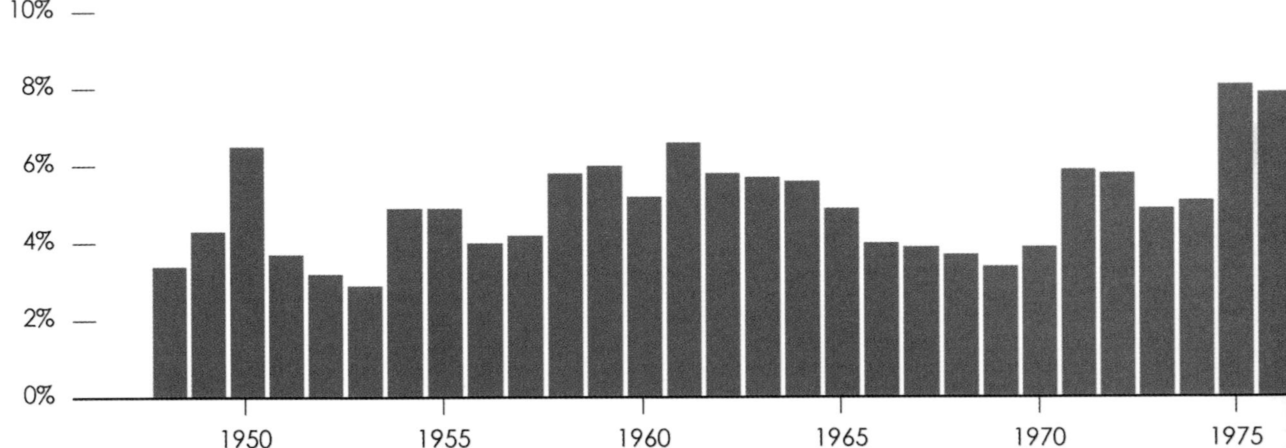

FIGURE 4-16 *Bar graphs for time series data, with monthly on top and annual on the bottom*

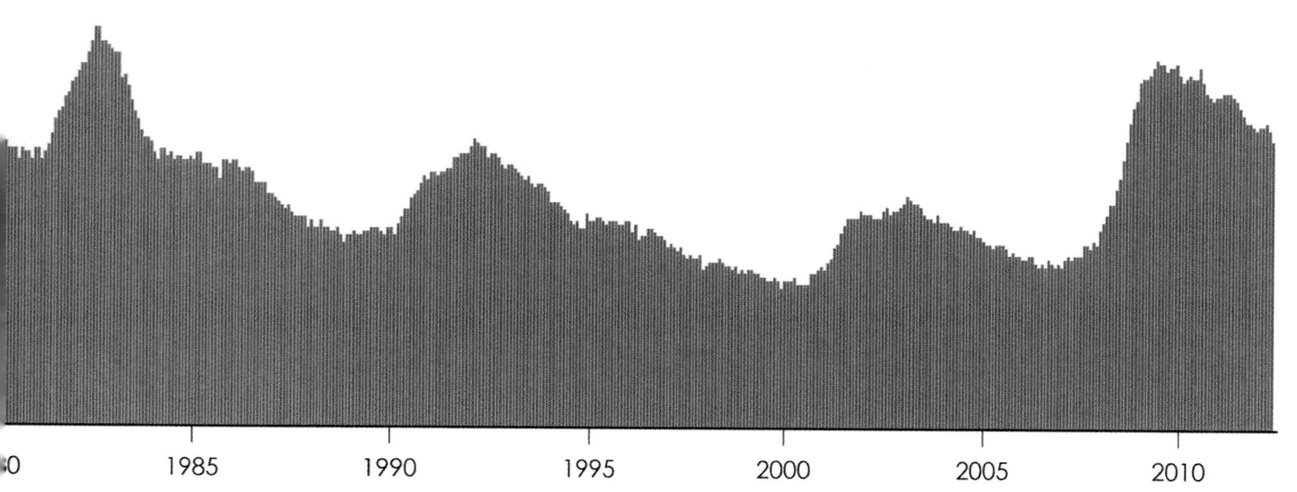
0
1985
1990
1995
2000
2005
2010

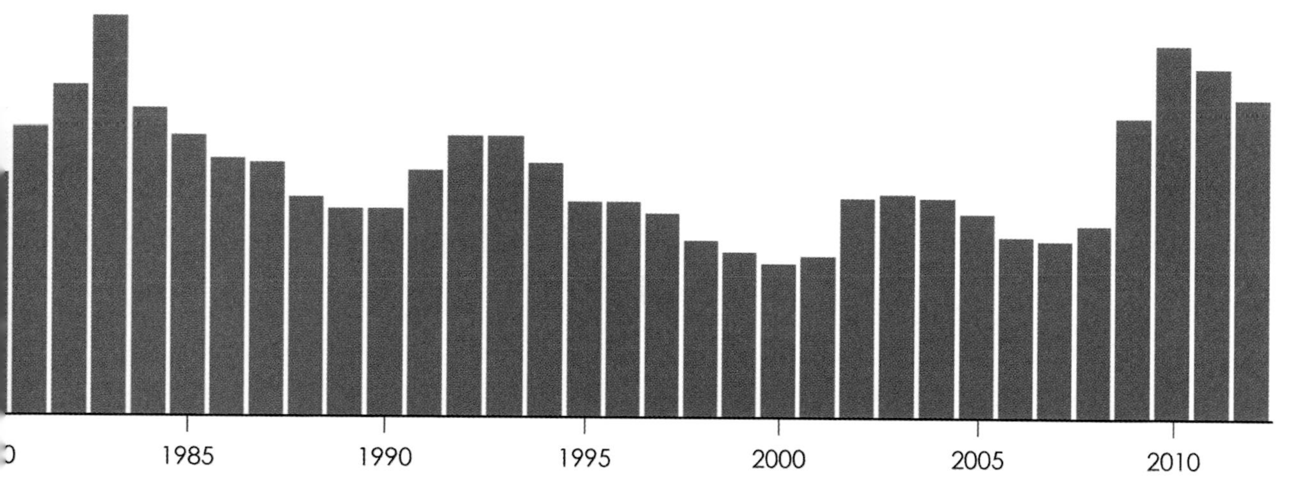
1985
1990
1995
2000
2005
2010

However, where the bars look like a continuous unit (refer to Figure 4-16), it's easier to distinguish changes because you can see the slope, or the rate of change in between points. It's even easier to see the slope when you use a continuous line, as shown in Figure 4-17. The line chart shows the same data as the bar graph on the same scale, but change is directly displayed via direction as a visual cue.

A dot plot can be used in the same way, as shown in Figure 4-18. Again, the data and axes are the same and the visual cue is different.

Like bar charts, dots put focus on each value, and trends can be harder to see. Although in this example, there are enough data points, so you don't need to mentally fill in the gaps. If the data were more sparse, such as in Figure 4-19, changes are less obvious.

When you connect the sparse dots with a line, as shown in Figure 4-20, the focus of the plot shifts again.

If you care more about an overall trend than you do about the more specific monthly variability, you can fit a LOESS curve to the dots, as shown in Figure 4-21, instead of connecting every dot. The closer you fit the curve to the dots, the more it resembles Figure 4-17.

Note: *LOESS* (or *LOWESS*) stands for *locally weighted scatterplot smoothing*. It's a statistical technique created by William Cleveland, which fits a polynomial function to a subset of the data at different points. When combined, they form a continuous line.

Of course, the chart style you choose depends on your data, and although it might seem like a grab bag of options at first, you get a feel for what type of chart to use with practice. It's not an exact science (or computers could do all the work) and options can vary a lot even if you have datasets that look similar.

For example, the previous charts on unemployment rate provide a historical view of the past few decades. You can see peaks and valleys, periods of recession such as in 2001 and from 2007 to 2009, and an overall picture of changing rates. If you were only interested in the five highest peaks and what happened immediately after them, the exploration would take a different route.

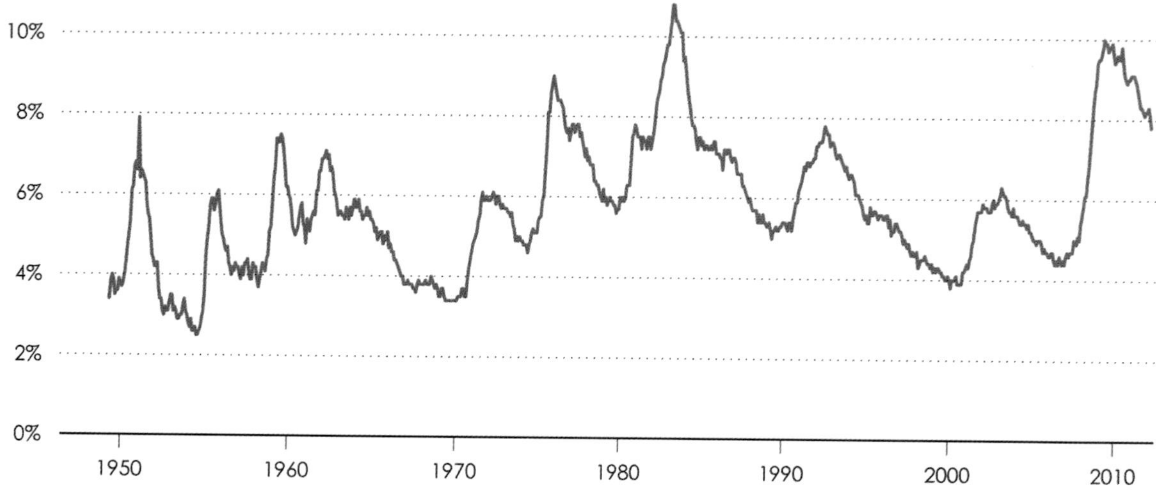

FIGURE 4-17 *Line chart to show time series*

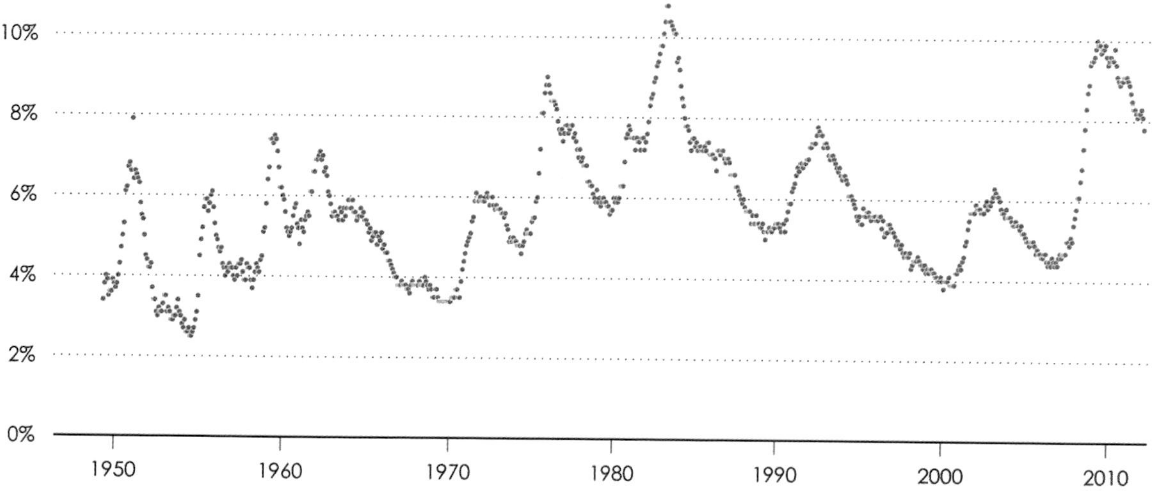

FIGURE 4-18 *Dot plot to show time series*

Unemployment rate

10%
8%
6%
4%
2%
0%
1950
1960
1970
1980
1990
2000
2010

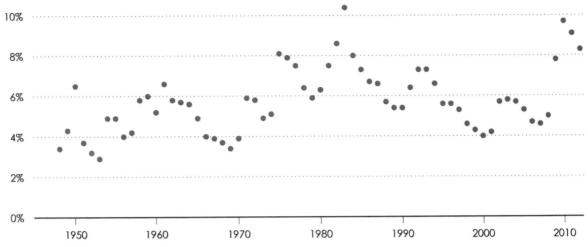

FIGURE 4-19 *Sparse dot plot*

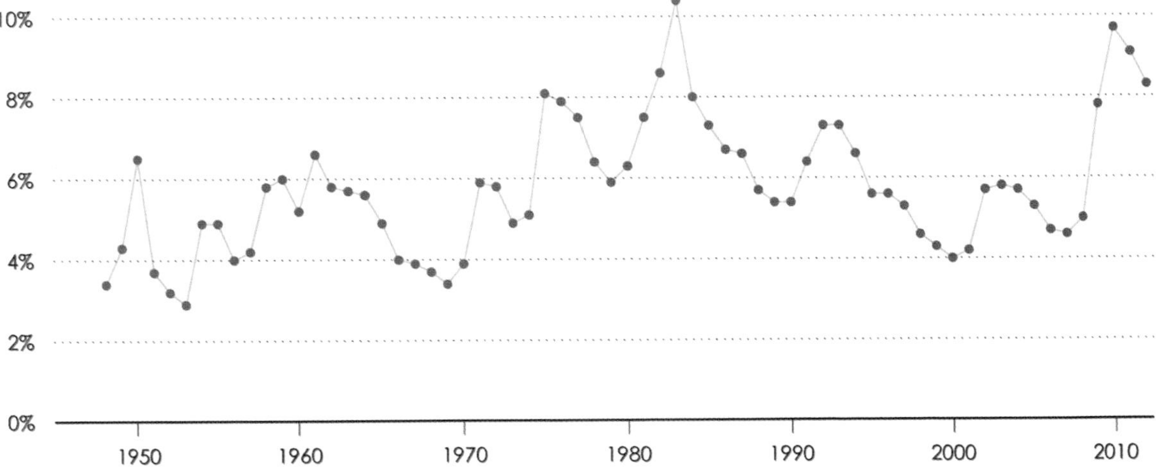

FIGURE 4-20 *Sparse dot plot with connecting line*

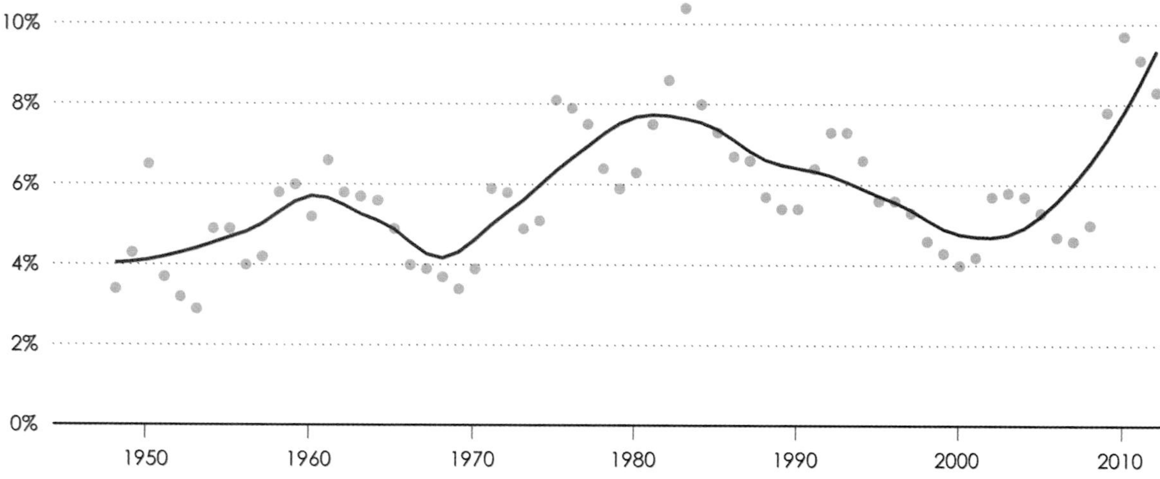

FIGURE 4-21 *Fitted LOESS curve*

CYCLES

A number of factors feed into the economy and affect the unemployment rate, so there aren't regular intervals in between significant increases. For example, the data doesn't suggest that unemployment goes up to 10 percent every 10 years. However, there are a lot of things that repeat themselves on regular intervals. Students get summer breaks and people often take summer vacations, and lunch is typically around noontime, so the restaurant around the corner that makes burritos bigger than your face usually has a longer line during that hour.

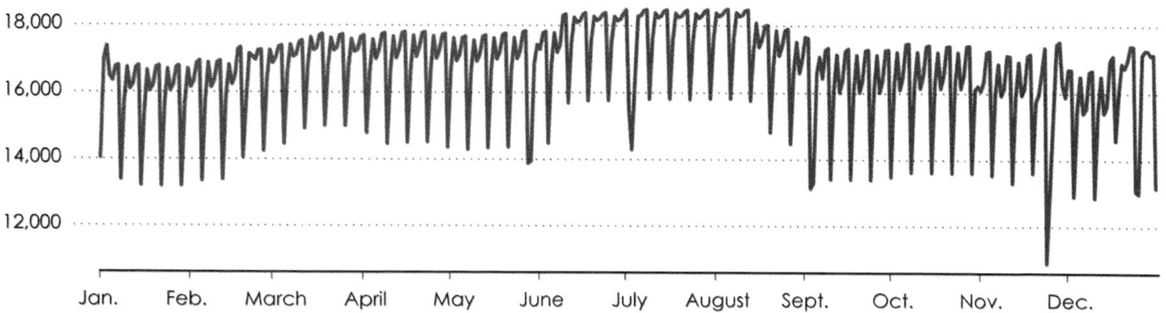

FIGURE 4-22 *Weekly cycle*

You saw repetition in the crashes data: More people travel during the summer months; more people leave work around 5 in the afternoon and head home; and more accidents occur on Saturday than any other day of the week. This information can be used to make sure there are enough people staffed during each day of the week and when to allot vacation times.

Flight data from the Bureau of Transportation Statistics shows a similar cycle, as shown in Figure 4-23. The chart shows a weekly cycle, with the fewest flights on Saturdays and typically the most flights on Fridays (a contrast to car crashes).

You can see the same pattern if you switch to a polar axis, as shown in the star plot in Figure 4-23. The data starts at the top, and you read the chart clockwise. The closer to the center a point is, the lower the value, and greater values move further away.

Note: The *star plot* is also commonly referred to as a *radar chart*, *radial plot*, and *spider chart*.

Because the data repeats itself, it makes sense to compare like days of the week to each other. For example, compare all Mondays. It's hard to do this when time is visualized as a continuous line or loop, but you can split the days

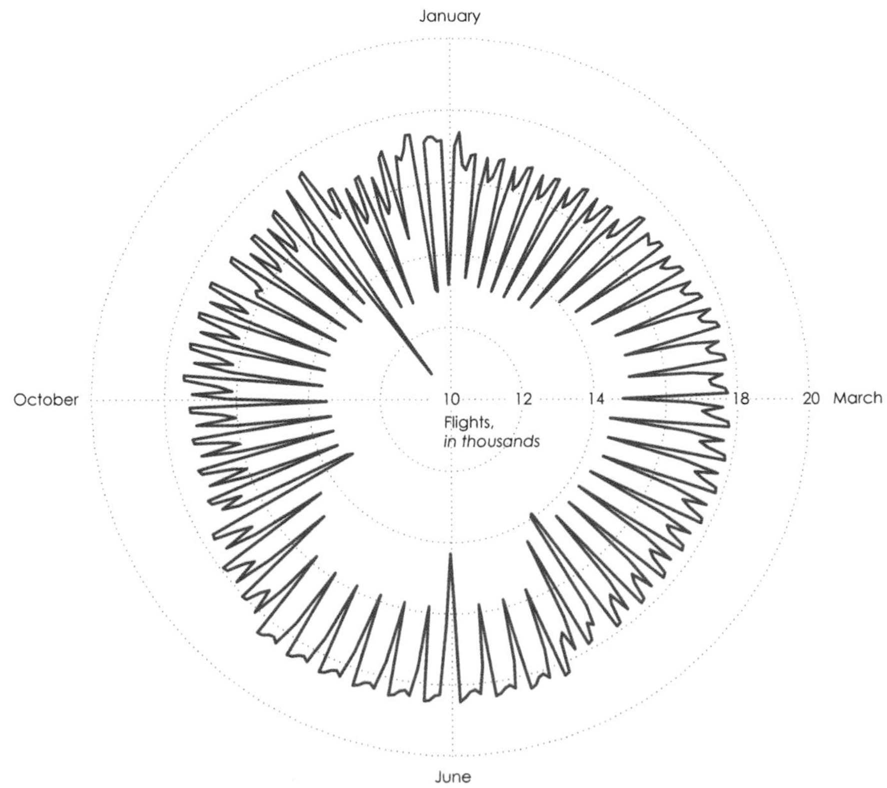

FIGURE 4-23 *Star plot to show time series data*

into weekly segments so that you can directly compare cycles, as shown in Figure 4-24, with both the line chart and star plot.

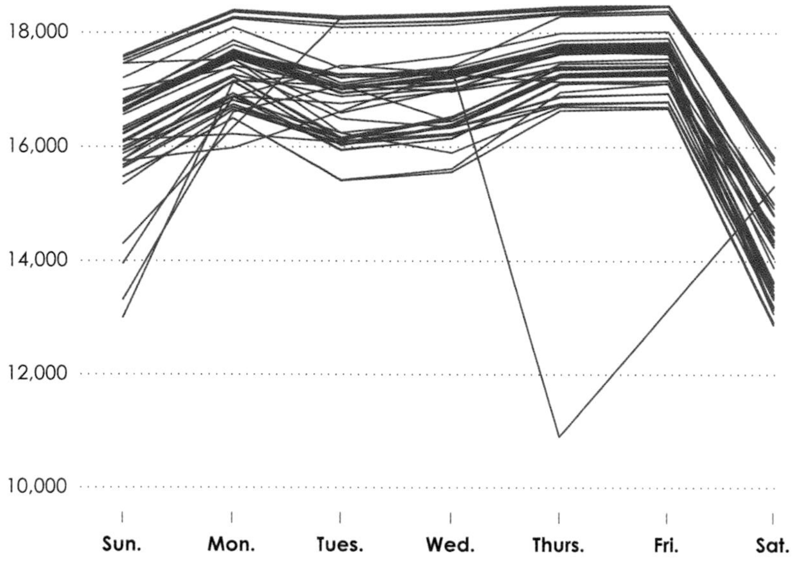

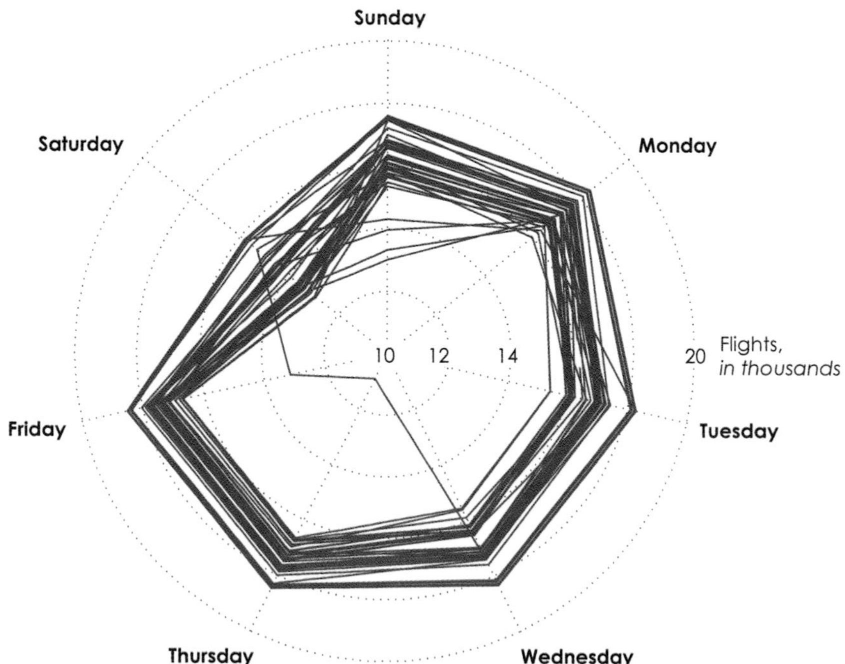

FIGURE 4-24 *Line chart and star plot to show overlapping cycles*

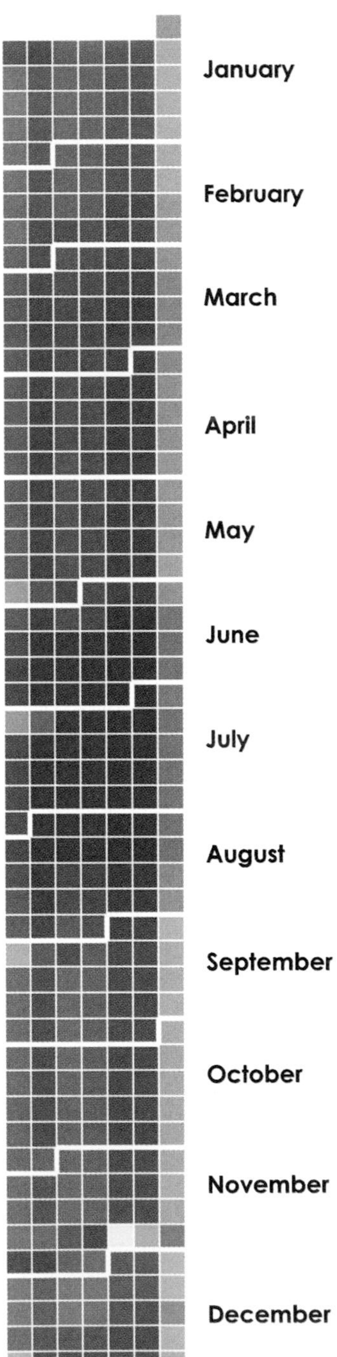

FIGURE 4-25 *Calendar heat map*

There's clearly a regular day-to-day pattern. The thin lines that overlap give the appearance of thicker bands.

There are also some obvious outliers. Thursdays and Fridays are much lower than the rest of the year, and four Sundays also appear lower. Take a moment to think about why travel might be lower than usual on those days.

The obvious way to figure out the dates of the outliers is to go back to the data and look at the minimums for each day. That's always there for your reference. You can also refer to the linear view in Figure 4-17 for a rough idea, but is there a way to directly get the dates and see the data? Figure 4-25 shows the data in a familiar calendar format. The first column is Sunday, the second is Monday, and so on, to Saturday at the end.

The advantage of the calendar heat map over the line chart is that, along with seeing cycles as you scan top to bottom, it's easy to see specific days in rows and columns, so it's easier to reference what day of the year each value is for. The first three low-volume Sundays precede holidays that fall on a Monday in the United States: Memorial Day on May 30, Independence Day on July 4, and Labor Day on September 5. The last Sunday was Christmas. As for the low-volume Thursday and Friday, that was Thanksgiving weekend in November.

A disadvantage of the calendar is that color is the visual cue, and it can be hard to see small differences. It's easier to compare positions on the line chart. So there are trade-offs between different views, but then again, there's no harm in looking at your data from all angles.

Note: The calendar heat map is an intuitive layout that you see often in your everyday life, but is underused as an exploratory tool. It can come in handy.

WHAT TO LOOK FOR

Generally speaking, look for changes over time. More specifically, note the nature of the changes. Are the changes relatively a lot or are they small? If they're small, is the change still significant? Think of possible reasons for what you see over time or sudden blips and if they make sense. The change itself is interesting, but more importantly, you want to know the significance of a change.

VISUALIZING SPATIAL DATA

Spatial data is easy to relate to because at any given moment—as you read this sentence—you have a sense of where you are. You know where you live, where you've been, and where you want to go.

There is a natural hierarchy to spatial data that allows, and often requires, you to explore at different granularities. Far out into space, Earth looks like a small, blue dot with little to see, but as you zoom in, you see land and large bodies of water. There are continents and oceans. Zoom in again, and you get countries and seas, then provinces and states, counties, districts, cities, towns, neighborhoods, all the way down to an individual household.

Global data is often categorized by country and national data by states, provinces, or territories. However, if you have questions about variation across blocks or neighborhoods, such high-level aggregates won't do you much good. So again, the exploration route you choose depends on the data you have or the data you can get.

The most obvious way to explore spatial data is with maps, which place values within a geographic coordinate system. Figure 4-26 shows some of your options, of which there are many.

If you care only about individual locations, you can place dots on a map, as shown in Figure 4-27. The map simply shows the 30 busiest airports in the United States, based on the number of outgoing flights in 2011. As you might expect, the busy airports are in or near major cities such as Los Angeles, Washington, DC, New York, and Atlanta.

Note: A map isn't always the most informative way to visualize spatial data. Often, you can treat regions as categories, and a bar graph might be more useful than seeing a location.

Locations

A direct translation of latitude and longitude to two-dimensional space is straightforward and intuitive, but can pose challenges when there are a lot of locations.

Location map

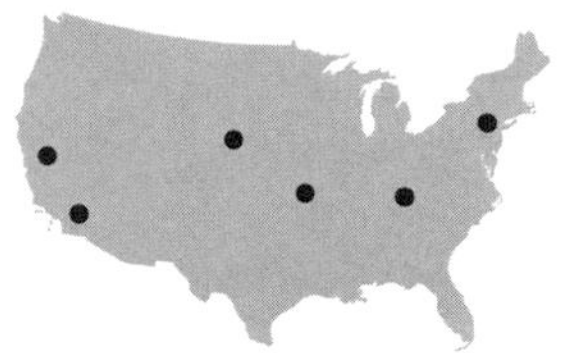

Points represent locations and can be scaled by metric

Connections

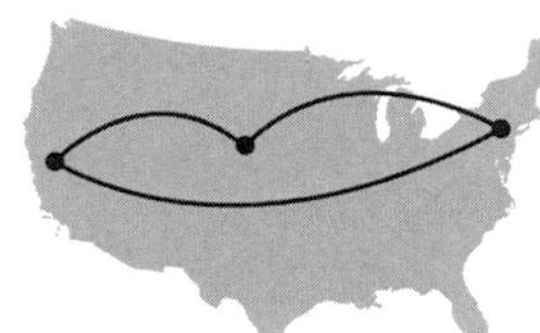

Points can be connected to show relationships between locations

Regions

Oftentimes the the density of individual points across regions is more informative than points on a map that can overlap.

Choropleth map

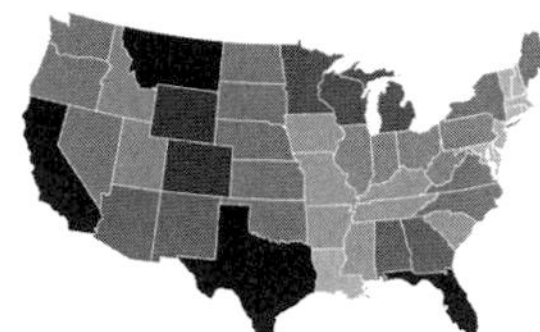

Defined regions colored by data and meaning can change based on scale

Countour map

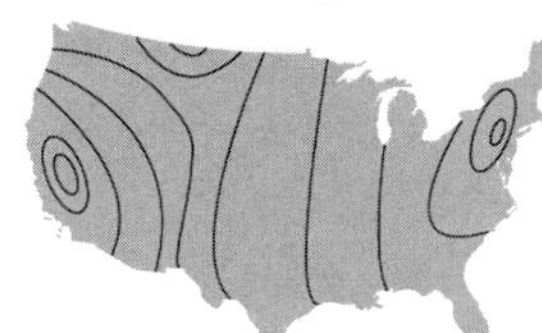

Lines show data continuously over geography, using density

Cartograms

Choropleth maps give large regions more visual attention, regardless of the data, so cartograms instead size regions by the data and ignore physical area.

Circular cartogram

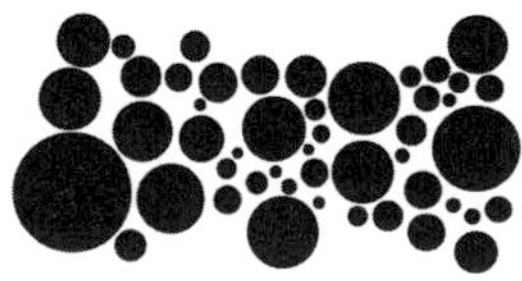

Entire regions sized by data instead of physical area using shapes

Diffusion-based cartogram

Regions sized by data but boundaries stay connected

FIGURE 4-26 *Visualizing spatial data*

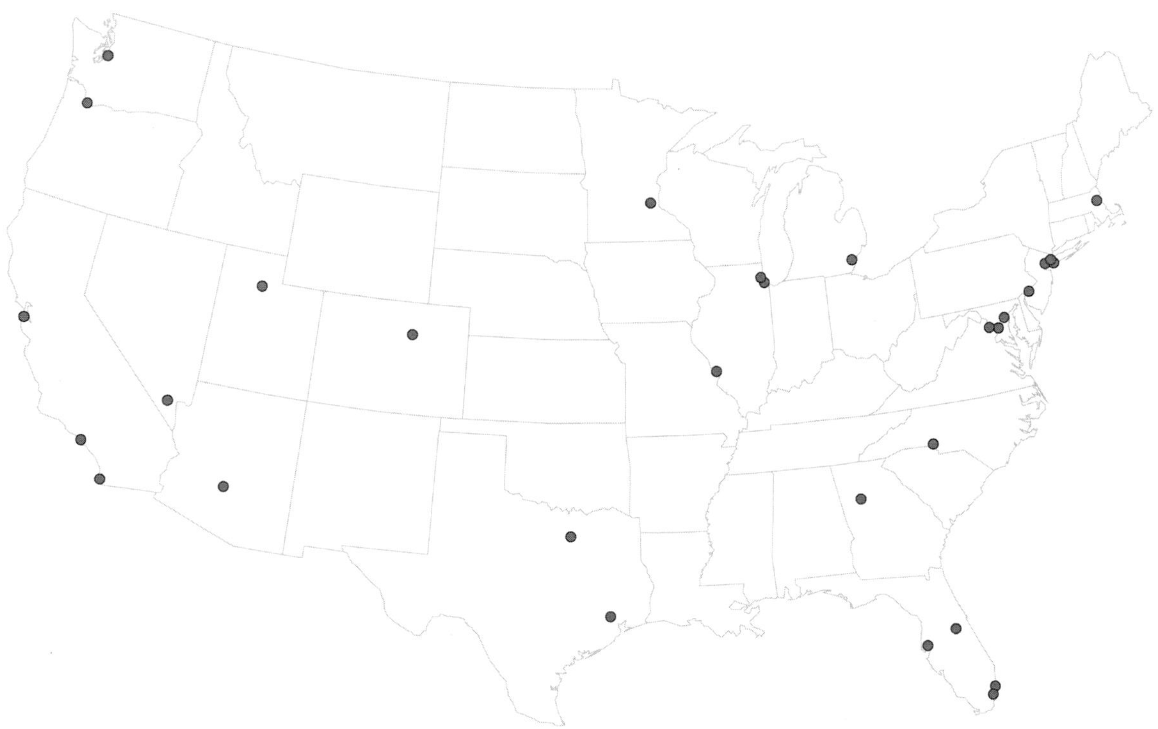

FIGURE 4-27 *Dots in a geographic coordinate system*

Figure 4-28 uses bubbles for the airports, sized by the number of outgoing flights. So, with the addition of an area as visual cue, you don't just see where the busiest airports are, but also how busy they are relative to each other. Atlanta International served the most outgoing flights in 2011, followed by Chicago O'Hare, Dallas-Fort Worth, Denver, and Los Angeles.

Rather than separate locations, you might want to explore connections between locations. For example, in recent years, people have visualized global friendships on social network sites such as Facebook and Twitter. It's one thing to see where people like to use the sites, but it's another to see how they interact.

With the flight data, you already saw counts for outgoing flights via bubbles on a map, but where did those flights go to? Each flight has an origin and a destination. Figure 4-29 shows these connections. The brighter a line is, the

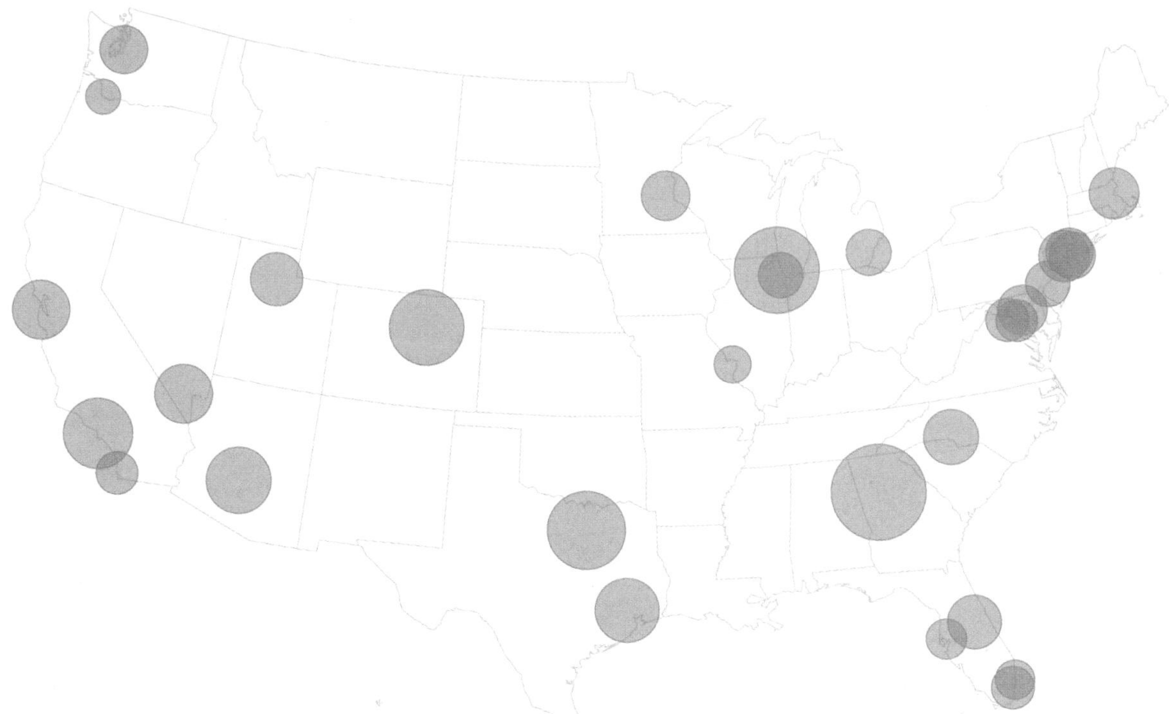

FIGURE 4-28 *Bubbles to show an additional metric on a map*

more flights that went to and from those two airports. Busier airports also appear where there is a higher density of flights.

It's fun to see patterns emerge when you plot a lot of data at once. The map represents more than 6 million domestic flights in 2011, and you gain a rough idea of where people flew to and from. But there's more you can take away from this data by splitting it into categories. For example, map flights by airline, as shown in Figure 4-30, and you see the data with a new dimension.

Note: When you have a lot of data, it is often to your benefit to split it into groups so that you can see details more clearly.

Hawaiian Airlines flies only from the west coast to the islands; Atlantic Southeast Airlines is true to its name; Southwest stays within the contiguous United States; and Delta flies to a number of places, but you can see their major hubs in Atlanta, New York, Detroit, and Salt Lake City.

FIGURE 4-29 *Connections between locations*

REGIONS

To maintain the privacy of individuals and to keep personal addresses hidden, it's common to aggregate spatial data before releasing it. Sometimes it's not possible to make estimates at a higher granularity because it would be too big of an undertaking. For example, it's rare to see global data more than country-specific because it's difficult to get a big enough sample in every country for such high detail.

FIGURE 4-30 *(following page) Categorizing data for more specific views*

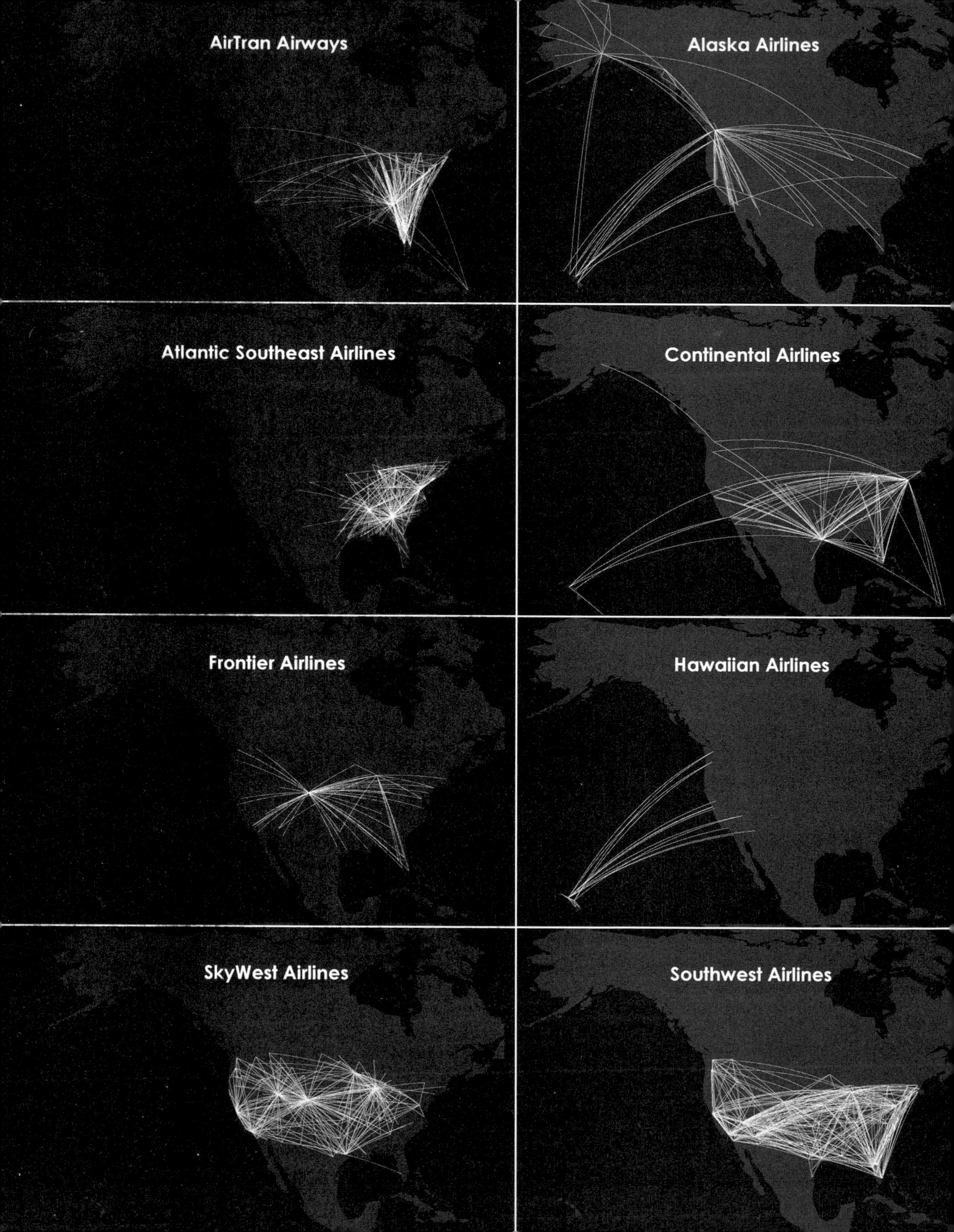
AirTran Airways
Alaska Airlines
Atlantic Southeast Airlines
Continental Airlines
Frontier Airlines
Hawaiian Airlines
SkyWest Airlines
Southwest Airlines

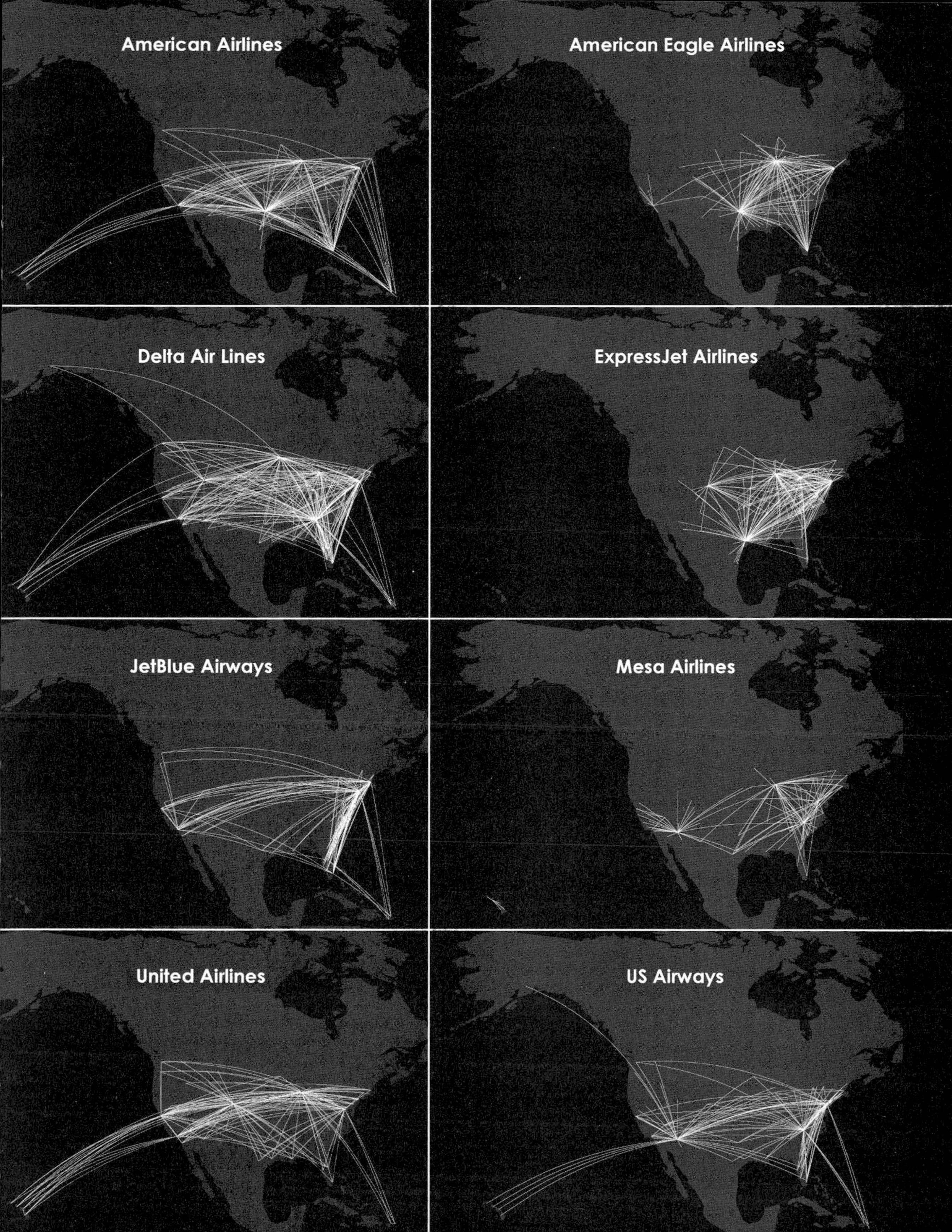
American Airlines
American Eagle Airlines
Delta Air Lines
ExpressJet Airlines
JetBlue Airways
Mesa Airlines
United Airlines
US Airways

Why not combine studies if they estimate the same thing? If methodology is different, it can be hard to make a case that the results are comparable.

Other times it just makes sense to provide data in aggregate because people want to compare regions. For example, if you work with open data, you often see estimates by country, state, or county. Although less specific, you can still extract information from aggregated data.

Choropleth maps are the most common way to visualize regional data in a spatial context. The method uses color as its visual cue, and regions are filled based on the data. Higher values are typically represented with higher saturation and lower values with lower saturation, such as the map in Figure 4-31.

The map shows estimated national gas prices around the world. The darker the shade of brown, the higher the price per gallon. Gray indicates that there was no data available for that country. Prices are relatively high in Europe and Africa, compared to that of the United States.

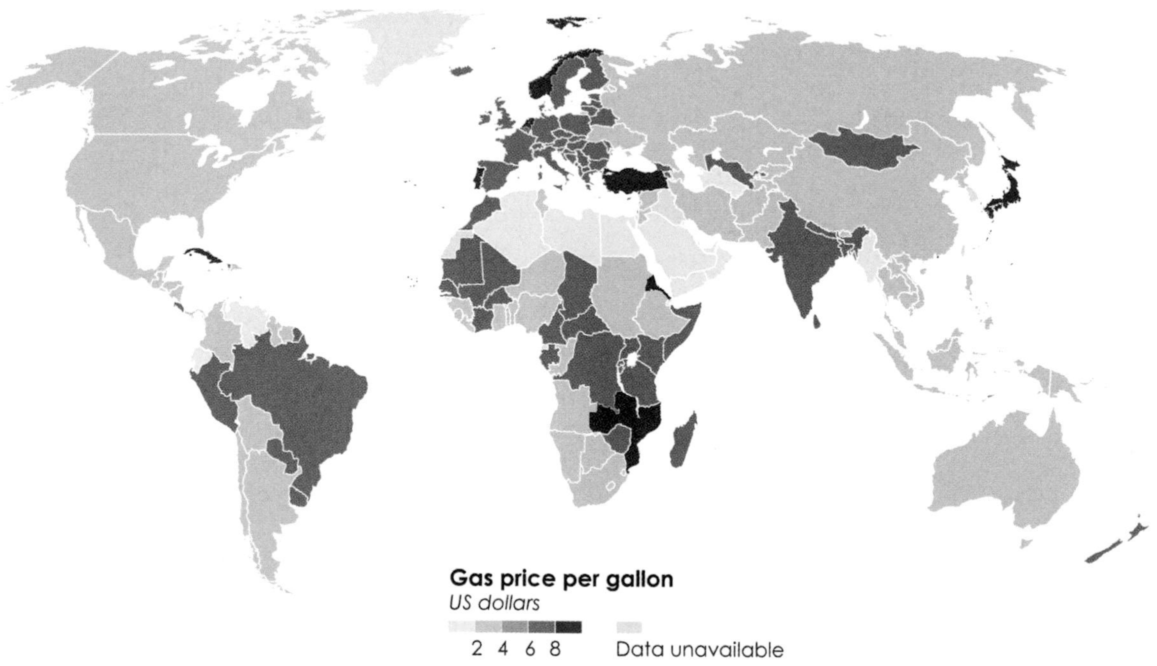

FIGURE 4-31 *Choropleth world map*

How deeply can you read into the data, though? Look at gas prices across your country, and there's variation. Heck, look at two gas stations within a few blocks of each other, and there can be a big difference. So although you can see general patterns, you shouldn't be too quick to judge as you explore. This data in particular comes from a variety of sources, such as government databases and newspaper articles, and from different years.

On the other hand, some sources use well-established methodologies and have done so for a long time. For example, the Bureau of Labor Statistics estimates the unemployment rate every month. You saw the national estimate over time in Figure 4-17, but you can also see the data by county, as shown in Figure 4-32. The map shows unemployment rate by county during August 2012. You can see high rates on the West Coast and in the Southeast and lower unemployment in the Midwest.

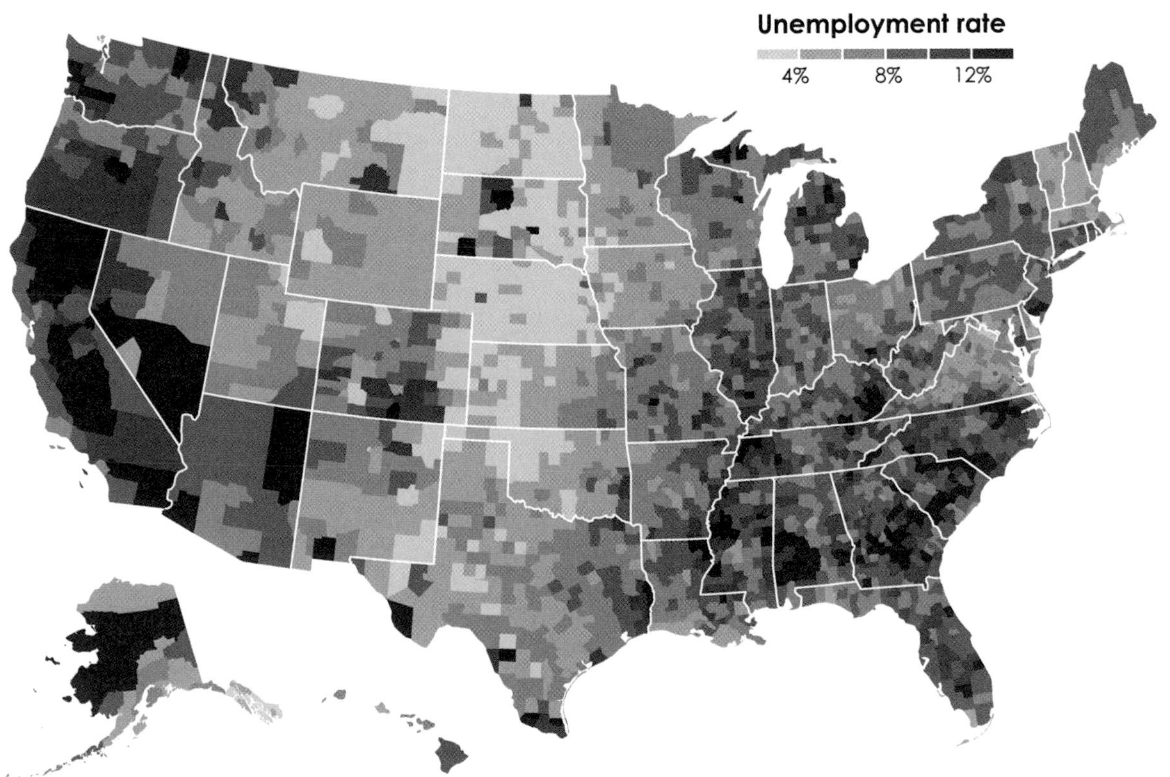

FIGURE 4-32 *County choropleth map*

There are also times when your spatial data actually does contain specific locations, but you're more interested in the aggregates. You might have a dataset with a lot of locations, and there are a lot of points in metropolitan areas. So when you map everything, points overlap, and it's difficult to tell how many observations there are in the dense areas.

For example, Figure 4-33 shows all recorded UFO sightings between 1906 to 2007, according to the National UFO Reporting Center. In areas where there were a lot of sightings (which curiously are where a lot of major airports are located), you just see a black blob, and it's hard to tell how many sightings there were when there is too much overlap.

Figure 4-34 shows the same data, but as a filled contour map. A color scale is used to show sightings density, where white means more sightings and black means none, and varying shades of red are for everything in between.

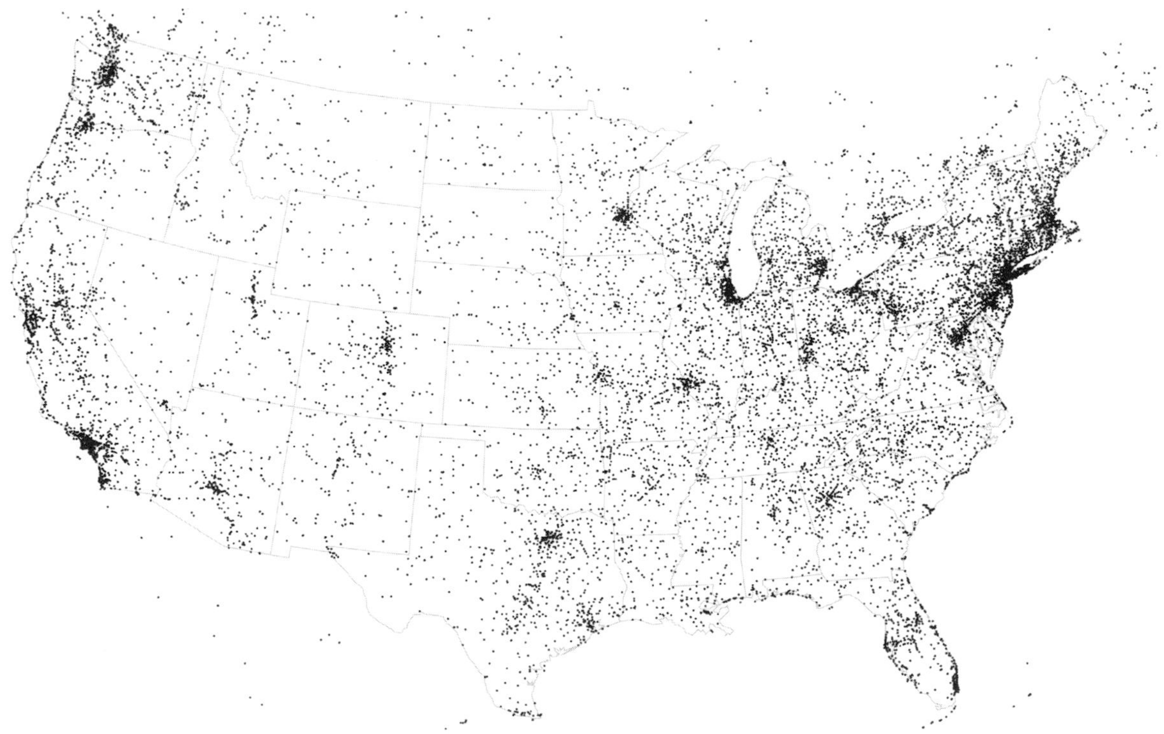

FIGURE 4-33 *Overlapping points on a map*

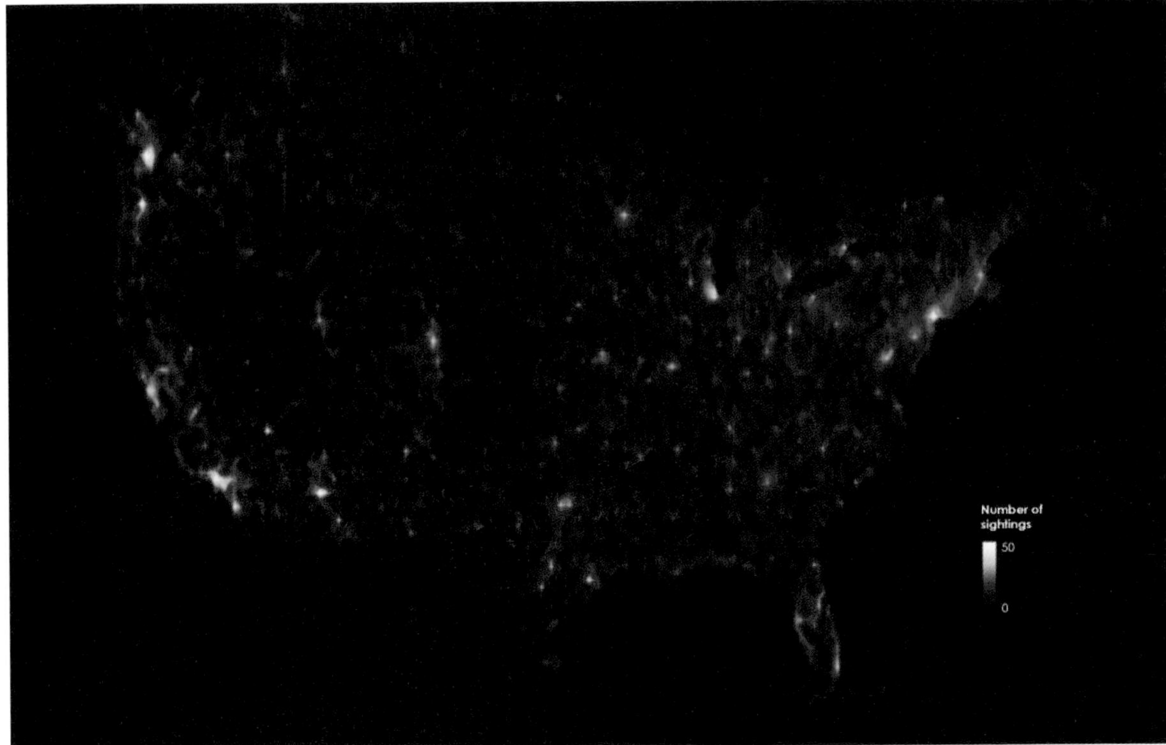

FIGURE 4-34 *Filled contour map*

CARTOGRAMS

A challenge with mapping regions, the choropleth map in particular, is that larger regions always get more visual attention regardless of the data. They take up more space in the physical world and on the computer screen. Cartograms are one way to remedy this. Location is somewhat preserved, but geographical areas and boundaries are not.

For example, a diffusion-based cartogram preserves boundaries but stretches them out so that the area of regions match the data. For example, Figure 4-35 shows the UFO sighting data as a cartogram. Notice the shrinking of Texas and swelling of California.

Obviously, the upside of cartograms is that areas fill the appropriate amount of space, but the trade-off is less geographic accuracy. When your data is for larger regions, with a wide range of sizes, this trade-off is worth it, but when regions are uniform in size, a choropleth map is most likely a better fit.

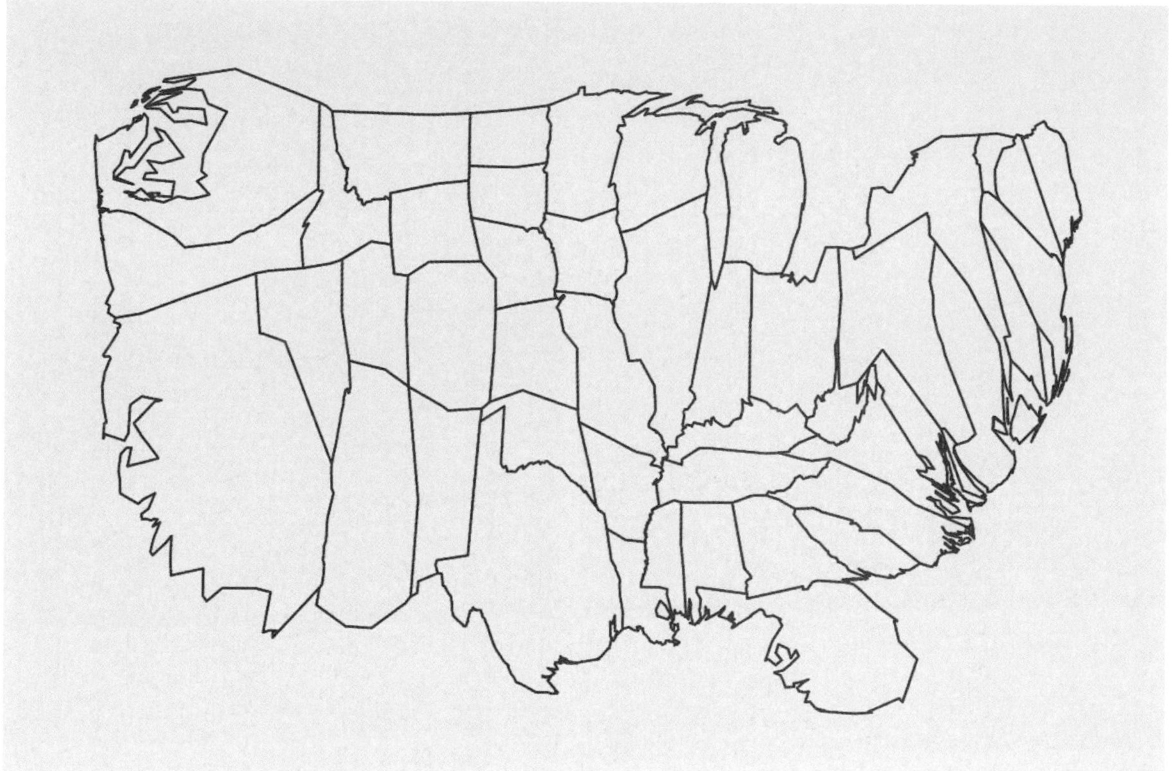

FIGURE 4-35 *Diffusion-based cartogram*

WHAT TO LOOK FOR

Spatial data is a lot like categorical data, but with a geographic component. You should know the range of the data to start with, and then look for regional patterns. Are there higher or lower values clustered in a certain area of a country or continent? Because a single value only tells you about a small part about a region filled with people, think about what a pattern implies and look to other datasets to verify hunches.

MULTIPLE VARIABLES

Data often comes in table form, with multiple columns, and each column represents a variable. You might have response data from a poll with multiple questions, results from an experiment that measured multiple aspects of a system, or demographic data on countries that includes multiple bits of information on each.

Some visualization methods let you explore multivariate data in one view. That is, all your data might fit onto a screen, and you can interpret relationships between variables and explore trends in individual ones.

Often though, the relationships between variables aren't straightforward. There isn't always a clear increasing or decreasing trend. In these cases, multiple views using more straightforward charts and graphs can help a lot. As usual, your approach depends on the data you have.

A FEW VARIABLES

With time series data, you look for how a variable changes when another variable, time, does. Similarly, when you have two metrics about people, places, and things, you might want to know how one metric changes, given the other does. Do cities with higher burglary rates also have higher homicide rates? What is the relationship between housing prices and square footage? Do people who drink more soda per day tend to weigh more?

You can visualize relationships similarly to how you look for them with time series data. Whereas the dot plots in this chapter placed time on the horizontal axis and a variable on the vertical axis, a scatter plot replaces time with a different variable, so you have two variables plotted against each other, as shown in Figure 4-36.

Each dot represents a player during the 2008–2009 NBA basketball season. Usage percentage, an estimated percentage of possessions that a player is involved in while on the court, is plotted on the horizontal axis, and points per game is plotted on the vertical axis. As you might expect, those who spend more time with the ball tend to score more points per game.

This statistical relationship between variables is called *correlation*. As one variable increases, the other one usually does, too. In this example, the correlation is strong and obvious in the chart, but the correlation strength can vary, as shown in Figure 4-37.

For a more defined view of how two variables are related, you can fit a line through the points, as shown in Figure 4-38. You saw the same method used with time series data in Figure 4-21. The increasing curve rounds off as points per game approaches zero, but the line straightens out, showing a linear relationship. (It'd be a different story if the line resembled a sine wave.)

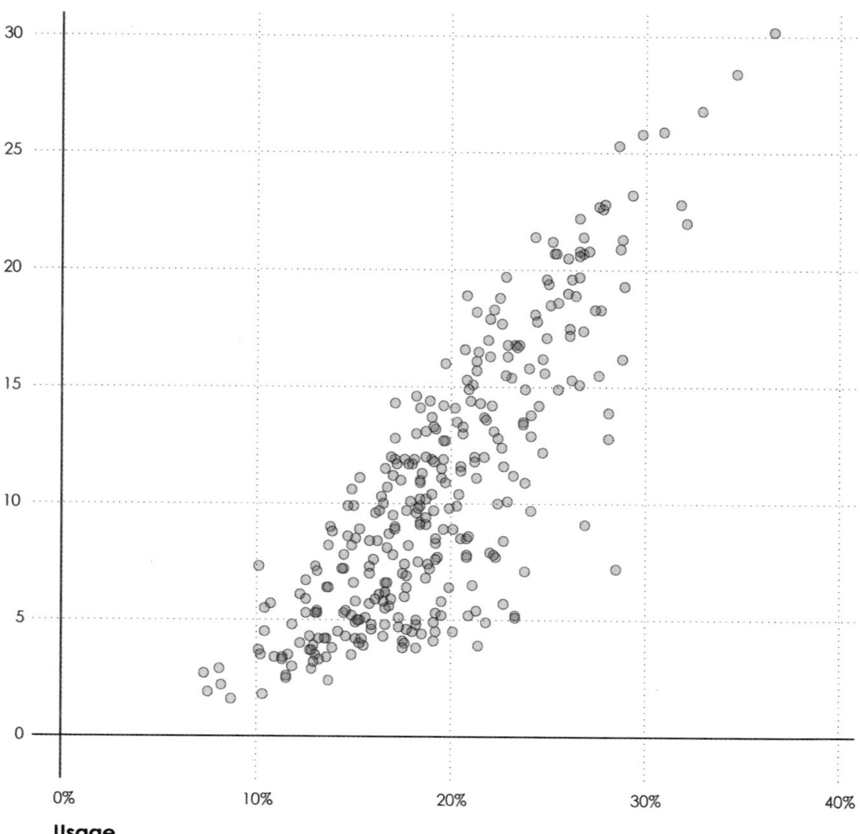

FIGURE 4-36 *Scatter plot to compare two variables*

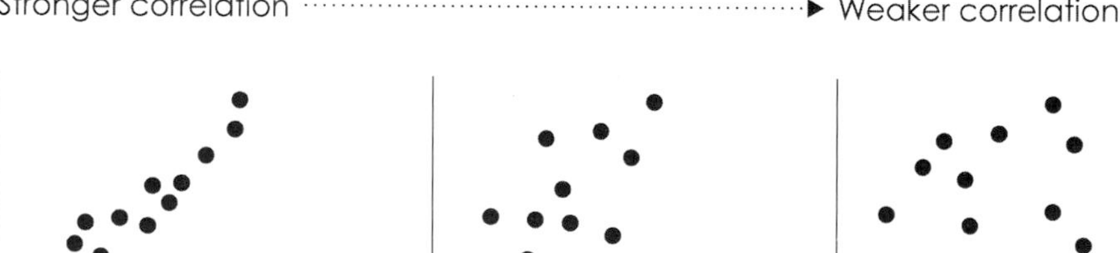

FIGURE 4-37 *Varying correlation strength*

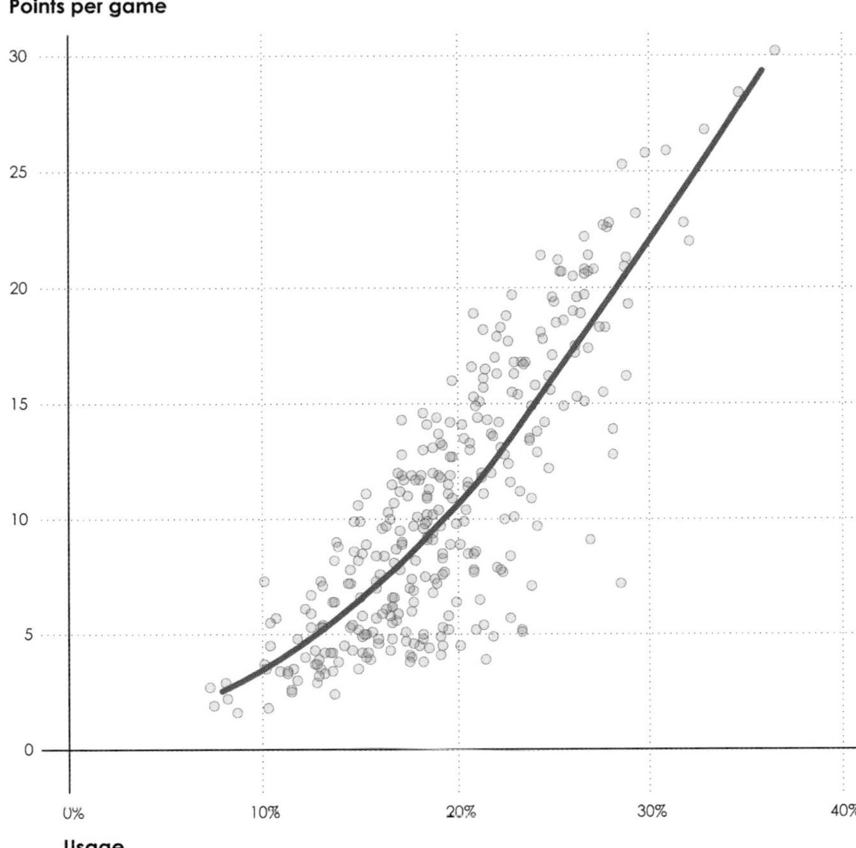

FIGURE 4-38 *Fitted LOESS curve on a scatter plot*

As you explore relationships between variables, don't confuse correlation with causation. Visualization-wise, a correlative and causal relationship between two variables will look similar, if not the same, but the latter usually requires rigorous statistical analysis and context from subject experts.

Obviously, some causal relationships are easy to interpret, such as when you place your hand over an open flame, you burn yourself. That's why you don't walk around in fire. On the other hand, the price of both milk and fuel has increased over the years. If you want to decrease the cost at the pump, should you just decrease the price of milk? Do basketball players score more points because they handle the ball more often, or do they handle the ball more often because they are good at scoring points and the coach runs more plays for such players?

Warning: Don't mix up causal and correlative relationships. They look the same when you visualize them, but the former is more difficult to prove than the latter.

Figure 4-39 shows two ways to incorporate a third variable in a scatter plot. The symbol plot on the left should look familiar because it was used with spatial data on a geographic scale earlier in the chapter. The area of a circle represents assists per game. The scatter plot on the right uses color instead of area to show the same thing. The darker the shade, the more assists per game.

The hope is that you'll see larger circles or darker shades clustered in an area of the scatter plot. In Figure 4-40, you see assist leaders closer toward the right corner of higher usage percentage and points, but there's high variability and there isn't a clear trend. There are players with a lot of assists per game who don't score that many points, and there are others who score a lot of points, have high usage percentage, and a lot of assists. It is clear however that those who don't score many points and have lower usage percentage typically don't have many assists either.

You can also double up on encodings, using both size and color to represent a third variable, as shown in Figure 4-40. The redundant visual cues help reinforce what might be more of a challenge to see with just one visual cue.

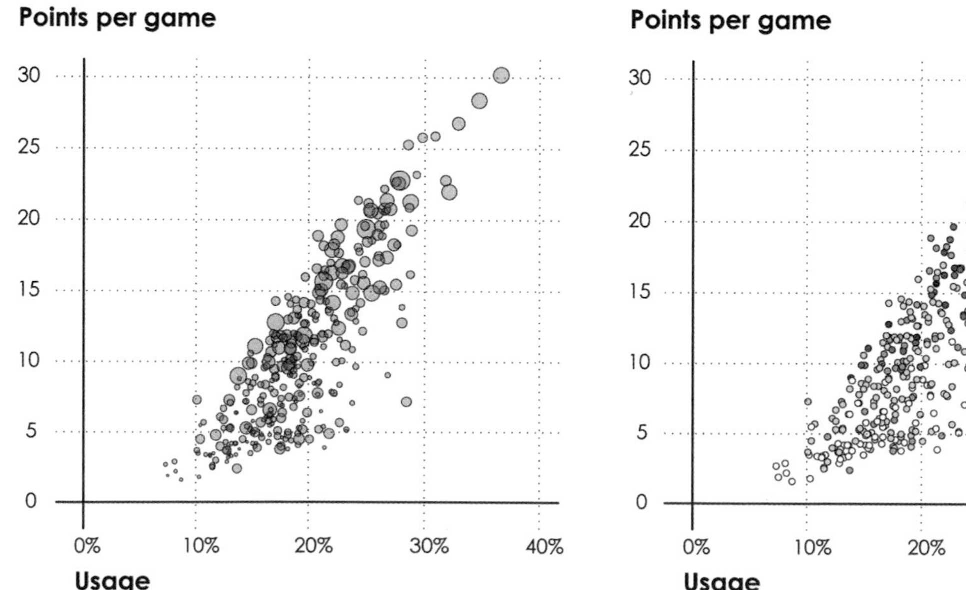

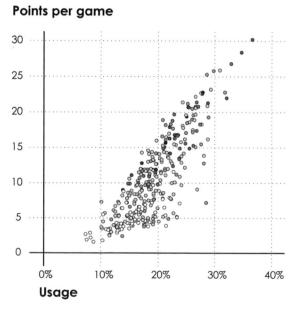

FIGURE 4-39 *Symbol plot on the left and colored scatter plot on the right*

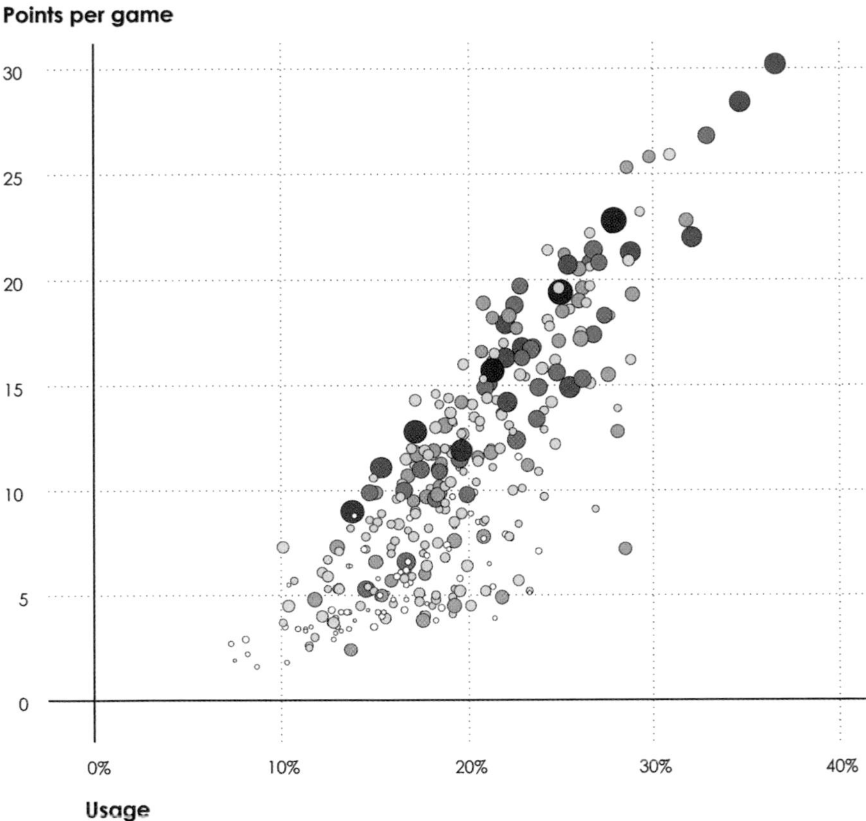

FIGURE 4-40 *Using redundant visual cues*

For example, if you were to go the opposite direction and use area and color to represent two separate metrics, the plot could be difficult to read. Figure 4-41 shows the same values on the axes, usage percentage and points per game, but uses area for rebounds and color for assists. Compare this to the previous figure, and it's clear that the additional encodings don't make anything clearer.

MANY VARIABLES

You might show four variables with a scatter plot, but what about five variables? Ten variables? There's only so much space in a scatter plot for so many visual cues. Unlike the scatter plot, there are views that are more conducive to comparing multiple variables at one time.

Note: You might be looking for a rule about how many encodings you can use at the same time before a visualization becomes useless, but I'd be overgeneralizing. It depends on the data. And the visualization. Experiment.

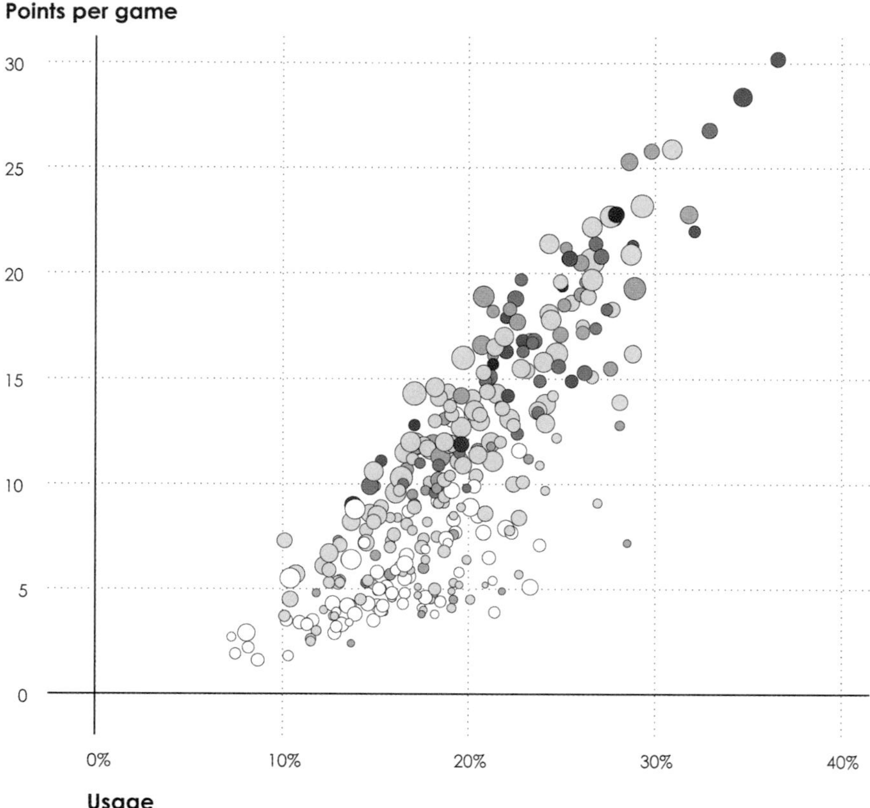

FIGURE 4-41 *Multiple encodings with a scatter plot*

A heat map, as shown in Figure 4-42, can be used to translate a table to a set of colors. It shows the same basketball player data, in addition to several other variables, including number of games played, field goal percentage, and three-point percentage. Each row represents a player, and darker shades represent relatively higher values.

With players sorted alphabetically, it's hard to see patterns, but if you sort by a column, say, points per game, as shown in Figure 4-44, relationships are easier to see. For example, usage percentage and minutes are roughly dark to light. On the other hand, the turnover rate appears to indicate a negative correlation because it goes from light to dark, and games played, field goal percentage, and three-point percentage look scattered, indicating a weak correlation, if any.

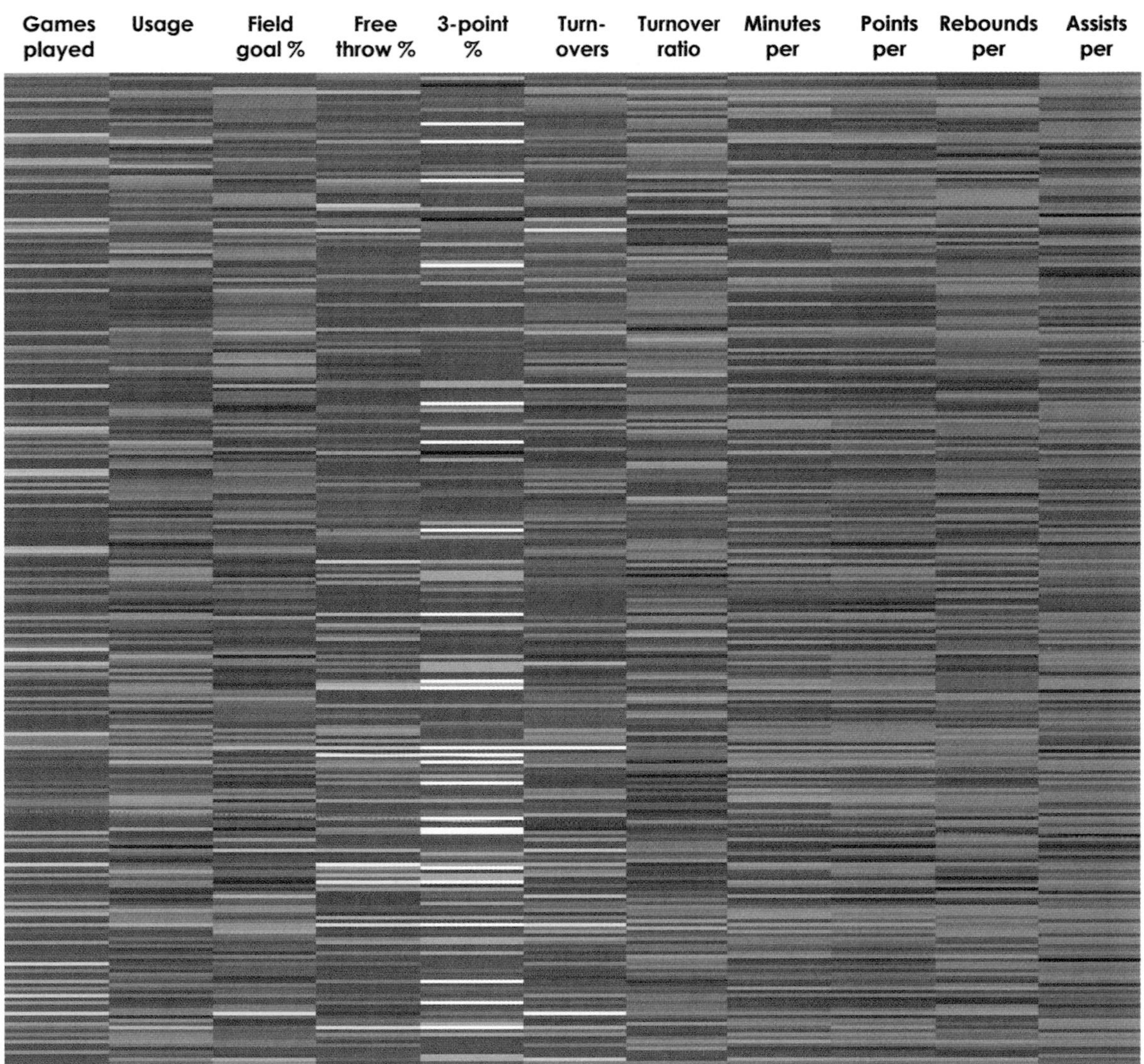

FIGURE 4-42 *Heat map to show multiple variables*

Parallel coordinates plots also arrange variables horizontally, but instead of using color like a heat map, you use vertical position, as shown in Figure 4-44. Each vertical axis represents a variable that typically ranges from the minimum and maximum of that variable, so the highest value is plotted at the top and the lowest at the bottom. Then lines are drawn left to right, positioned by the variables of each observation.

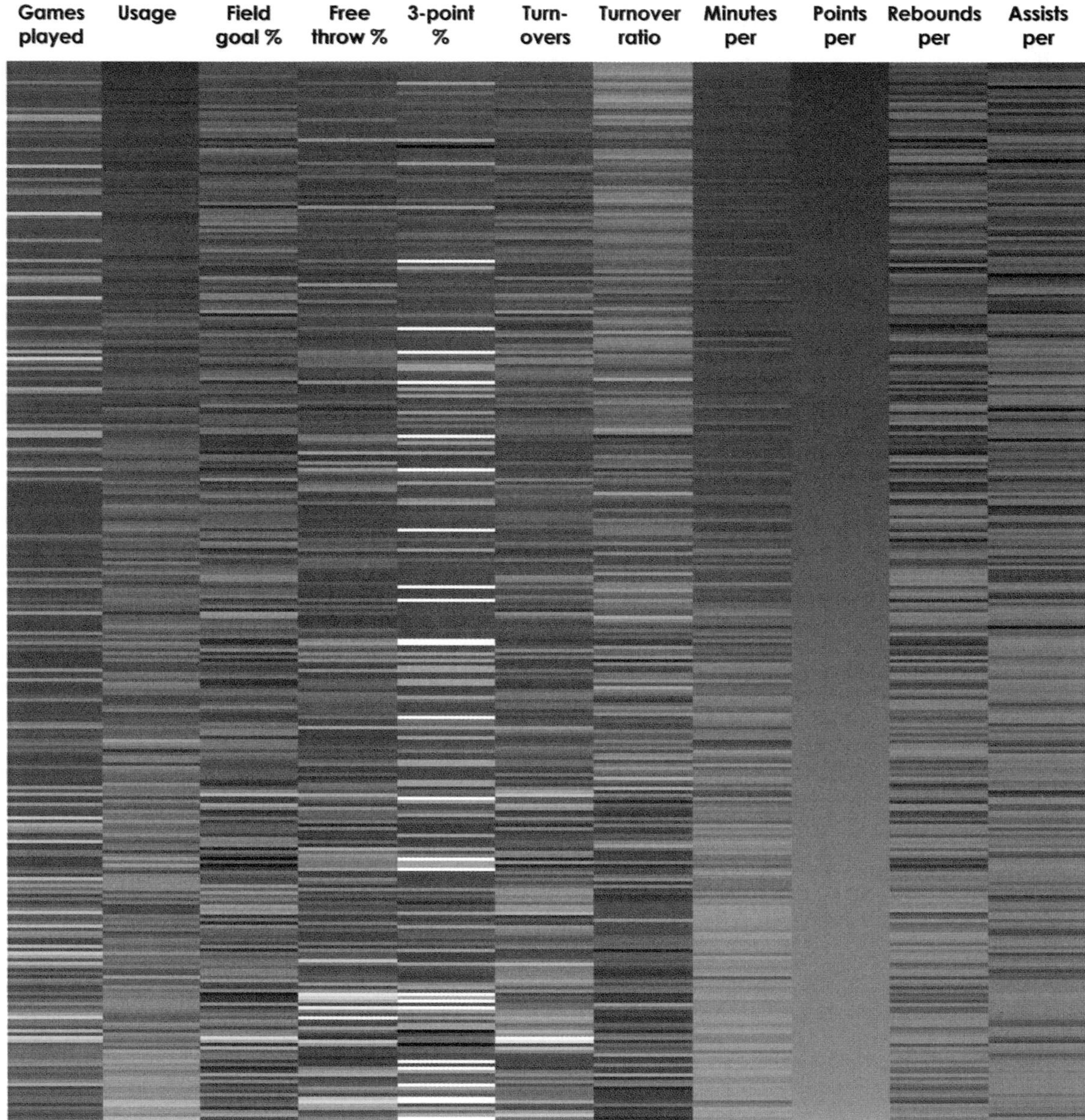

FIGURE 4-43 *Relationships with heat map*

For example, to plot a player, you start on the left, look up how many games he played, and start a line in the corresponding spot on the first vertical axis. Draw the line to the spot on the next axis that corresponds to the player's usage percentage. Do that for all the variables and all the players, and that's the parallel coordinates plot.

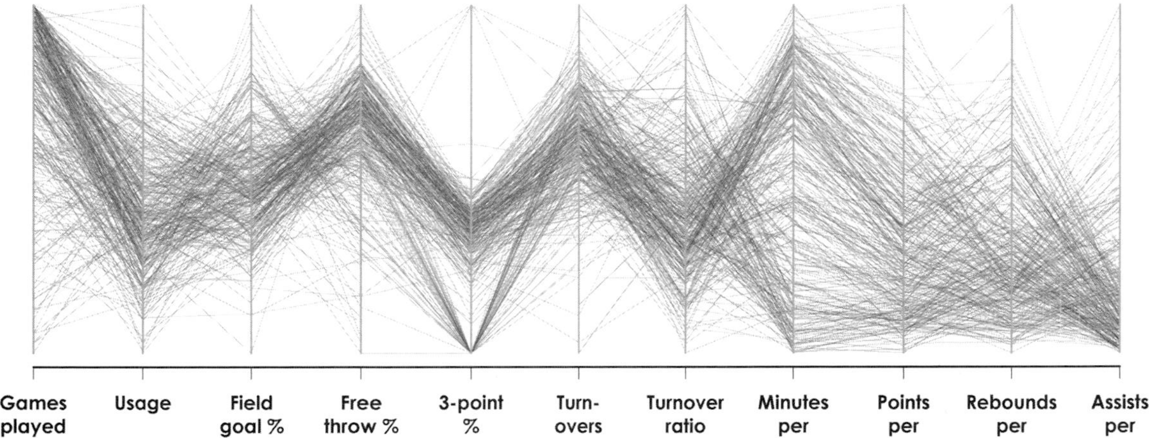

FIGURE 4-44 *Parallel coordinates plot*

If all the variables had strong positive correlations (which almost never happens), all lines would run straight across. If two variables were negatively correlated, you'd see lines on the top of one variable connect to the bottom of the axis for the other variable. Figure 4-45 shows a few more relationships.

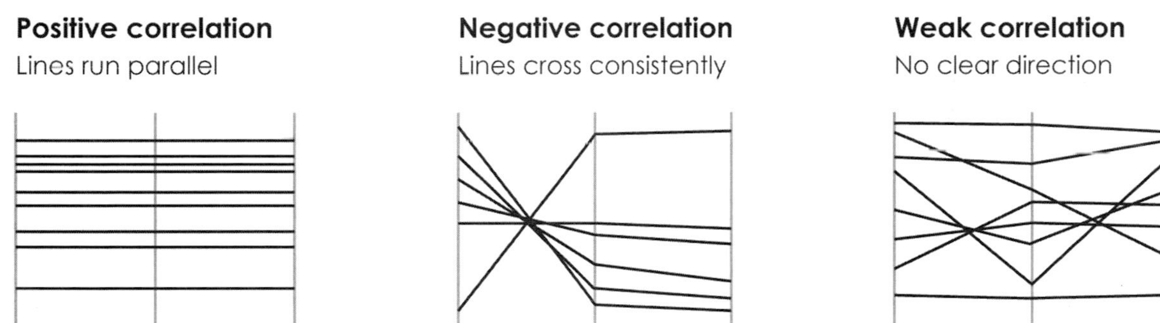

FIGURE 4-45 *Relationships with parallel coordinates plot*

When there aren't clear relationships across the board, it can be hard to see patterns. There's high variability from player to player in Figure 4-45, so you end up with a jumble of lines. You can however, highlight data based on criteria for a better view.

For example, if you highlight players who averaged five assists or more and gray out everyone else, as shown in Figure 4-46, it's easier to see how these type of players perform in other categories. Assist leaders play in more games, play more minutes, and tend to rebound less, but still vary in terms of points and field goal percentage.

Whereas the heat map and parallel coordinates plot provide an overview of the data, you might also want to look at individual data points more closely. Star plots, as you saw with time series data and shown again in Figure 4-47, present data separately. That is, you represent each row of data with its own plot.

The time series example uses the angle portion of the polar coordinate system for time. This example uses multiple variables. So in the same way that the star plot can be a polar coordinate version of a time series chart, it can also be a polar coordinate version of a parallel coordinates plot.

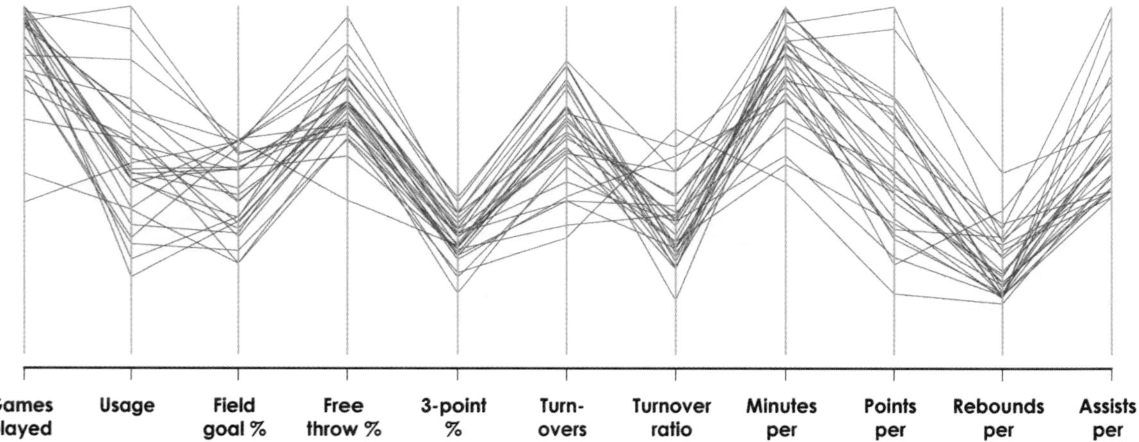

FIGURE 4-46 *Highlight data for easier reading*

USING MULTIPLE VIEWS

Note: As you gain more experience visualizing data, you'll notice that you can use many of the same type of charts for different types of data.

There's an inclination to show all the data at once, in a single view. When you have categorical data, you make a bar chart, and when you have time series data, you use a line chart. So if you have multiple variables that might be categorical, temporal, and spatial, you might also want to put it all in a single chart.

However, it can and often is better to use multiple charts instead because it lets you see the data from more angles.

You can, for example, make a lot of the same type of chart on multiple dimensions, such as the maps in Figure 4-30. The flight data is actually spatial, categorical, and temporal, so you get natural breaks in the data and a hint of places to look. Figure 4-48 explores the time series component of the flight data. Each line represents flight volume for an airline.

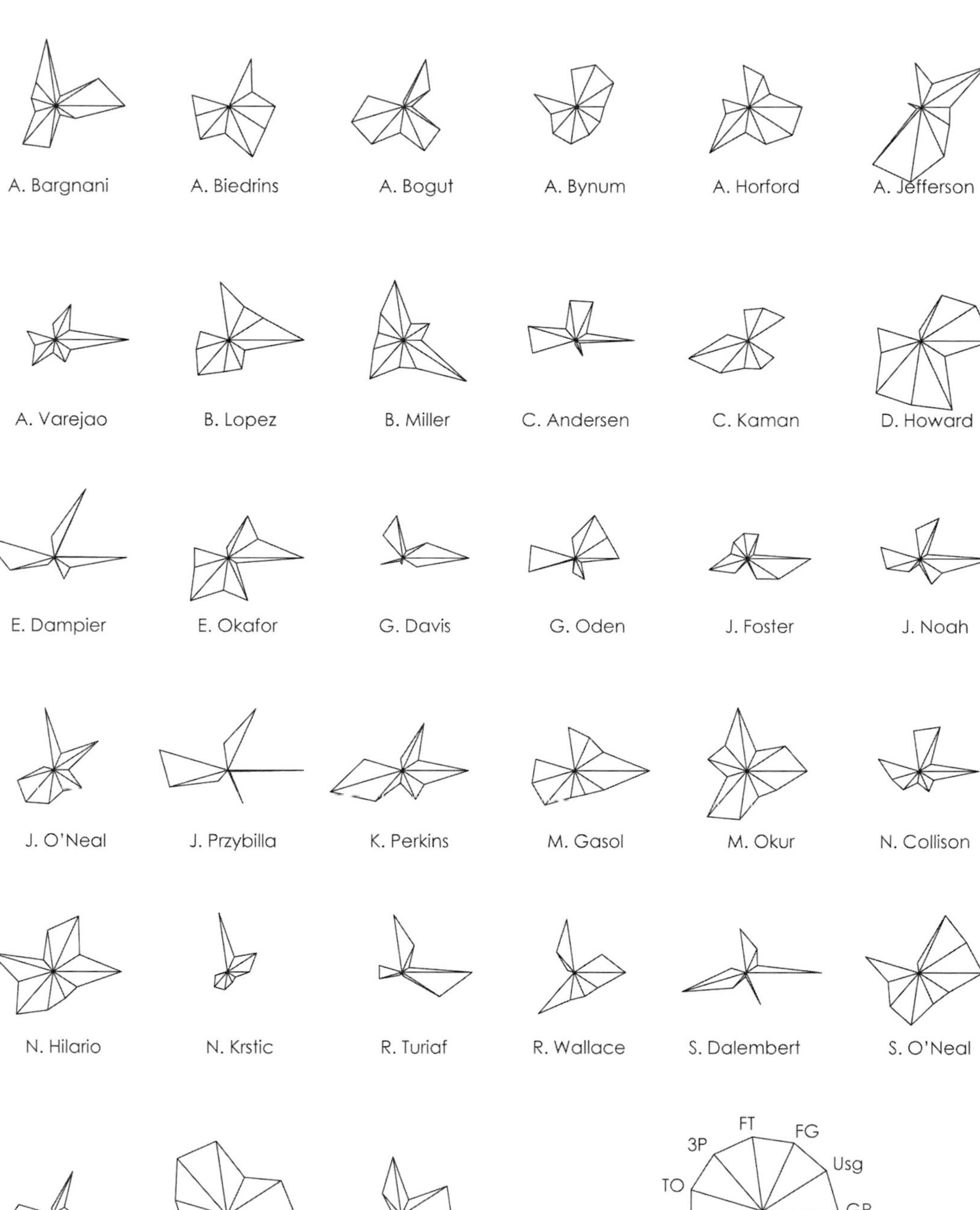

FIGURE 4-47 *Star plots*

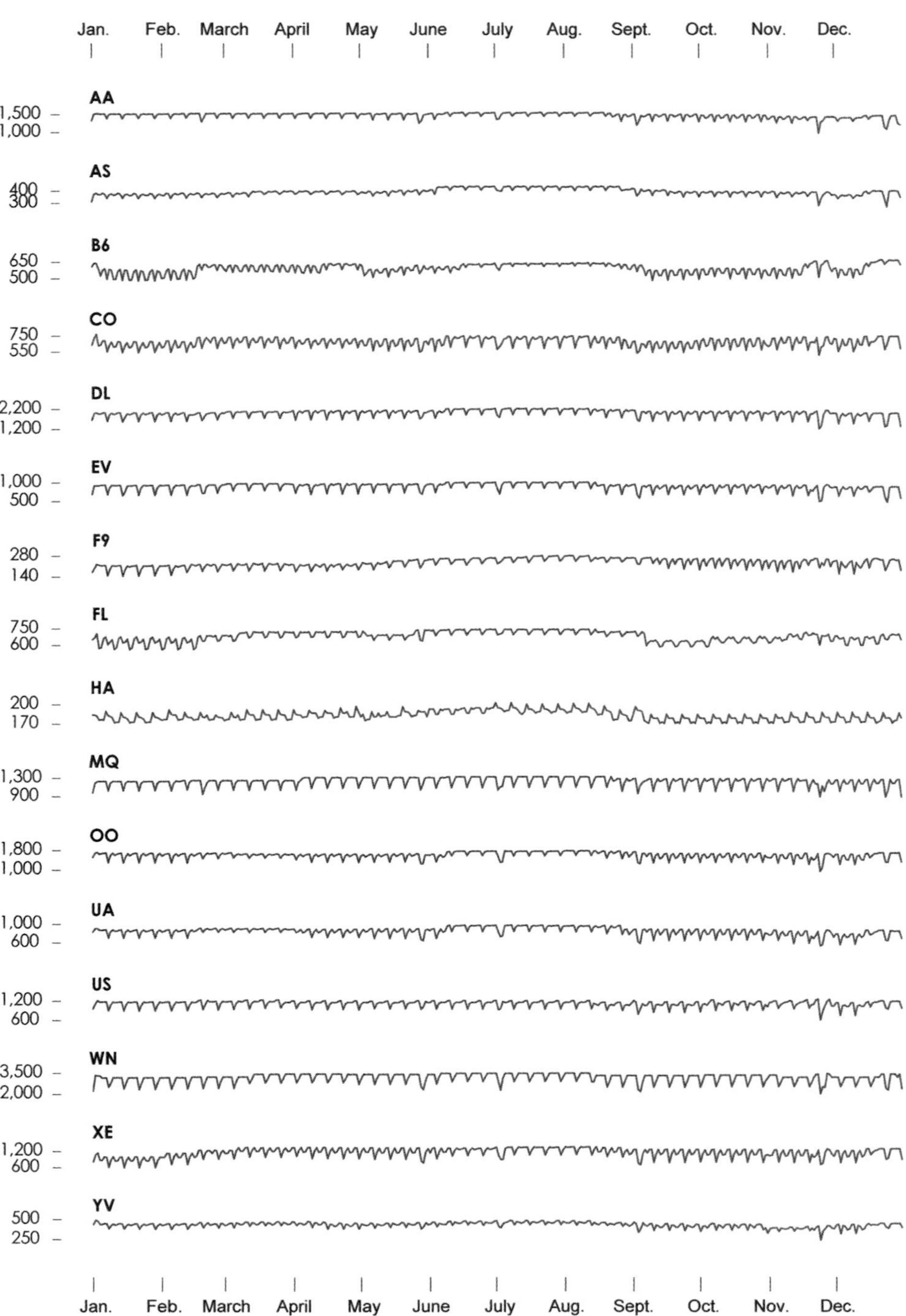

FIGURE 4-48 *Multiple time series charts for categories*

An alternative to the parallel coordinates plot, a scatter plot matrix, as shown in Figure 4-49, can show similar relationships. The relationships between variables are often easier to see in the matrix than with parallel coordinates because you can compare pairwise correlations instead of trying to decipher relationships between multiple variables at once. The latter is often complicated and hard to see.

It's also often useful to look at data with different views at the same time. For example, Figure 4-50 shows data as a heat map, bar chart, and star plot, for several players. The heat map provides detailed information about where the players shoot from; the bar graph provides an overview of the aggregates; and the star plot shows values for additional variables. Together, the views represent the playing style of several individuals, or more generally speaking, a detailed overview of several categories.

WHAT TO LOOK FOR

There are a lot of visualization methods that help you explore various aspects of your data, whether it is categories, time, space, or a combination of these. You can visualize the data all at once, but you can also make use of simpler, more straightforward views, which can help extract relationships. Sometimes the relationships are straightforward between two variables, but usually the relationship is complex, especially when you introduce more than two variables. Don't make assumptions as you explore relationships, and keep in mind there are variables not captured in the data that might contribute to changes. Finally, when it comes to correlation and causation, you need to take in all the context you can before you assign the latter.

DISTRIBUTIONS

You often hear or read about means and medians. They're used to describe a group of people, places, or things, and these measurements typically imply what is "normal" or "average," and anything that is far away from these measurements is abnormal or above or below average. However, what qualifies as extremely above average or just slightly below average? Is something 10 percent greater than the median a lot or a little? To answer these questions, you must know more about the data than just where the middle is. You have to know the spread.

FIGURE 4-49 *(following page)* *Scatter plot matrix*

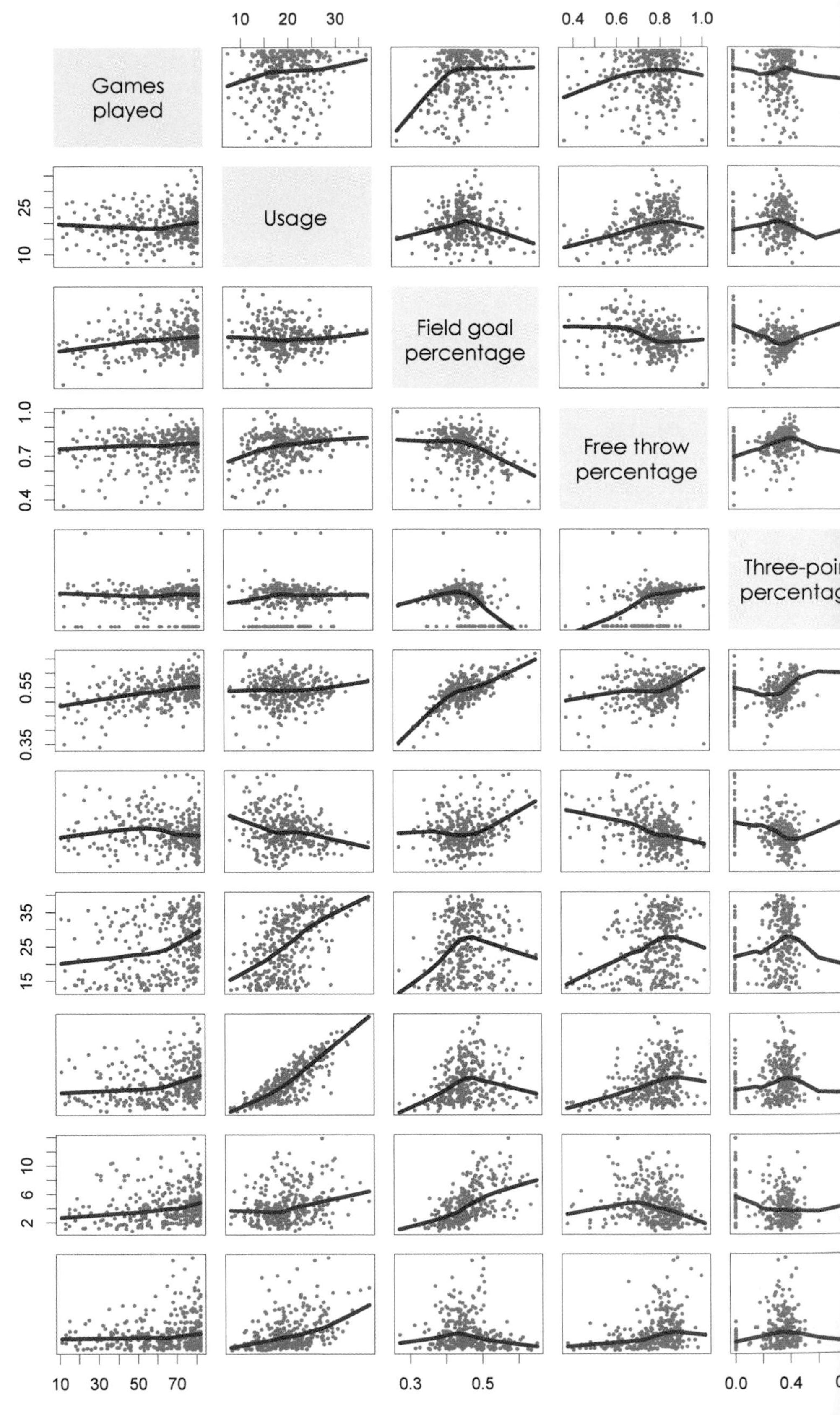

Games played
Usage
Field goal percentage
Free throw percentage
Three-poir percentag
10 20 30
0.4 0.6 0.8 1.0
25
10
1.0
0.7
0.4
0.55
0.35
35
25
15
10
6
2
10 30 50 70
0.3 0.5
0.0 0.4

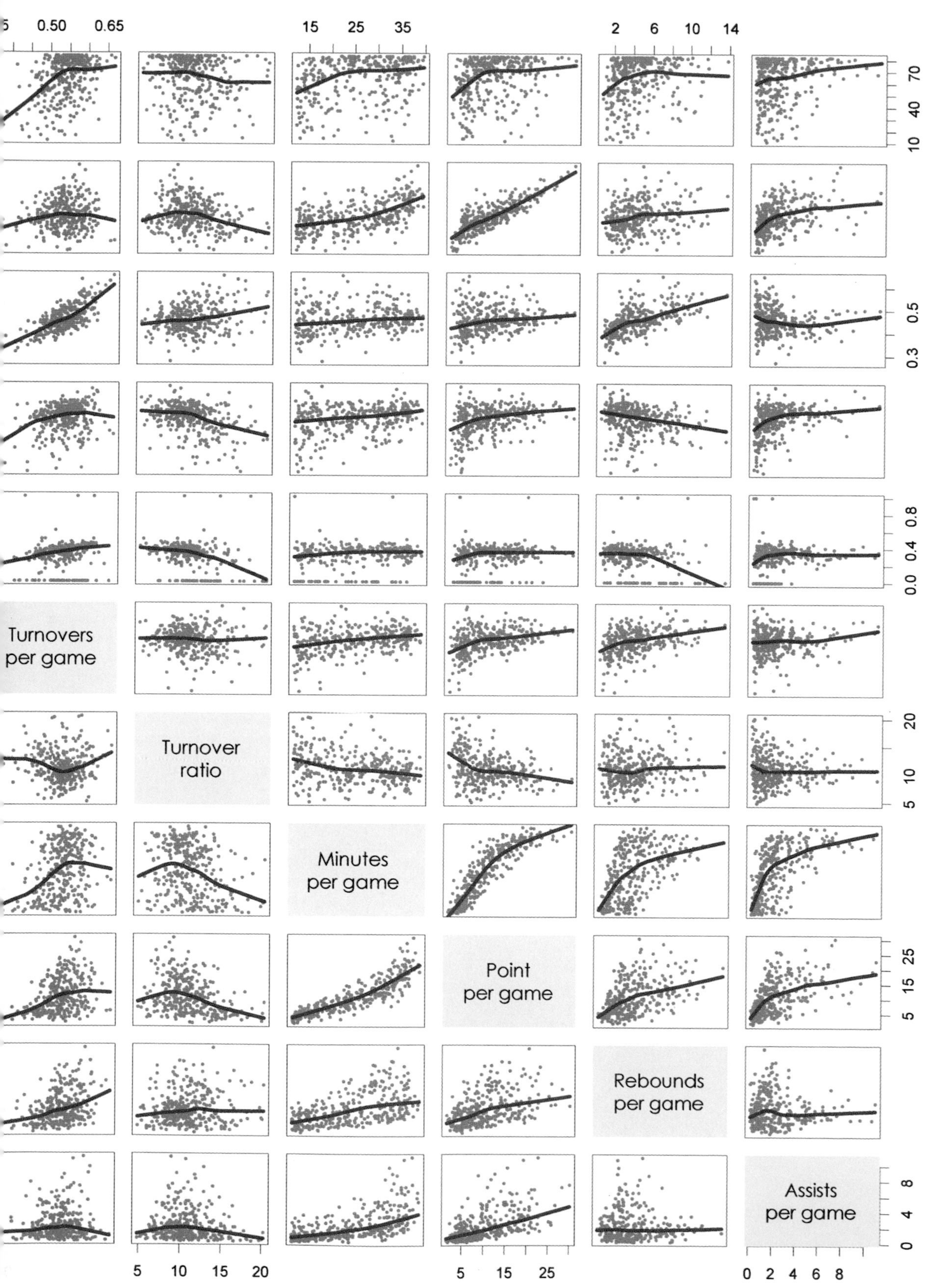

Turnovers per game
Turnover ratio
Minutes per game
Point per game
Rebounds per game
Assists per game

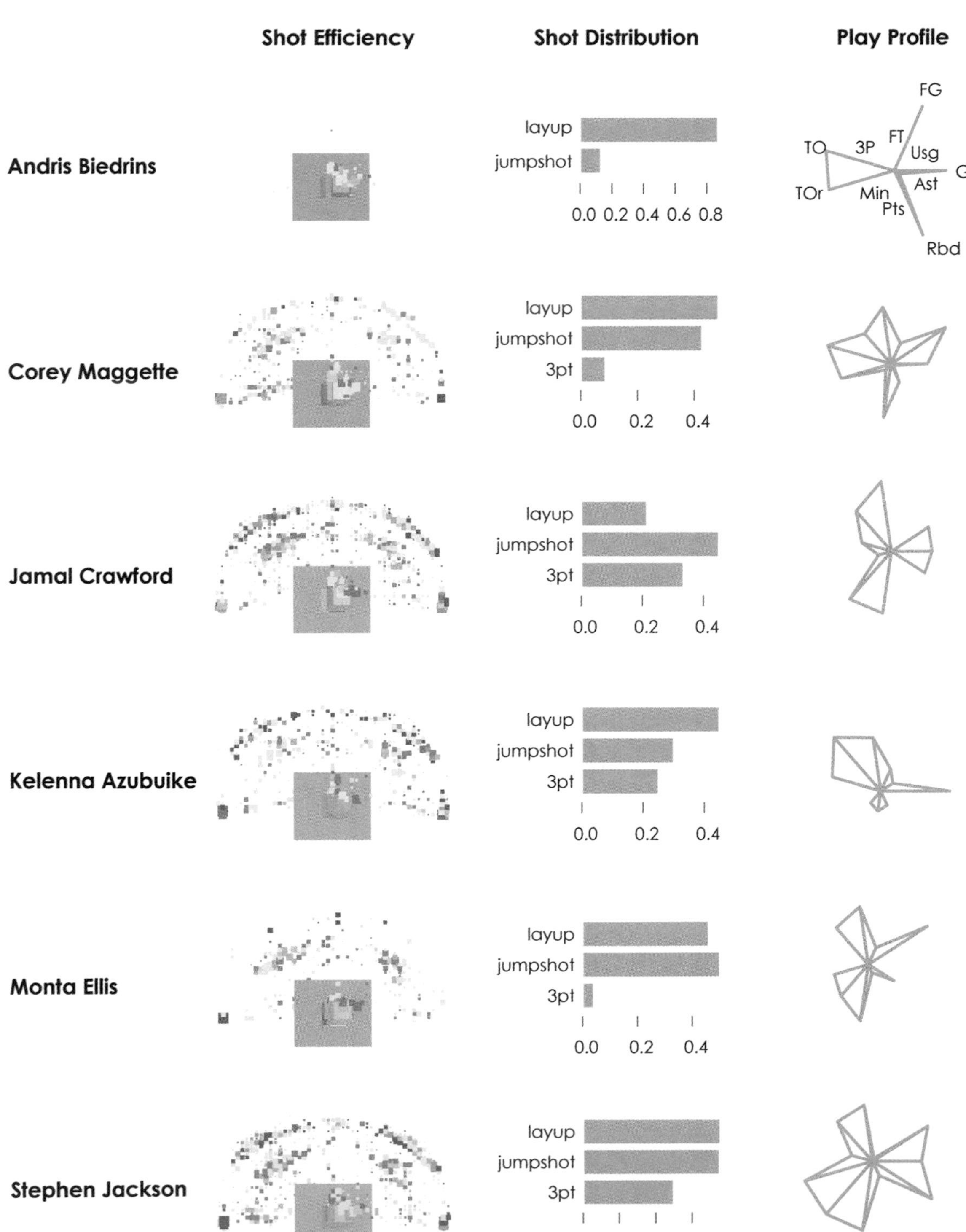
Shot Efficiency
Shot Distribution
Play Profile
Andris Biedrins
layup
jumpshot
0.0 0.2 0.4 0.6 0.8
FG
FT
TO
3P
Usg
GP
Ast
TOr
Min
Pts
Rbd
Corey Maggette
layup
jumpshot
3pt
0.0 0.2 0.4
Jamal Crawford
layup
jumpshot
3pt
0.0 0.2 0.4
Kelenna Azubuike
layup
jumpshot
3pt
0.0 0.2 0.4
Monta Ellis
layup
jumpshot
3pt
0.0 0.2 0.4
Stephen Jackson
layup
jumpshot
3pt
0.0 0.1 0.2 0.3

Now look at a classic example. Imagine there are 100 adults in a room. These 100 people have different heights, as shown in Figure 4-51. They range from 4 feet and 10-inches tall to 6½-feet tall, and the average height for the group is 5 feet and 4 inches.

FIGURE 4-50 *(facing page) Using multiple visualization methods to explore different dimensions*

It's hard to determine how many people there are in various height ranges without counting each dot, but you can get a better idea if you sort everyone from shortest to tallest, as shown in Figure 4-52. There are a few relatively tall people and a few short people, but most heights are around the 5 to 6-foot range. The median line at 64 inches is in the middle, where 50 people are shorter and 50 people are taller.

You get a better sense of the heights in the room, but there's a better way to see the distribution. You can group them into height categories or bins, such as those in between 4 feet and 4½- feet, as shown in Figure 4-54.

Now it is easy to see where most people are centered and to see the spread across a range. However, the dot plot can take a lot of space, especially if you had a lot more heights to show. So instead of dots, you could use bars, as shown in Figure 4-54. This chart is called a *histogram*, which you'll see more of soon. This counting and binning process is the basis for visualizations used to explore distributions.

As shown in Figure 4-55, you can visualize distributions with varying levels of granularity. Some views show only summary statistics, such as median, whereas other views, such as the histogram, show distribution in greater detail.

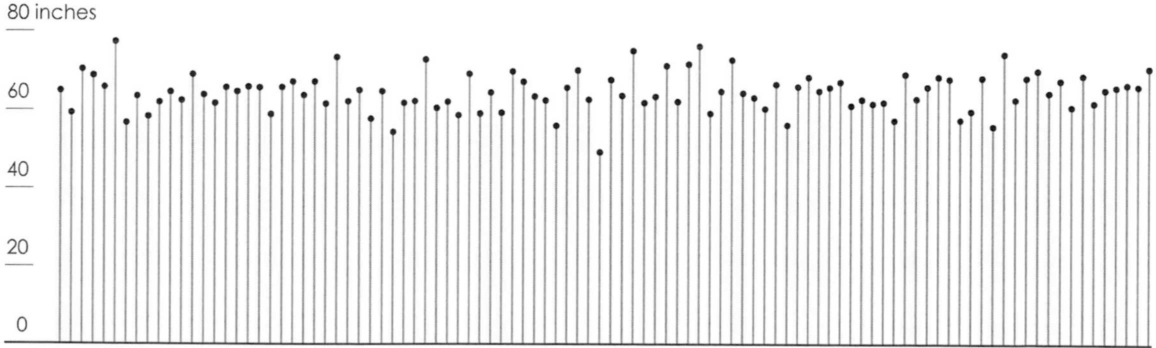

FIGURE 4-51 *Heights of 100 imaginary people*

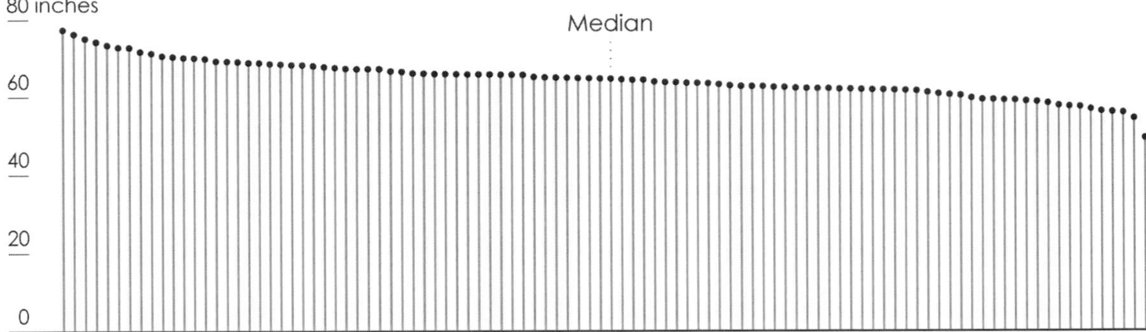

FIGURE 4-52 *Heights of imaginary people, sorted from shortest to tallest*

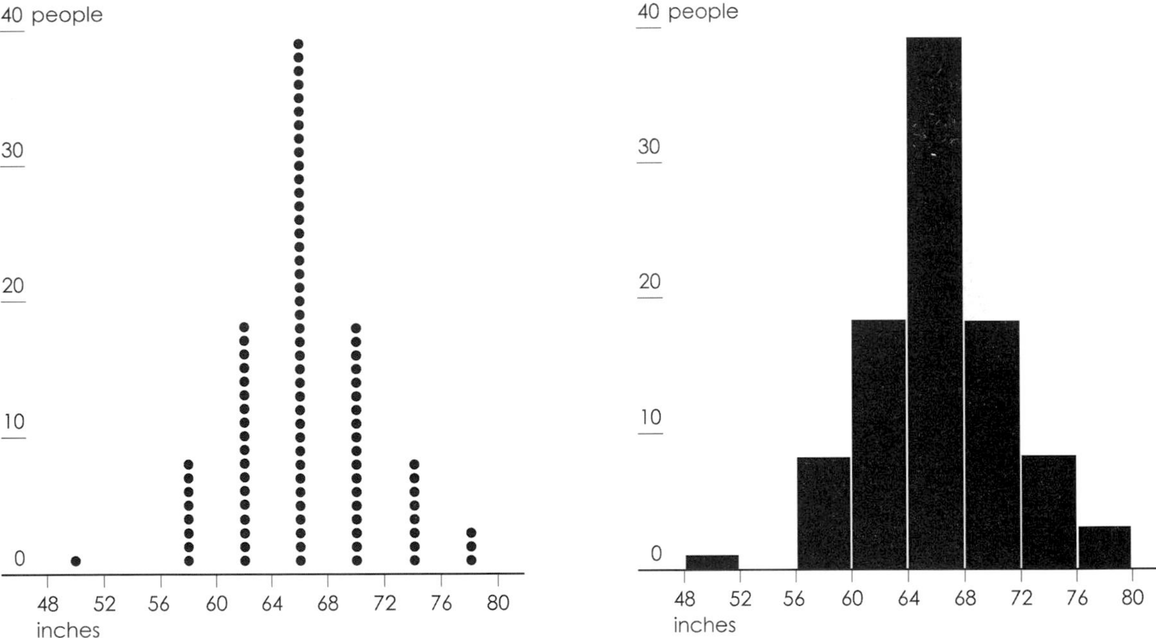

FIGURE 4-53 *Heights arranged in bins*

FIGURE 4-54 *Histogram showing heights*

The box plot, as shown in Figure 4-56, is an overview visualization that provides a general sense of distribution. The box in the middle is defined by the lower and upper quartiles. That is, whereas the median (the line in the middle) represents the halfway point, the lower quartile represents where one-quarter of the values are lower, and the upper quartile represents where one-quarter of the values are higher.

Distribution Summary

You can visualize data at different granularities with the charts above. These show key values for a less specific view of distributions.

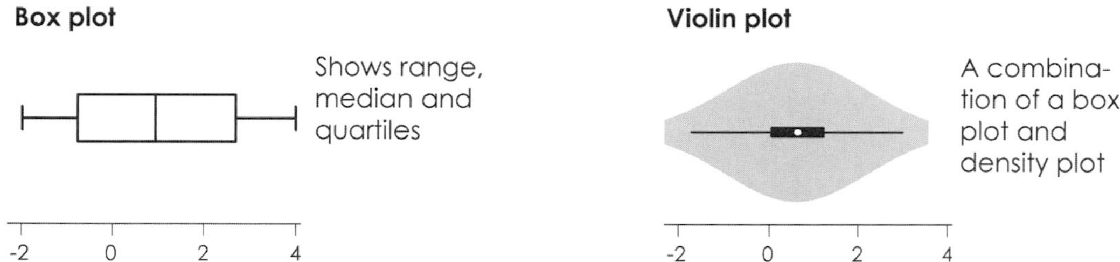

Distribution of one variable

You can see where data is clustered and see any outliers by keeping track of where they sit on a value axis.

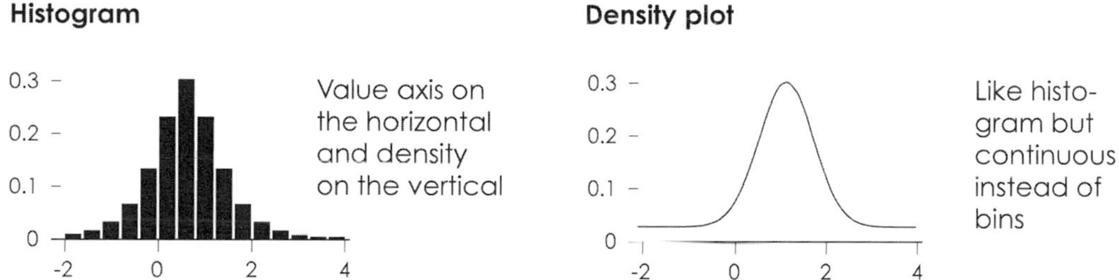

Distribution of multiple variables

Sometimes values come as pairs, and it makes sense to show both values at the same time.

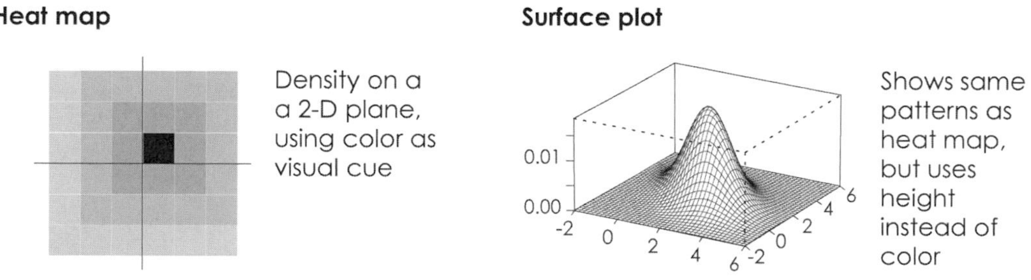

FIGURE 4-55 *Visualizing distributions*

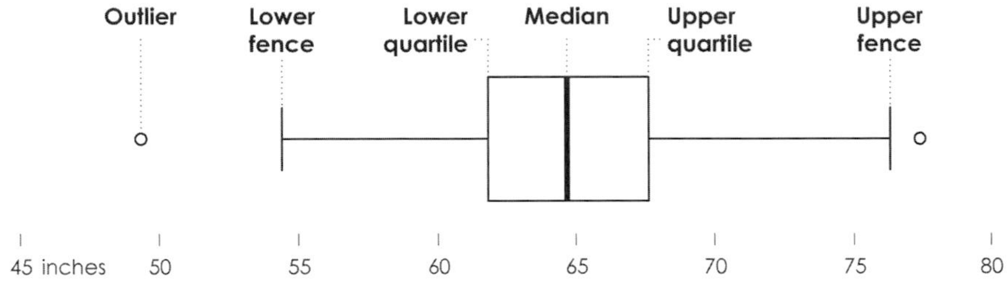

FIGURE 4-56 *Box plot*

The range in between the upper and lower quartiles is called the *interquartile range*. The outer lines are the lower and upper fences, defined by subtracting and adding 1½ times the interquartile range from the lower and upper quartiles, respectively. If the maximum and minimum values are within the upper and lower fences, the outlines are only drawn to the extremes. Otherwise, dots are used to represent any points that fall outside the upper and lower fences and are considered outliers.

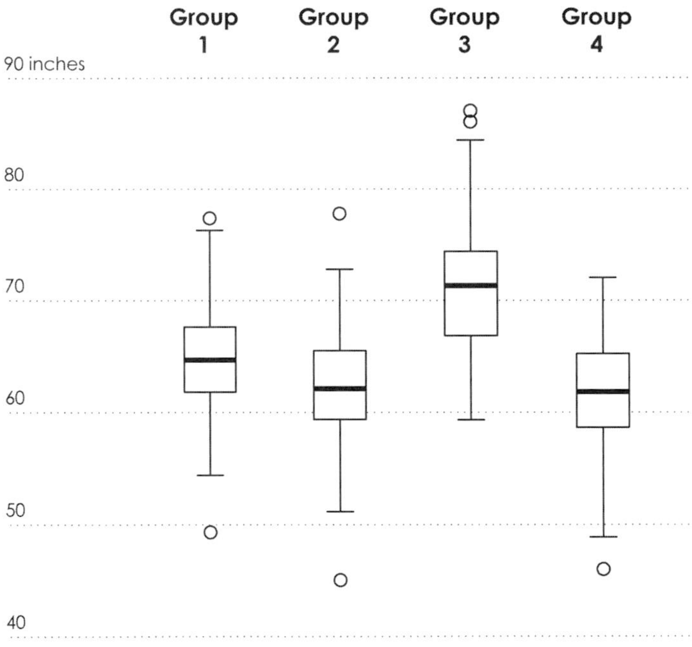

FIGURE 4-57 *Multiple box plots for comparison*

That said, the terminology makes the chart more confusing than it actually is. The main point: You can see a general distribution with a box plot. You can also use multiple box plots to compare distributions, as shown in Figure 4-57.

The histogram provides a more detailed view, which you saw in Figure 4-54. Bar height represents the proportion of values within a corresponding range, and when you change bin sizes, you change how much variability is visible. Figure 4-58 shows how the same height data can be represented with different bins.

Like box plots, you can also use multiple histograms to compare distributions. In one last return to the flight data, Figure 4-59 shows the distributions of arrival delays for major airlines. Delays of more than 15 minutes are highlighted in orange.

Note: Bin size changes by dataset, but you want it to be big enough so that you can see the variability over the range of values, but not so small that the histogram is too noisy to interpret.

One-inch bins

Small bins shows variations at higher granularity.

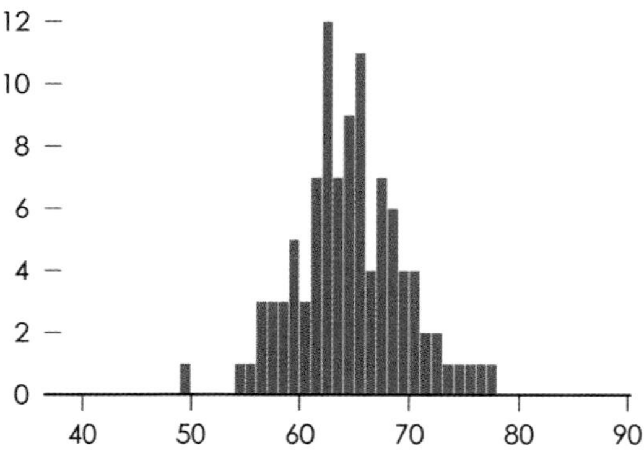

Two-inch bins

You see less variation, but the distribution around the median is more obvious.

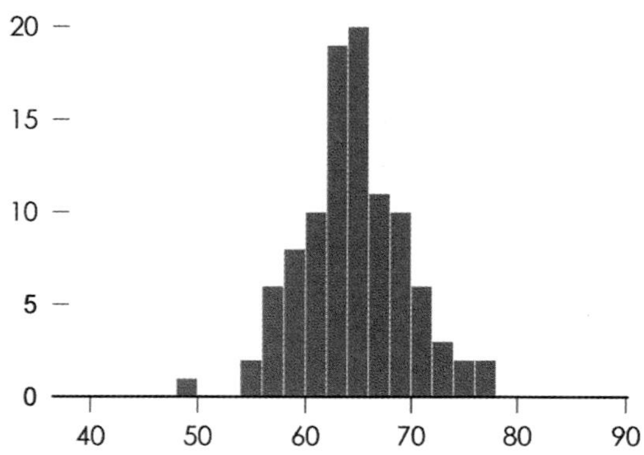

Half-foot bins

You can see distribution around the median, but you can only see some variation.

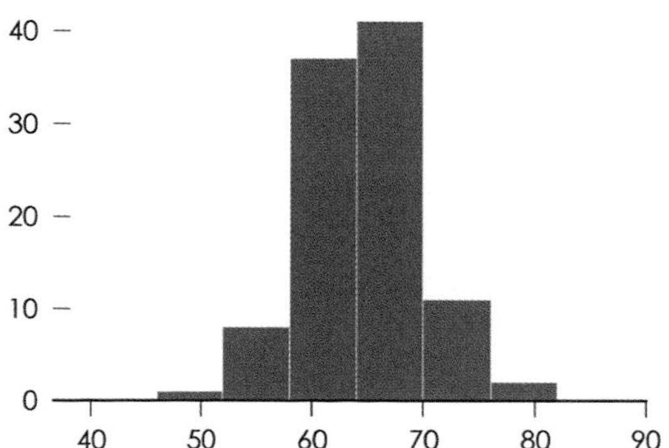

One-foot bins

The spread of the data isn't as obvious. because the larger bins show less detail.

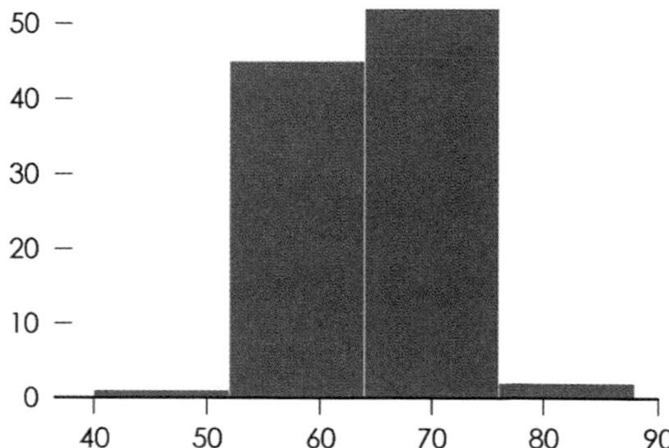

FIGURE 4-58 *Varying bin sizes with histograms*

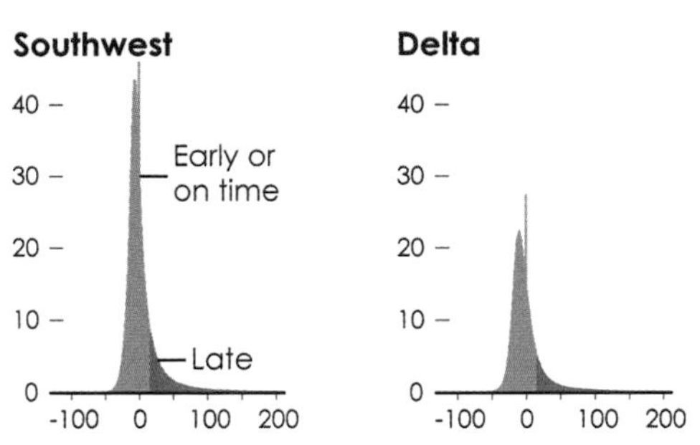

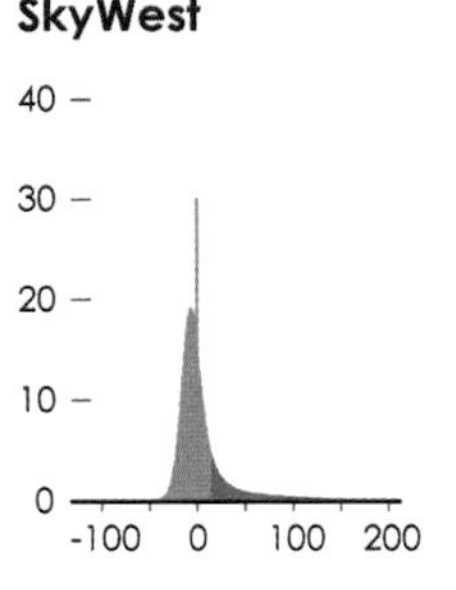

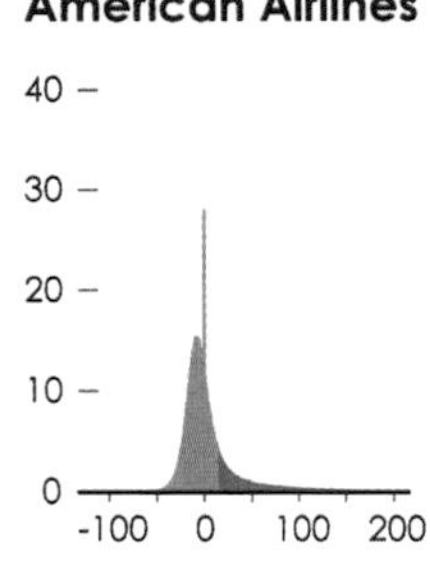

American Eagle

40
30
20
10
0
-100 0 100 200

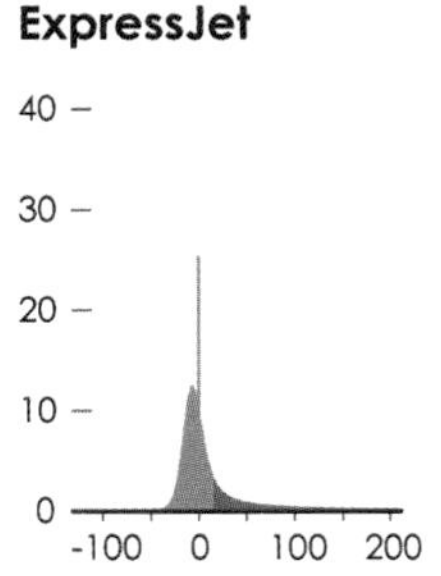

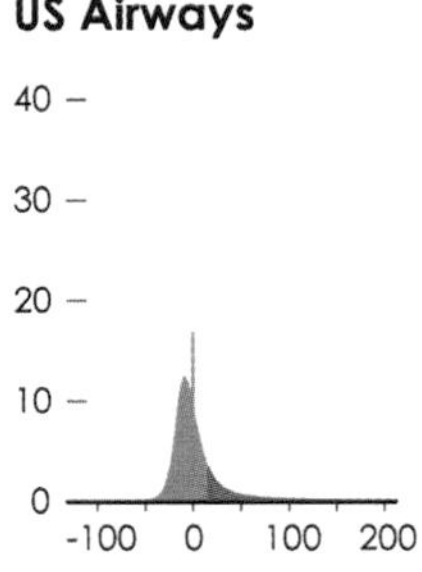

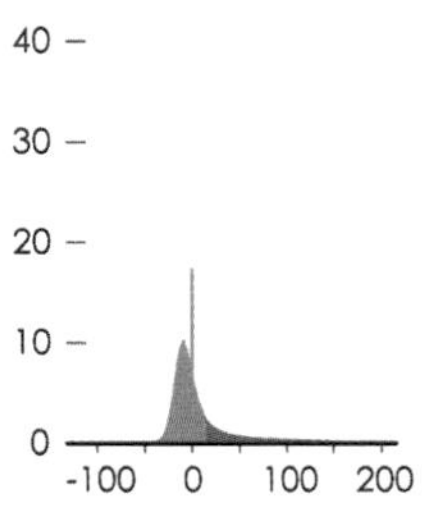

United Airlines

40
30
20
10
0
-100 0 100 200

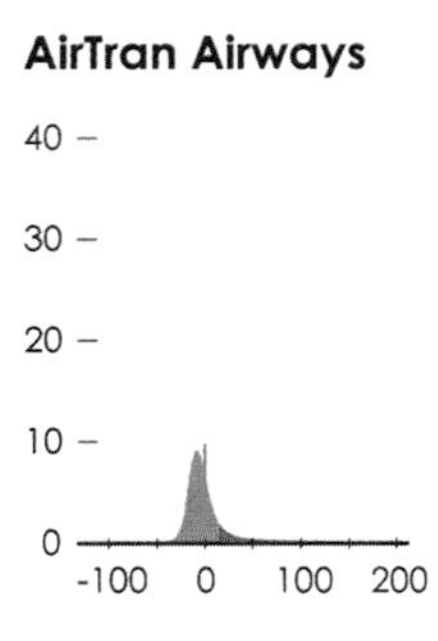

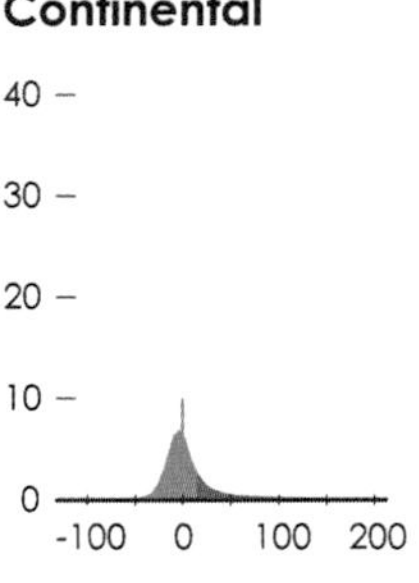

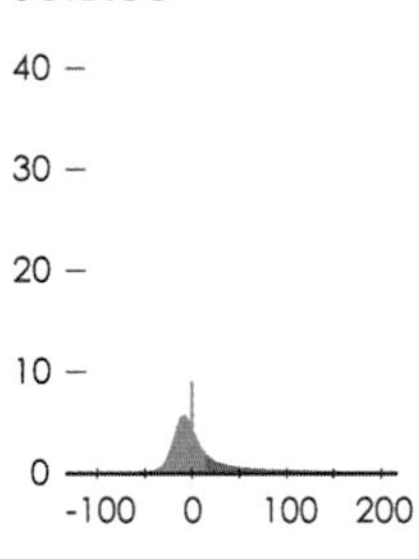

Mesa

40
30
20
10
0
-100 0 100 200

Alaska

40
30
20
10
0
-100 0 100 200

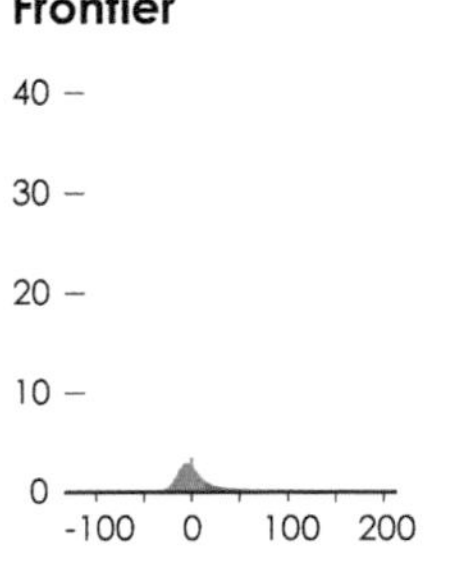

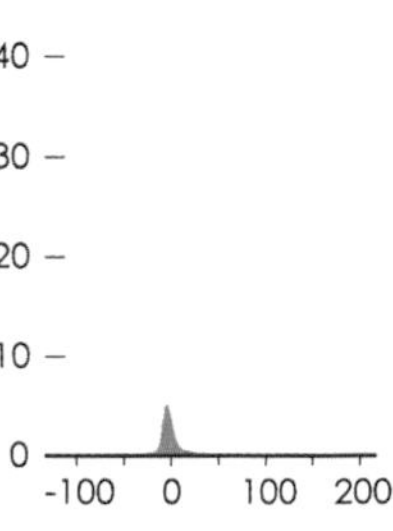

FIGURE 4-59 *Multiple histograms to compare distributions*

Notice the spike at the zero-minute mark, where airlines aim to be right on time? It looks like airlines either arrive right on time or whoever records the data rounds to on-time arrivals. You wouldn't see that with just means and medians.

WHAT TO LOOK FOR

Regardless of the type of visualization you use to explore distributions, look for peaks and valleys, range, and the spread of your data, which tell you a lot more than just the mean and median would. The visual analysis of raw data and the variation in between the summary statistics are almost always more interesting, so make use of the opportunity when you get it.

WRAPPING UP

Visualization can be a great tool to explore your data, and with advancing technologies, computers are less of a limiting factor than they were just a few years ago. So the key to getting the most out of your data—to understand what it represents and what it means—isn't so much about finding the right software than it is to learn how to use the tools you have and to know what questions to ask.

Consider what data you have and what you can get, where the data is from, how it was derived, and what all the variables mean, and let that extra information guide your visual exploration. If you use visualization as an analysis tool, you must learn as much as you can about your data. Even if your goal is to visualize data for presentation, exploration can lead to unexpected insights, which makes for better graphics.

Visualizing with Clarity

During the exploration phase, you get to look at your data from a variety of angles and browse various facets, without having to dwell on charting standards and clarity. You understand a chart better because you know more about the data after you examine lots of other quickly generated charts. However, when you use graphics to present results to other people, you must make your graphics readable to those who don't know your data as well as you do.

A common mistake is that all visualization must be *simple*, but this skips a step. You should actually design graphics that lend clarity, and that clarity can make a chart "simple" to read. However, sometimes a dataset is complex, so the visualization must be complex. The visualization might still work if it provides useful insights that you wouldn't get from a spreadsheet.

As an effort toward clarity, people often preach removing all elements of a graphic that don't help you interpret the data. When you "let the data speak," you have done your job. This is fine, but it assumes the only goal of visualization is quick analytical insight, which is a small subset of what you can get out of data. It's okay to ponder and reflect, and elements that are not helpful in one situation might be helpful in another.

That said, whether it's a custom analysis tool or data art, make graphics to help others understand the data that you've abstracted, and try your best not to confuse your audience. How do you do this? Learn how we see data, and use that to your advantage.

VISUAL HIERARCHY

When you look at visualization for the first time, your eyes dart around trying to find a point of interest. Actually, when you look at anything, you tend to spot things that stand out, such as bright colors, shapes that are bigger than the rest, or people who are on the long tail of the height curve. Orange cones and yellow signs are used to alert you on the highway of an accident or construction because they stand out from the monotony of the black pavement. In contrast, Waldo is hard to find right away because he doesn't stand out enough to stick out in a sea of people.

You can use this to your advantage as you visualize data. Highlight data with bolder colors than the other visual elements, and lighten or soften other elements so they sit in the background. Use arrows and lines to direct eyes to the point of interest. This creates a visual hierarchy that helps readers immediately focus on the vital parts of a data graphic and use the surroundings as context, as opposed to a flat graphic that a reader must visually rummage through.

For example, Figure 5-1 is the scatterplot from the previous chapter that shows NBA players' usage percentage versus points per game. The dots, fitted line, grid, border, and labels are of the same color and thickness, so there is no clear visual focus. It's a flat image, where all the elements are on the same level.

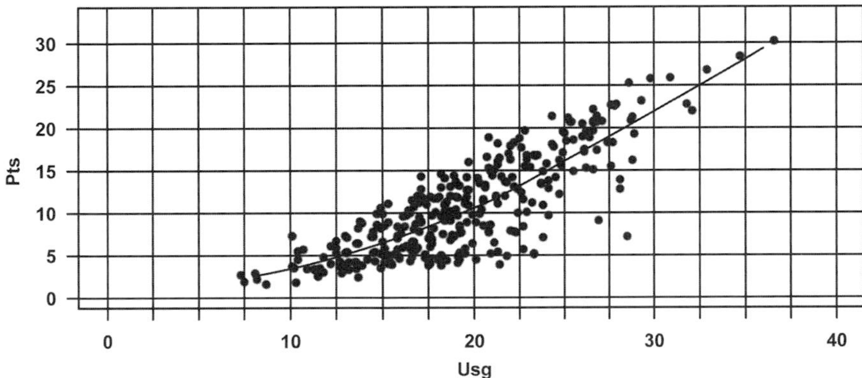

FIGURE 5-1 *All visual elements on the same level*

This is easily remedied with a few small changes. In Figure 5-2, the line width of the grid lines is reduced so that they are no longer as thick as the fitted line. In this example, you want the data to stand out. The grid lines also alternate in width so that it is easier to see where each data point lies in the coordinate system, and there's no imaginary blur that you get in the original chart.

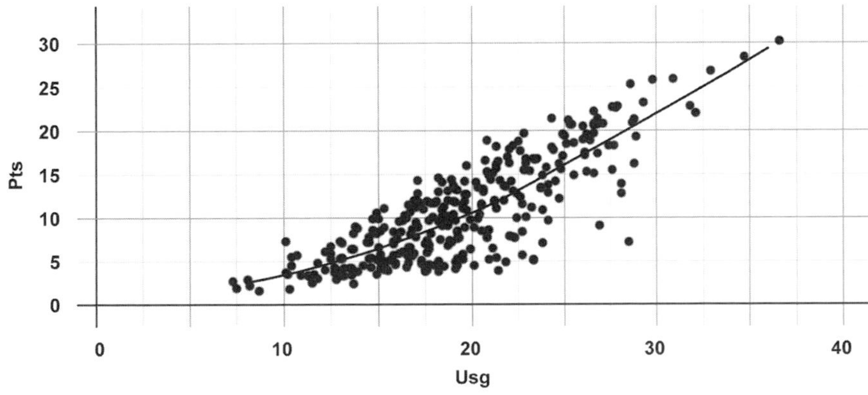

FIGURE 5-2 *Width of grid lines reduced to fit in background*

Still though, the fitted line is obscured by all the dots, because (1) it's thin compared to the radius of each dot and (2) it still blends in with the grid behind it. Figure 5-3 changes the color to blue to make the data stand out more, and the width of the fitted line is increased so that it clearly rests on top of the dots.

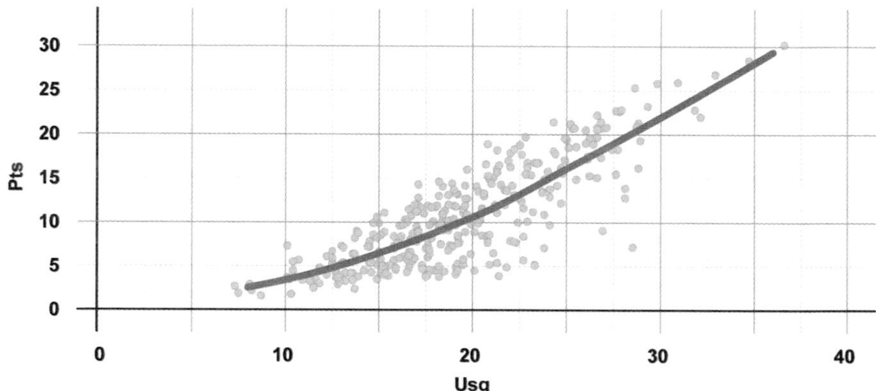

FIGURE 5-3 *Focus of chart shifted to fitted line with color and width*

The chart is a lot more readable now, but if you imagine people viewing the graphic like they would a body of text—from top to bottom and left to right—more descriptive axis labels and less prominent value labels can help, as shown in Figure 5-4. The text within the chart works similar to how it does in an essay or a book. Headers are often printed bigger and in a bold font to provide both structure and a sense of flow. In this case, the bolder labels provide immediate context for what the chart is about. Also, notice fewer and less prominent gridlines, which directs focus further to the upward trend.

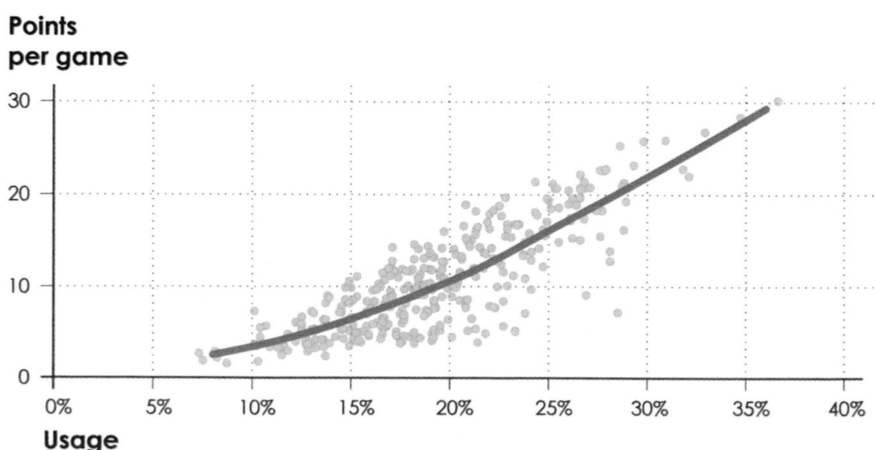

FIGURE 5-4 *Grid and value labels adjusted and fewer, less prominent gridlines*

Even if a graphic is exploratory or made to show an overview rather than a specific point or story in the data (such as a trend line), you can still use a visual hierarchy to provide structure. A presentation of a lot of data at the same time can be visually intimidating, but a breakdown by category helps readers browse visually. For example, Figure 5-5 shows 2,000 films across 20 genres over 100 years.

FIGURE 5-5 *(following page)* The History of Film *(2012) by Larry Gormley, http://historyshots.com*

Each layer alternates in color to separate genres, which makes the chart easier to read left to right, even if the name of the genre is not within view. Font and color separate genres (medium and red) and film titles (small and black), and the timeline on the bottom shows a division of film eras with tick marks. Had the same colors and fonts been used throughout, as in the scatterplot in Figure 5-1, it'd be a headache to browse this poster.

Sometimes visual hierarchy is used to show process or reflect how you might explore a dataset. Imagine you generate a lot of charts during the data exploration phase. You make a few graphs to see an overall picture, and in that summary, you note specifics and then make charts that focus on those. You can design your graphics to follow this same logic, basically taking readers on a tour of your analysis.

The bottom line: Graphics that follow a visual hierarchy are easier to read and can be used to guide readers toward points of interest. In contrast, flat graphics that lack flow make it harder for readers to interpret results and discourages closer looks. You don't want that.

READABILITY

An author who uses words to describe a world or character interactions makes abstractions so that a reader can picture what's going on. Poor descriptions and little character development challenge readers to make sense of what seem like obscure clues. If readers can't connect the dots and understand what the author tries to describe, the words lose their value.

Similarly, you encode data with visual cues when you visualize it, and then you or others have to decode the shapes and colors to draw insights or to understand what a visualization represents, as shown in Figure 5-6. If you don't describe the data clearly, which makes a data graphic readable, then the shapes and colors lose their value. The connection between the visual and the underlying data is broken, and you end up with a geometry lesson.

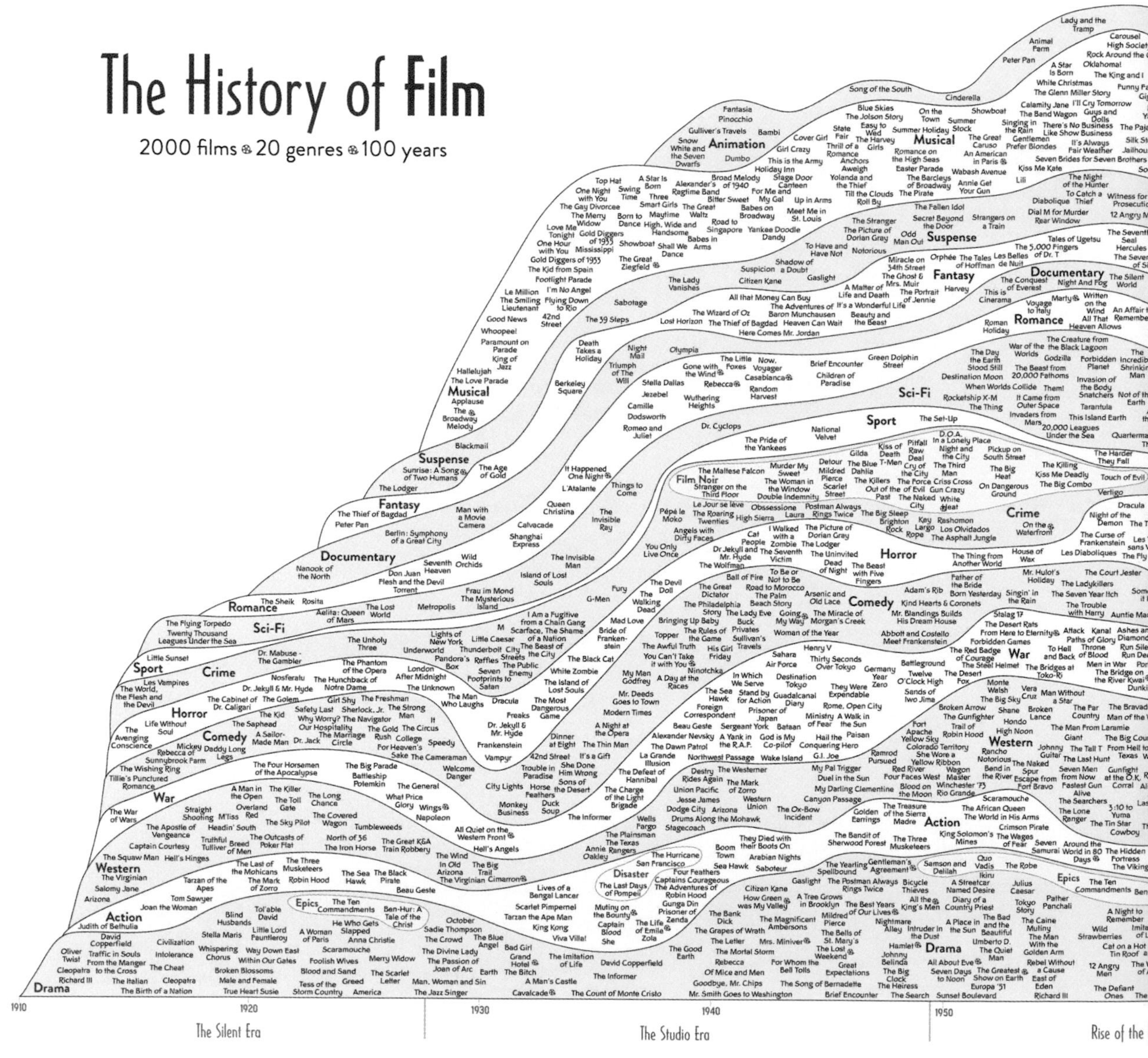

The History of Film
2000 films • 20 genres • 100 years
Animation
Musical
Suspense
Fantasy
Documentary
Romance
Sci-Fi
Sport
Film Noir
Crime
Horror
Comedy
War
Western
Action
Epics
Disaster
Drama
1910
1920
1930
1940
1950
The Silent Era
The Studio Era
Rise of the

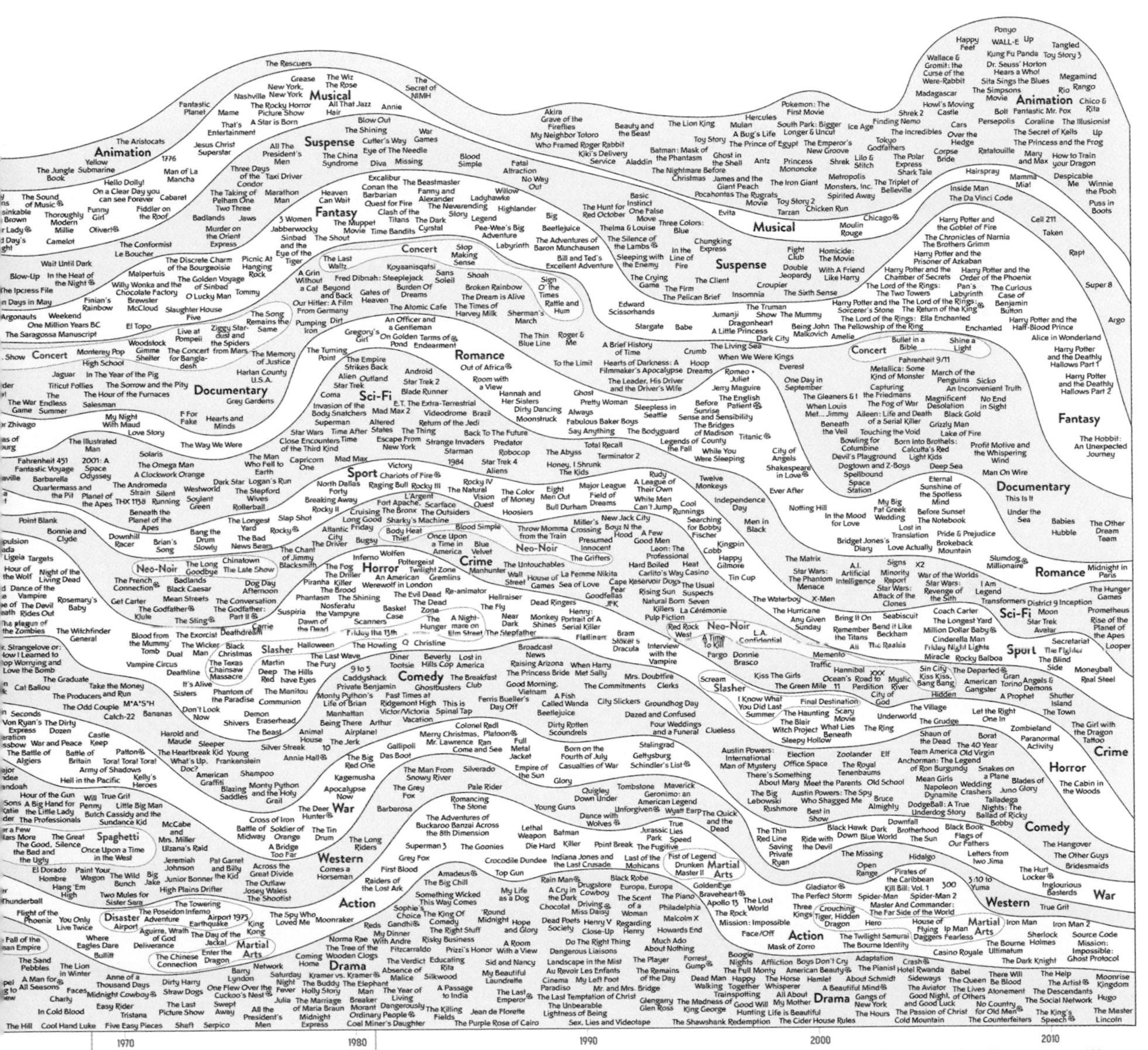

New Hollywood Era

More than 2000 of the most important feature-length films are mapped into 20 genres spanning 100 years. Films that won the best picture Academy Award are marked with ®

The History of Film by Larry Gormley ©2012

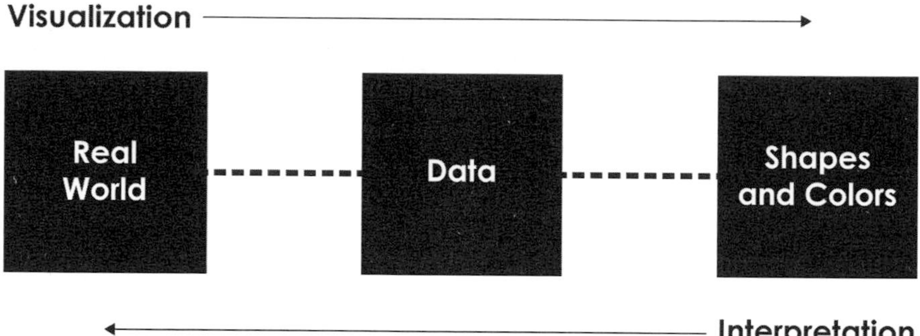

FIGURE 5-6 *Connecting visual cues to what data represents*

You must instead maintain the connection between visual cues and data because the data is what connects a graphic to the real world. So readability is key. Allow comparisons, consider the context of your data and what it represents, and structure shapes and colors (and the space around them) for clarity.

ALLOW COMPARISONS

Allowing comparisons across points is the main purpose of visualizing data. In table form, you can compare only point by point, so you place data in a visual context to see how big one value is relative to the rest and how all the individual data points relate to each other. As a way to better understand data, your visualization isn't useful if it doesn't fill this basic requirement. Even if you just want to show that values are equal across the board, the key is still to allow that comparison and conclusion to be made.

Traditional graphs, such as the bar, line, and dot plots that you've seen throughout this book were designed to make comparisons as straightforward and obvious as possible. They abstract the data into basic geometric shapes so that you can compare length, direction, or position. However, as shown in Figure 5-7, you can apply small variations to these charts that can make a graphic more or less challenging to read.

You saw how area should be used as a visual cue in Chapter 3, "Representing Data." When area is used to indicate values, determine the size of shapes such as bubbles and squares by their total area rather than the length of radius or side length. Essentially, the size of shapes are based on how people interpret them visually.

However, also keep in mind that it can be harder to see small changes between two-dimensional shapes than it is to see differences between position or length. This is not to say to avoid area as a visual cue. Instead, area is more useful when there are exponential differences between values. When small differences are important, look to a different visual cue, such as position or length.

Harder to Compare **Easier to Compare**

Narrow color scale

Colors look washed out and pattern is less obvious.

vs.

Wide color scale

Greater contrast between bins makes pattern obvious.

Showing data points only

It's harder to compare positions as you scan across.

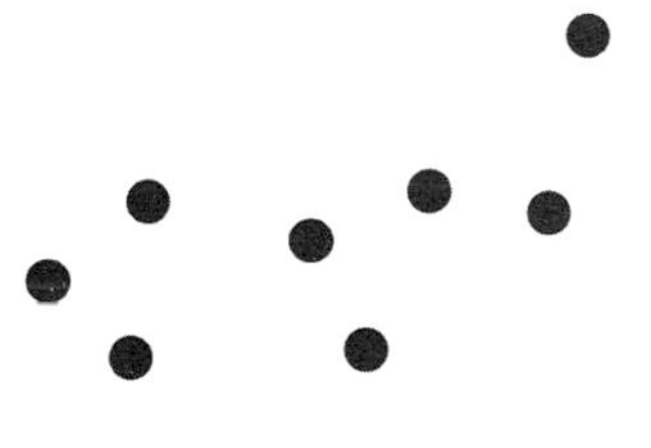

vs.

Additional visual elements

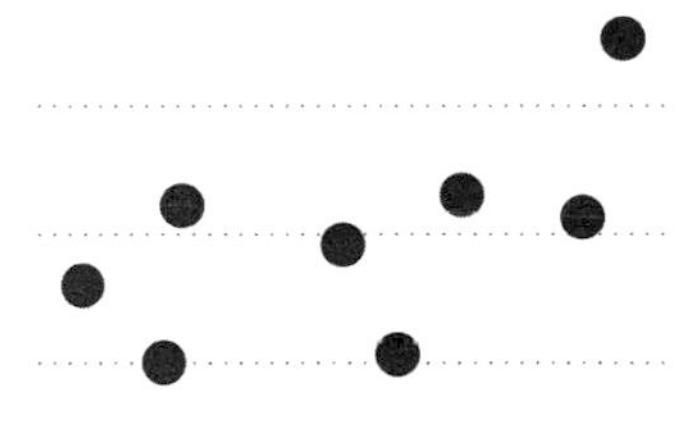

Line increments make comparison quicker.

Using area as visual cue

Although area has its merits, it can be hard to see small differences.

vs.

Using length as visual cue

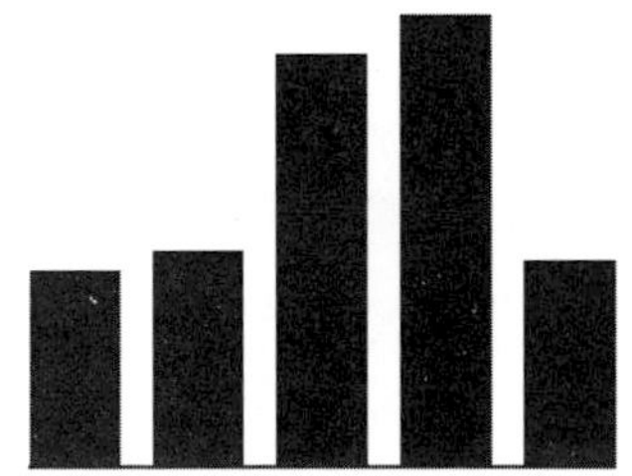

Small differences are easier to see without a square root transform.

FIGURE 5-7 *Allowing comparisons*

For example, Figure 5-8 shows a number of identified species of invertebrates and vertebrates. The bar chart on the left and the bubble chart on the right show the same data, but because there so many more identified species of insects than vertebrates, the bars for the latter are dwarfed. They are barely visible, and the bar for corals is also just a sliver.

On the other hand, the bubbles let you put large and small counts in the same space. The downside is that you can't visually compare values as accurately as a bar chart, but in this case, the bar chart doesn't even give you a chance to compare the values. So there's a trade-off.

Note: Area can also make data seem more tangible or relatable, because physical objects take up space. A circle or a square uses more space than a dot on a screen or paper. There's less abstraction between visual cue and real world.

The graphic in Figure 5-9, made in 1912 when the Titanic crashed into an iceberg in the Atlantic Ocean, also places information within a familiar geographic context.

Each layer from top to bottom represents the time it would take to travel across the ocean via a 17th century ship (40 days), the Titanic (4 days), and by a not yet realized airplane (1 day). If only you could fly the Atlantic! Grid lines separate the modes of transportation as well as provide estimated travel times. Grids can also improve readability in more traditional charts because they dictate spacing and reflect scale.

Introduce color as a visual cue, and there are additional considerations. For example, you saw how those who are color-deficient see shades of red and green. If you use red and green hues with the same saturation, the colors look the same to those who are color-deficient. Color options also change based on what scale you use for a chart or what you want to show. As shown in Figure 5-10, there are three main categories of color scales, with variation within each.

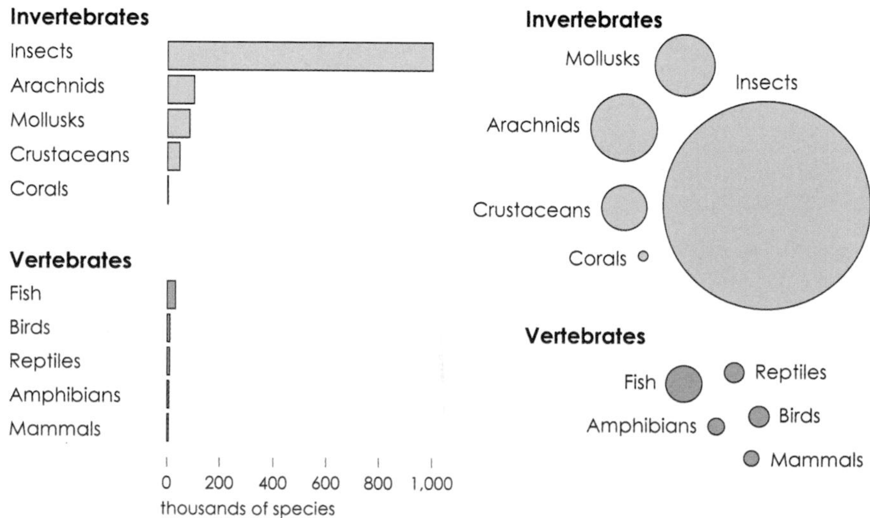

FIGURE 5-8 *Bar chart versus bubble chart to show large counts*

FIGURE 5-9 If only we could fly the Atlantic! *(1912) by Sid Treeby*

Sequential

The same or similar hues are used, and saturation varies for a single metric.

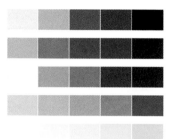

Diverging

Two hues are used to indicate a division, such as positive and negative values.

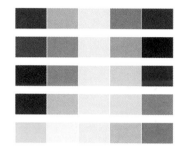

Qualitative

When data is non-numeric, contrasting colors are used for each category.

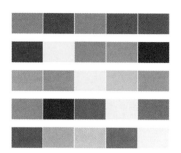

FIGURE 5-10 *Color scale options*

A sequential color scale is used to represent a single variable without a separation requirement (positive versus negative, for example). Darker shades typically represent higher values and lighter shades represent lower values. You essentially choose a saturated hue and then decrease the saturation in increments to create a scale. With the sequential scales in Figure 5-10, the saturated hues are on the right and saturation is decreased as you shift left.

When you do have a natural or defined split in the data, such as increases and decreases or political leanings toward two parties, you can use a diverging color scale. It's like a combination of two (or more) sequential color scales with a separator in between to indicate a neutral value, such as a change of zero or a balance of political favor.

Qualitative color scales are useful when your data is categorical or non-numeric. Each color might represent a category, and the varying shades should provide visual separation.

Regardless of the type of color scale you use, there should be enough variation between your choice of hues and saturation so that you can see differences. Choose shades that are too similar and it's a challenge to make comparisons. A narrow color span restricts the amount of difference between shades, as shown on the left of Figure 5-11, whereas a wider color span on the right makes it easier to see differences. This works in the opposite direction, too. A color span that's too wide can exaggerate differences, and if

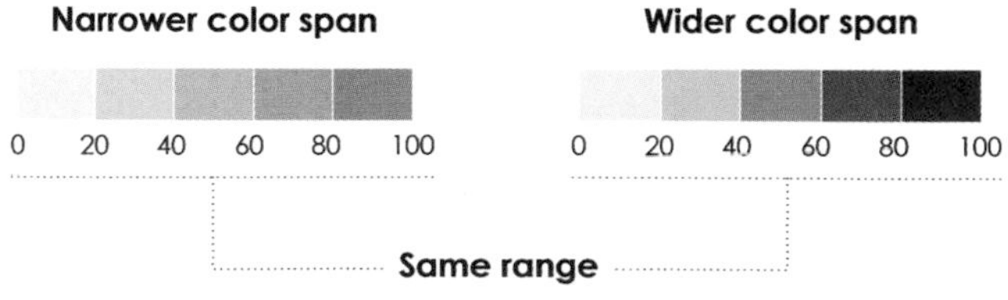

FIGURE 5-11 *Color scales that span the same range of values*

you don't pay attention to the context of the data, you might show patterns that look obvious but are not significant.

Figure 5-12 is the Cartesian equivalent. The space between each tick mark on the vertical scale is tiny, but because the span of the values is also small, the change in the line's position looks big.

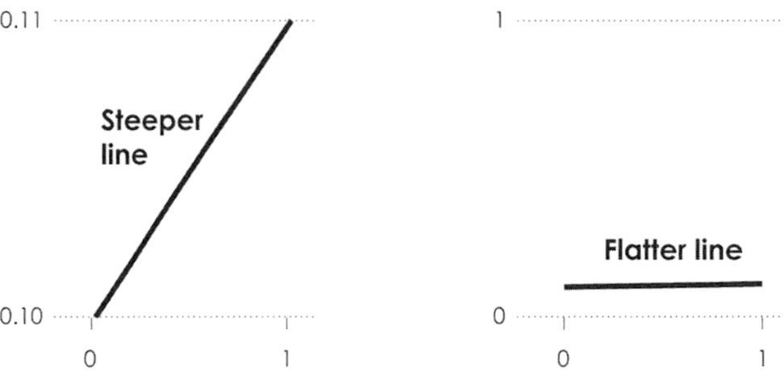

FIGURE 5-12 *Zoomed in on a Cartesian coordinate system*

Sometimes it makes sense to do this, and other times the zoom exaggerates what's actually there. A rule of thumb: Match the amount of visual change to the significance of the change in real life, and as always, represent the data fairly so that others can make fair comparisons.

REPRESENT CONTEXT

Context helps readers relate to and understand the data in a visualization better. It provides a sense of scale and strengthens the connection between abstract geometry and colors to the real world. You can introduce context through words that surround a chart, such as in a report or story, but you can also incorporate context into the visualizations through your choice of visual cue and design elements.

As shown in Figure 5-13, Stephen Von Worley showed the increased variety of colors in the Crayola crayon spectrum. In 1903, on the release of the first wax crayons under the brand name Crayola, there were just 8 colors. Over the years, Crayola inherited and created other colors in between the existing hues, and by 2010, there were 120 shades offered. In addition to red, there is now also bittersweet, brick red, mahogany, maroon, orange red, red orange, violet red, wild watermelon, radical red, razzmatazz, fuzzy wuzzy, and scarlet.

It makes sense to use the actual colors to represent the shades each year, to show the increase in diversity. A grayscale version would require a label for each shade and would quickly clutter by 1949.

Often your choice of visual cues changes based on the expectations of those you make a graphic for. A graphic that does not fulfill expectations can confuse readers. (I of course, mean this from a design perspective rather than a data one. Unexpected trends, patterns, and outliers are always welcome.)

Note: Choose geometry and color based on the context of your data. Software defaults are rarely, if not never, the best option.

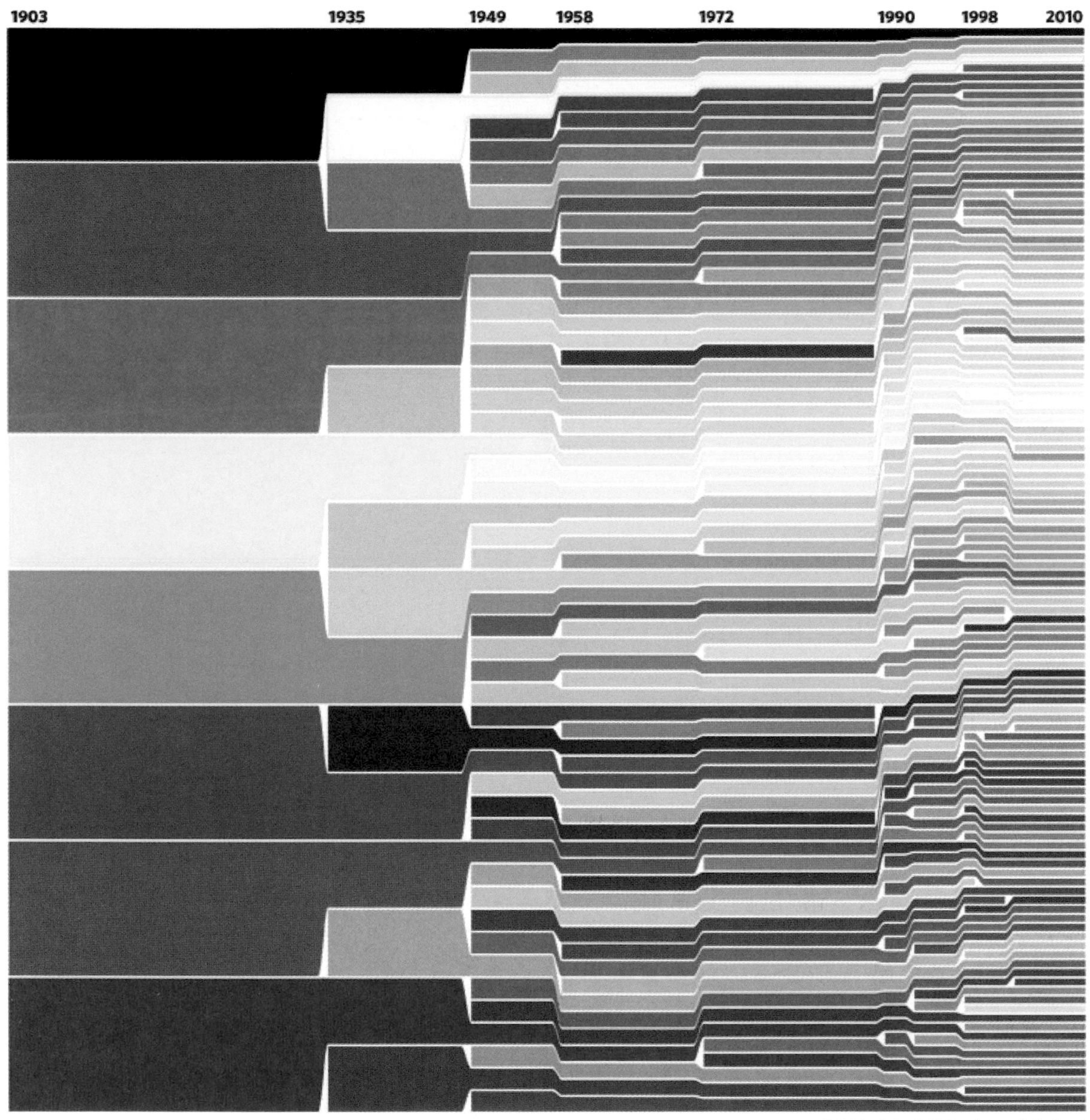

FIGURE 5-13 Crayola Color Chart 1903–2010 *(2010) by Stephen Von Worley, http://www.datapointed.net/visualizations/color/crayola-crayon-chart/*

For example, the United States has a two-party system, with Democrats and Republicans. Blue is the color of the Democratic party and red is the color of the Republican party. Therefore, a map, as shown in Figure 5-14, should reflect the party colors. Flip the colors, and the proportions between two groups would be the same, but because the party colors are so commonplace, it's probable that readers would misinterpret the results as a Barack Obama win in the Midwest and Southeast and a Mitt Romney in the West and Northeast.

The charts in Figure 5-15 show movie trilogy ratings from the review aggregation site Rotten Tomatoes. On the site, a ripe red tomato is used for movies that earn at least 60 percent positive reviews (fresh), whereas a splattered green tomato is used for movies below the 60 percent threshold (rotten). The graphs match the site's color scheme so that you can easily see which movies were fresh and which were rotten. The length of each bar provides a more exact value.

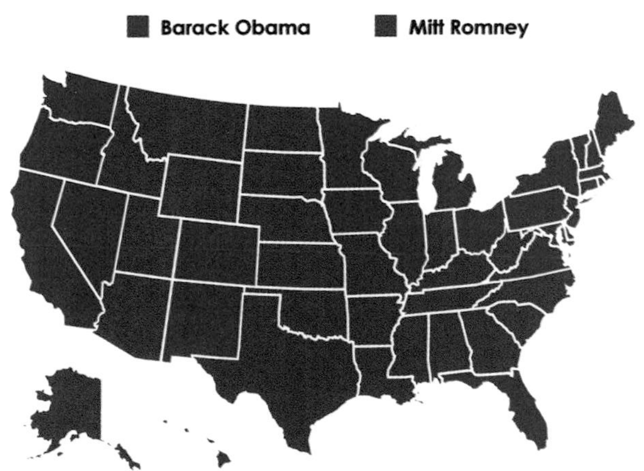

FIGURE 5-14 *Color by expectation*

Context can also affect your choice of geometry. For example, the Bureau of Labor statistics releases monthly estimates for number of jobs lost and gained. Figure 5-16 represents jobs lost between February 2008 and February 2010. More jobs were lost than gained every month during this period. The taller the bar is, the more jobs that were lost on the corresponding month.

The chart with values in the positive makes sense, but consider the context the chart is usually presented in. People expect to see bars in the positive for jobs gained and in the negative for jobs lost. However, the coordinate system in Figure 5-17 would put jobs gained in the negative. Negative jobs lost means new jobs.

So instead, it's more intuitive to frame jobs lost as negative values, as shown in Figure 5-17. It makes more sense to show something lost moving downward, when that something is looked at negatively. On the other hand, decreased weight, when the goal is actually to lose weight, might work better on the positive side of the axis.

Trilogies: Fresh originals and rotten finales

Movie reviews aggregator, Rotten Tomatoes, defines a movie as *fresh* if at least 60% of reviews are positive, and *rotten* otherwise. Sequels and finales usually don't fair well.

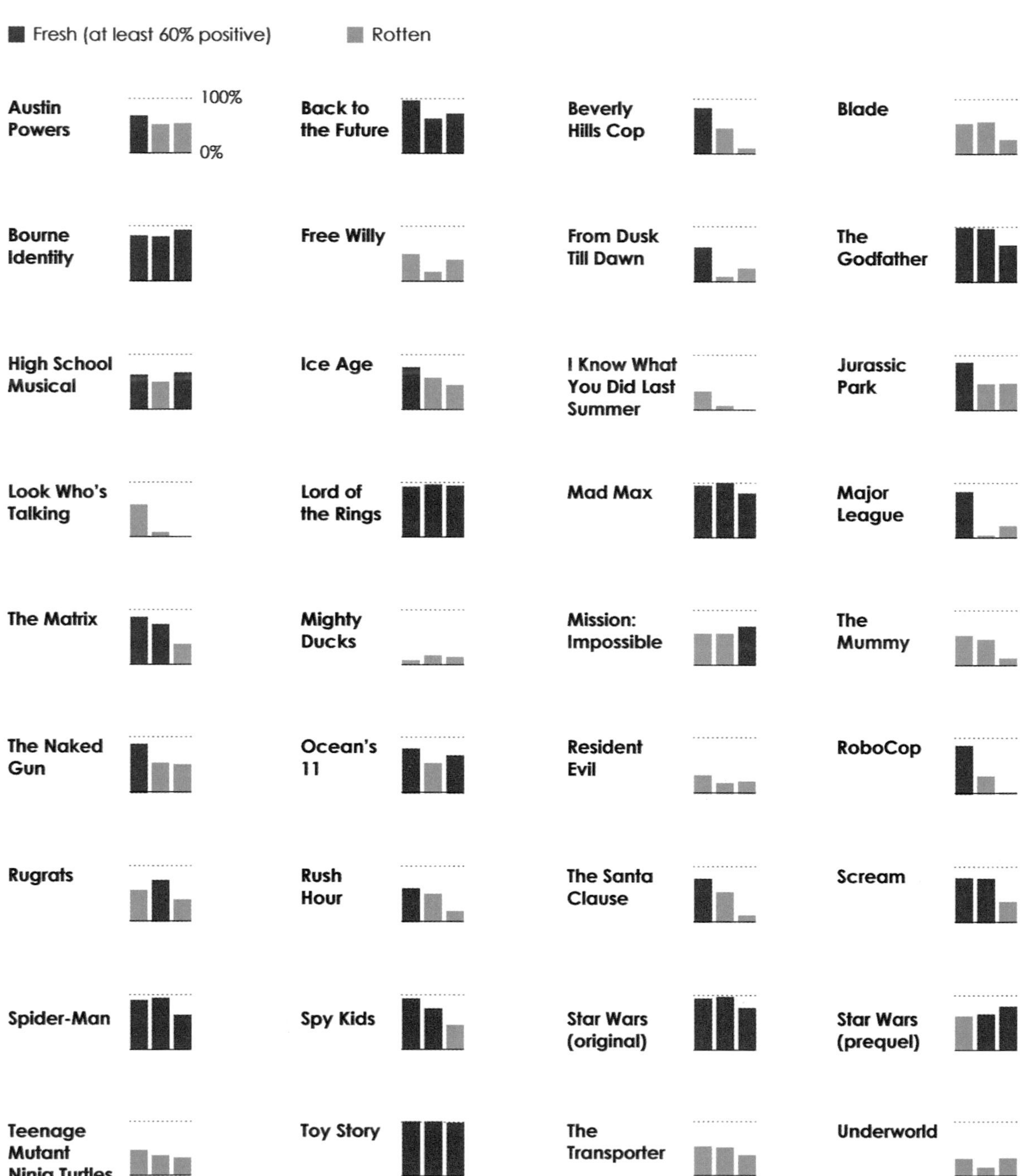

FIGURE 5-15 *Color based on where the data comes from*

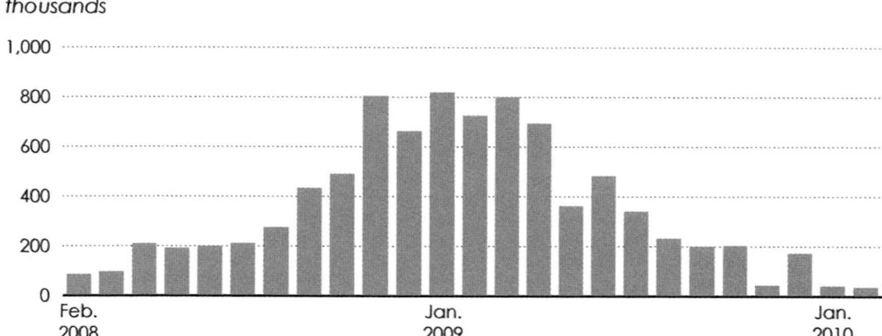

FIGURE 5-16 *Visualizing data generically*

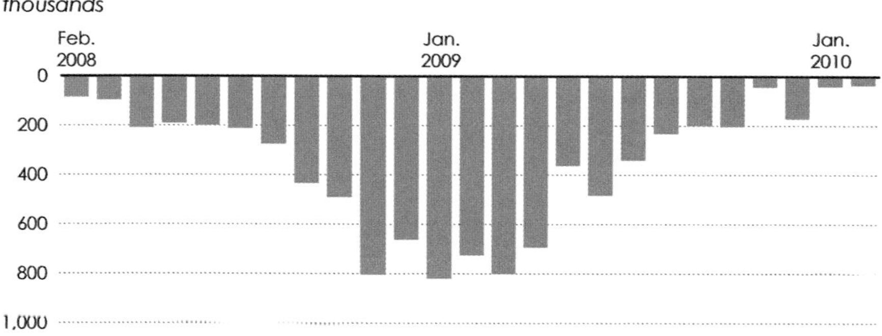

FIGURE 5-17 *Visualizing data in context*

NEGATIVE SPACE

Clutter is the enemy of readability. A lot of objects and words packed into a small area can make a visualization confusing and unclear, but put some space in between and it's often a lot easier to read. You can use space to separate clusters within a single visualization, or you can use space to divide multiple charts, so that they are modular and don't all run together. This makes a visualization easier to scan and mentally process piece-wise.

Figure 5-18 shows equally spaced rectangles, which appear to be in the same cluster, followed by ways to separate them with space and other elements, such as lines and contrasting colors. The space implies division (which you should keep in mind when you don't want to separate visual elements).

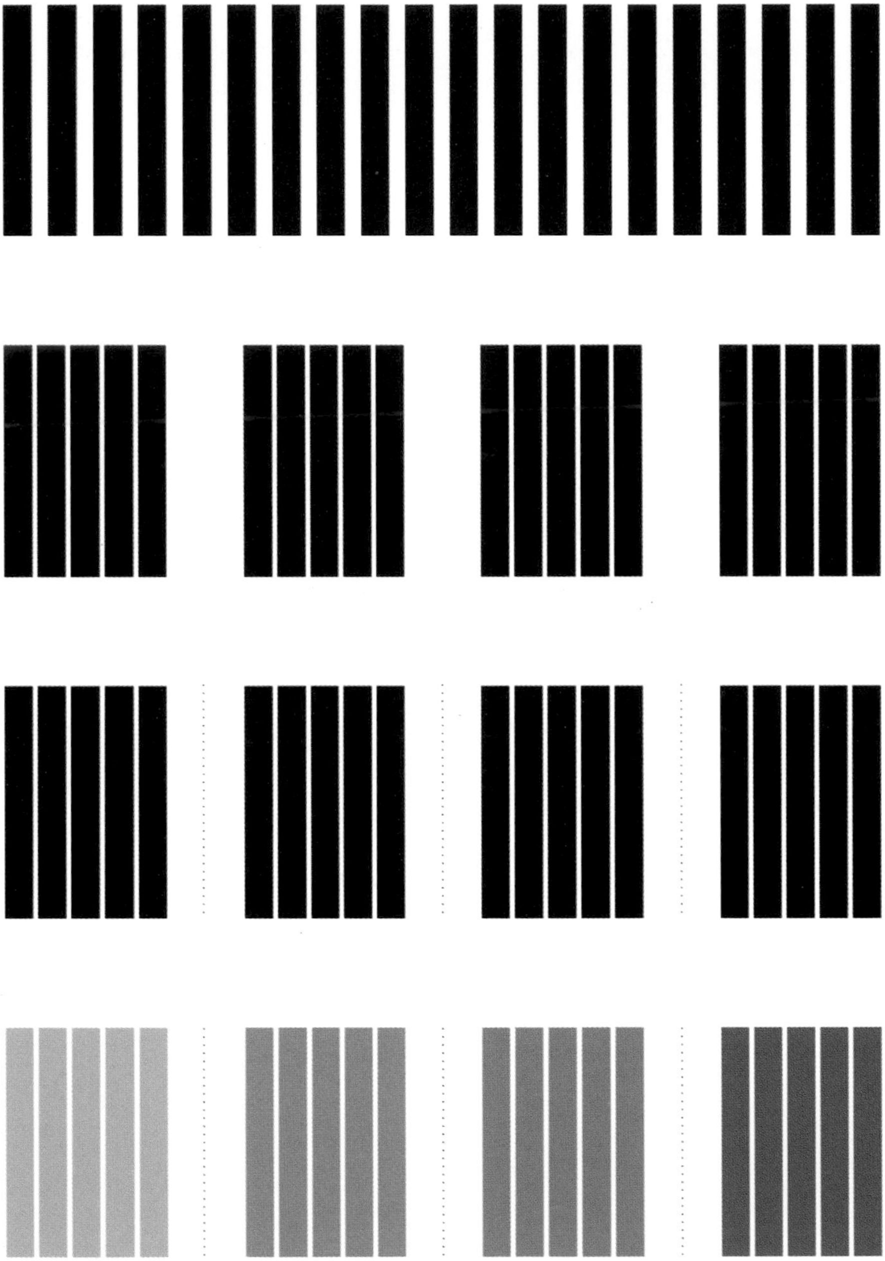

FIGURE 5-18 *Grouping shapes with space and other elements*

It's easy to see how this works in practice. Decrease the space between the labels and small charts (refer to Figure 5-15), and you get Figure 5-19. Although you can figure out which bars correspond to which labels by their positions, it is not immediately clear.

The same applies even if you don't want to show specific groups. Figure 5-20 is the map from Chapter 1, "Understanding Data," which shows fatal crashes in the United States. The top version uses small dots to show each accident, and the bottom version uses larger circles.

Because small dots are used in the top version, it's easier to see the pattern of roads and city centers. The negative space in between points help show where there are no roads or where fewer people drive cars. Places where there is no data is just as important as the places where there is data. On the other hand, the bottom version uses large circles that are relatively large compared to the size of the United States and the total number of crashes during the selected time period. There is practically no negative space, so roads and city centers are hidden by the data, and you only see country boundaries. Without a sufficient amount of negative space, the visualization is useless.

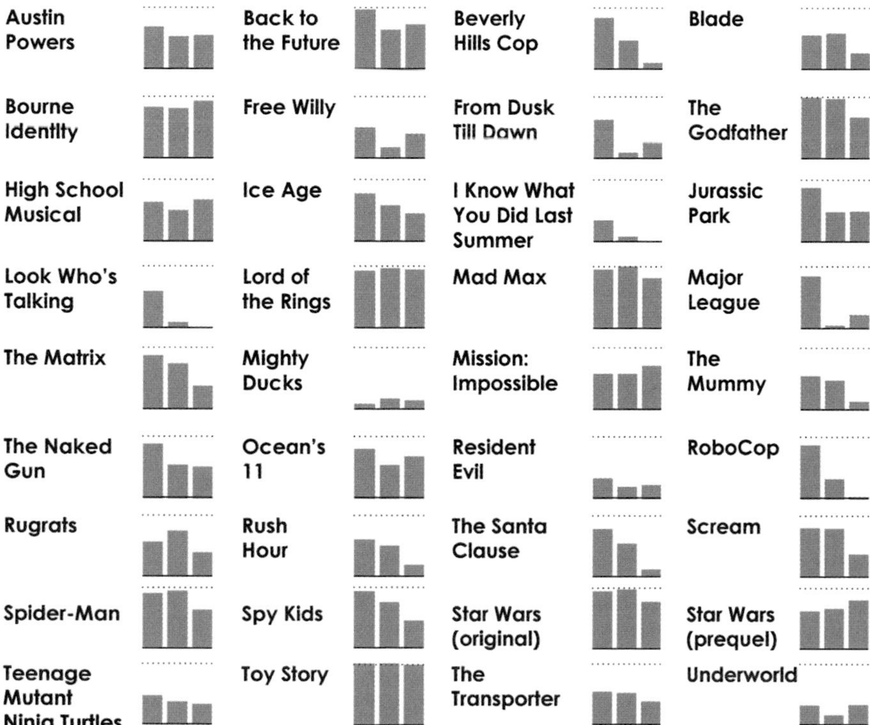

FIGURE 5-19 *Decreased negative space, decreased readability*

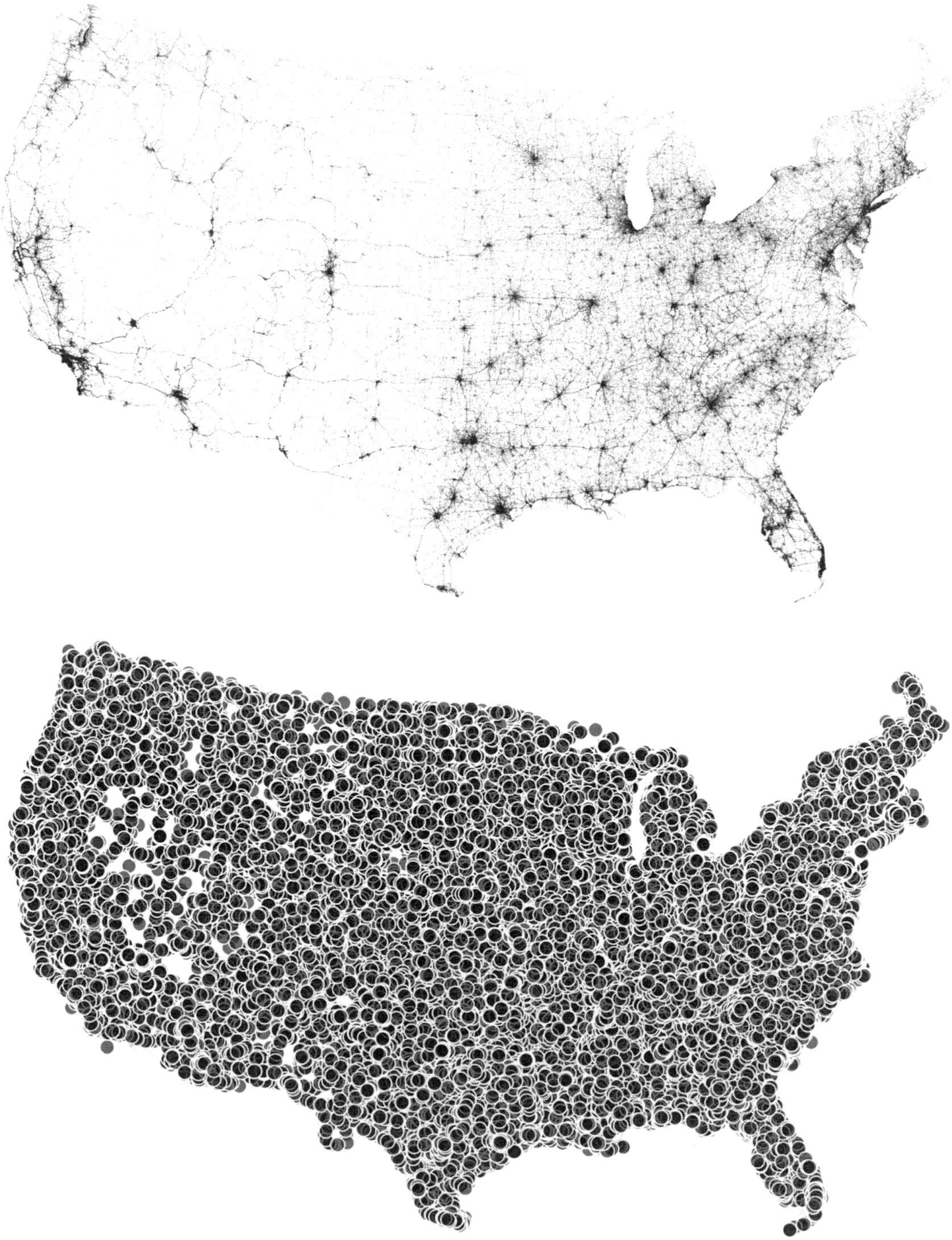

FIGURE 5-20 *Spacing between data points makes patterns more obvious*

You can also have too much negative space that interferes with the primary elements. Figure 5-21 shows a series of bars where the space in between each one is about the same as the width of a single bar. Look at the bars individually, and it's easy to see the separation between each gray bar, but look at the group as a whole, and you perceive a visual vibration between gray and white that almost makes the figure appear blurry. The equal negative space confuses your brain about which part to focus on.

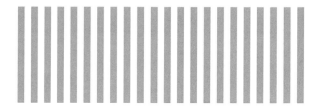

FIGURE 5-21 *Visual vibrations from equal negative space*

On the other hand, reduce the negative space and increase bar width, and the image appears less blurry, as shown in Figure 5-22. The bars clearly dominate focus, and the negative space serves as thin separators in between.

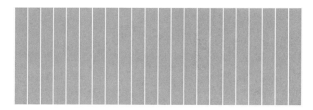

FIGURE 5-22 *Bars visually dominate with little negative space.*

As shown in Figure 5-23, the opposite direction works, too. The bars are thin strips and negative space is relatively large. Like in the earlier section on allowing comparisons between shapes and colors, contrast is the key.

FIGURE 5-23 *Less vibration with a contrast between bars and negative space*

With little differentiation between negative space and the elements of interest, a visualization is less clear, so experiment to find the right balance.

HIGHLIGHTING

Readability in visualization helps people interpret data and make conclusions about what the data has to say. Embed charts in reports or surround them with text, and you can explain results in detail. However, take a visualization out of a report or disconnect it from text that provides context (as is common when people share graphics online), and the data might lose its meaning; or worse, others might misinterpret what you tried to show.

Highlighting can guide readers through the data and direct eyeballs to the most important parts in a graphic. It reinforces what people might already see or draw attention to areas or data points that people should see.

To draw visual attention to a data point, you simply do what you would in real life. You make it stand out. Speak a little louder. Make it a little brighter. Edit an area or point in a visualization—while keeping the data, its visual cues, and readability in mind—to differentiate it from the rest. Use a brighter or bolder color, draw a border, thicken a line, or introduce elements that make the point of interest look different.

For example, Figure 5-24 shows how to use color to highlight a specific point. Most of the shapes are a neutral color, and the point of interest is purple, so attention immediately focuses on the parts that stand out.

Visualize time series data, and you might focus on specific years, such as in Figure 5-25. As you know, America loves their competitive eating, and no contest is more important than the annual hot dog eating contest on Coney Island. The top bar chart shows the number of hot dogs and buns that winners ate each year, but you can highlight bars to shift focus to years when someone broke a world record or when a certain person won.

On to more important matters: Figure 5-26 shows the world life expectancy chart from Chapter 2, "Visualization: The Medium," categorized by geographic regions. Each line represents a country's time series. The graphic shows all the countries that data was available for but shifts focus for each region. So the current point of interest is highlighted and brought to the front, and the rest are moved to the back and made a light gray, which remain for a sense of scale and context.

Again, to highlight elements, you make points of interest more visually prominent than the rest of a graphic. You place it higher in the visual hierarchy, so you either move the point of interest up or move everything else down. Elements on the same level get the same attention.

For example, Figure 5-27 shows the availability of movies that led at the box office, via streaming rental on iTunes, Amazon.com, and Vudu or the subscription-only Netflix. DVD availability is provided for reference. Availability is the point of interest, so a brighter color moves it up in the hierarchy, whereas neutral colors move other areas down. More specifically, rectangles highlighted yellow indicate that a movie was available via a service; an empty rectangle means not available; and a gray rectangle means the movie was only available for purchase.

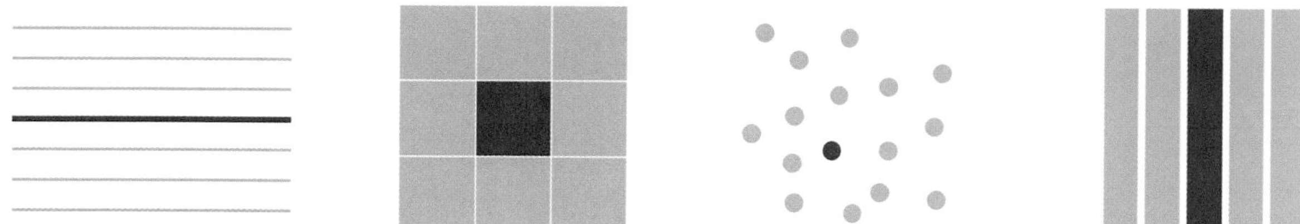

FIGURE 5-24 *Examples of highlighting with color*

Breaking hot dog eating records

70 Hot dogs and buns (HDBs)

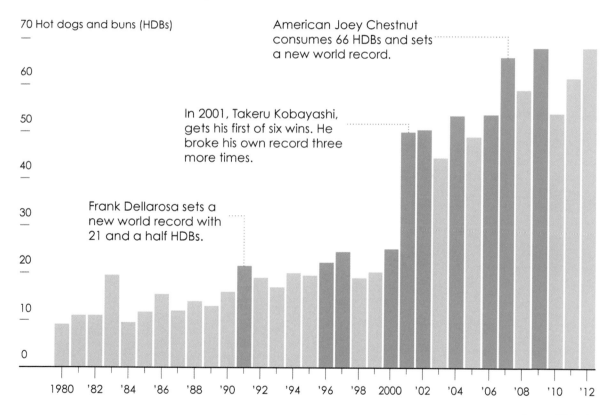

American Joey Chestnut consumes 66 HDBs and sets a new world record.

60

50

In 2001, Takeru Kobayashi, gets his first of six wins. He broke his own record three more times.

40

30

Frank Dellarosa sets a new world record with 21 and a half HDBs.

20

10

0

1980 '82 '84 '86 '88 '90 '92 '94 '96 '98 2000 '02 '04 '06 '08 '10 '12

Source: Wikipedia

FIGURE 5-25 *Placing focus on various aspects of the data*

Increasing Life Expectancy

According to data from World Bank, the number of years a person lives on average has been steadily increasing over the decades. However, as seen in some regions, war and economic turmoil can lead to sudden dips.

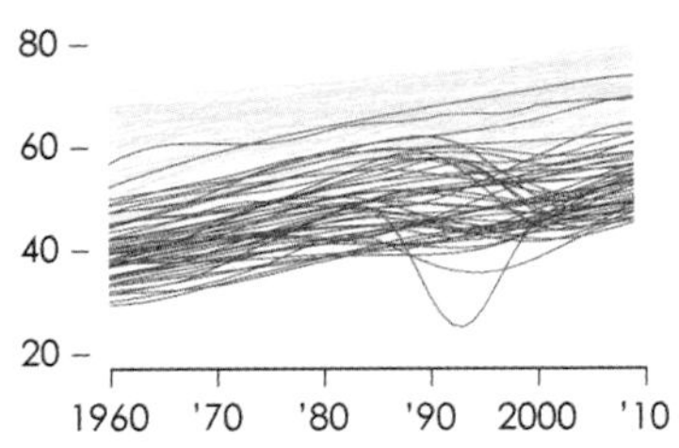

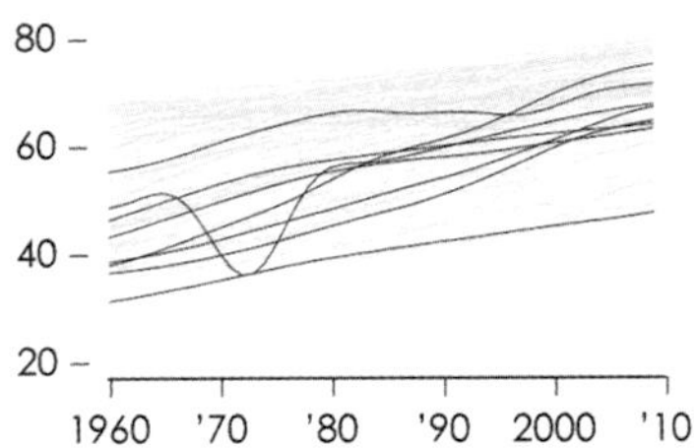

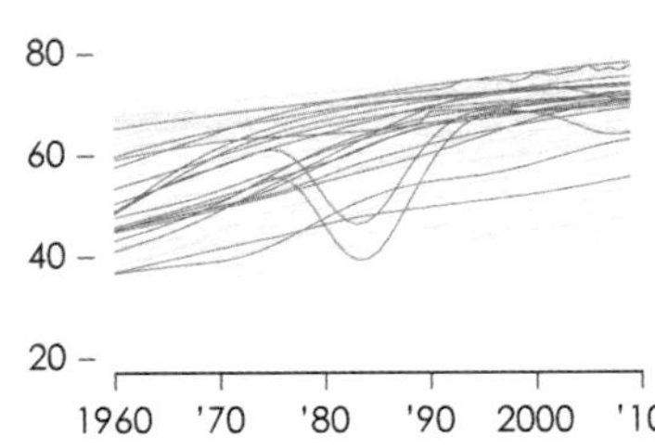

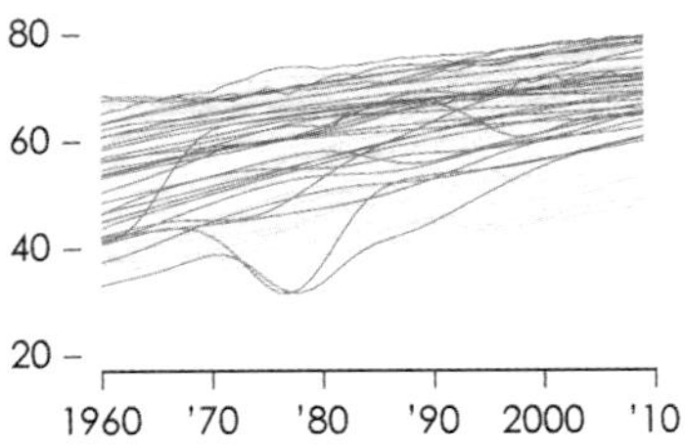

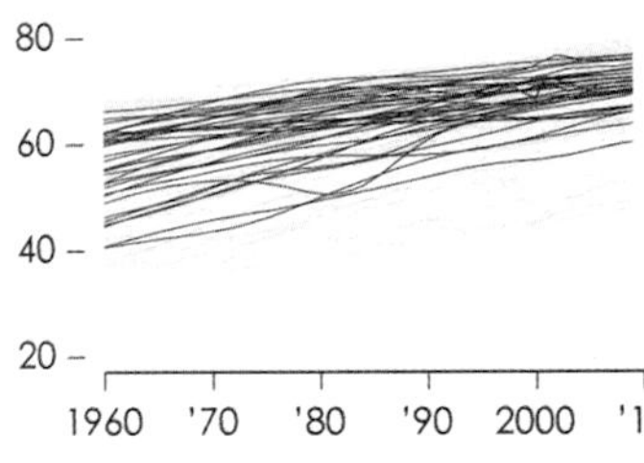

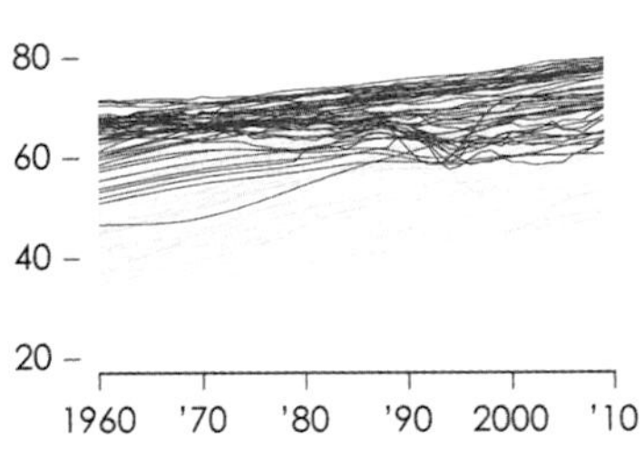

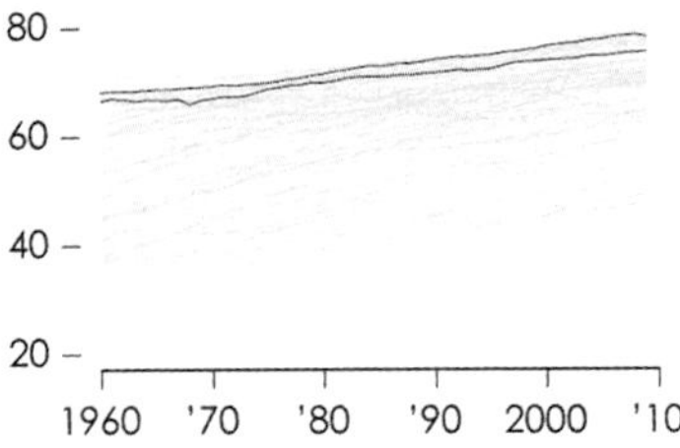

FIGURE 5-26 *Showing all data, but shifting focus with highlighting*

As shown in Figure 5-28, the focus easily shifts. A brighter color for nonavailable or shades of equal brightness take away focus from the main point of the graphic. A poor choice of color can also lead to misinterpretation, where it looks like Netflix has the most available streaming movies, even if the legend indicates otherwise.

Highlighting doesn't always have to be front and center. You can put it in the background, as shown in Figure 5-29. The unemployment time series data still keeps focus, but gray bars highlight periods of recession and provide information outside the primary dataset.

Streaming the Box Office

iTunes | Amazon | Vudu | Netflix | DVD

Top 50 in 2011

Streaming movies grows more common, but services – especially the subscription-only Netflix – still lack selection among more popular movies.

Available
Not available
Purchase only

1 Harry Potter and the Deathly Hallows Part 2
2 Transformers: Dark of the Moon
3 The Twilight Saga: Breaking Dawn Part 1
4 The Hangover Part II
5 Pirates of the Caribbean: On Stranger Tides
6 Fast Five
7 Cars 2
8 Thor
9 Rise of the Planet of the Apes
10 Captain America: The First Avenger
11 The Help
12 Bridesmaids
13 Kung Fu Panda 2
14 X-Men: First Class
15 Puss in Boots
16 Rio
17 The Smurfs
18 Mission: Impossible — Ghost Protocol
19 Sherlock Holmes: A Game of Shadows
20 Super 8
21 Rango
22 Horrible Bosses
23 Green Lantern
24 Hop
25 Paranormal Activity 3
26 Just Go With It
27 Bad Teacher
28 Cowboys & Aliens
29 Gnomeo and Juliet
30 The Green Hornet
31 Alvin and the Chipmunks: Chipwrecked
32 The Lion King (in 3D)
33 Real Steel
34 Crazy, Stupid, Love.
35 The Muppets
36 Battle: Los Angeles
37 Immortals
38 Zookeeper
39 Limitless
40 Tower Heist
41 Contagion
42 Moneyball
43 Justin Bieber: Never Say Never
44 Dolphin Tale
45 Jack and Jill
46 No Strings Attached
47 Mr. Popper's Penguins
48 Unknown
49 The Adjustment Bureau
50 Happy Feet Two

As of January 20, 2012
Source: Tristan Louis

FIGURE 5-27 *Highlighting the theme of a graphic*

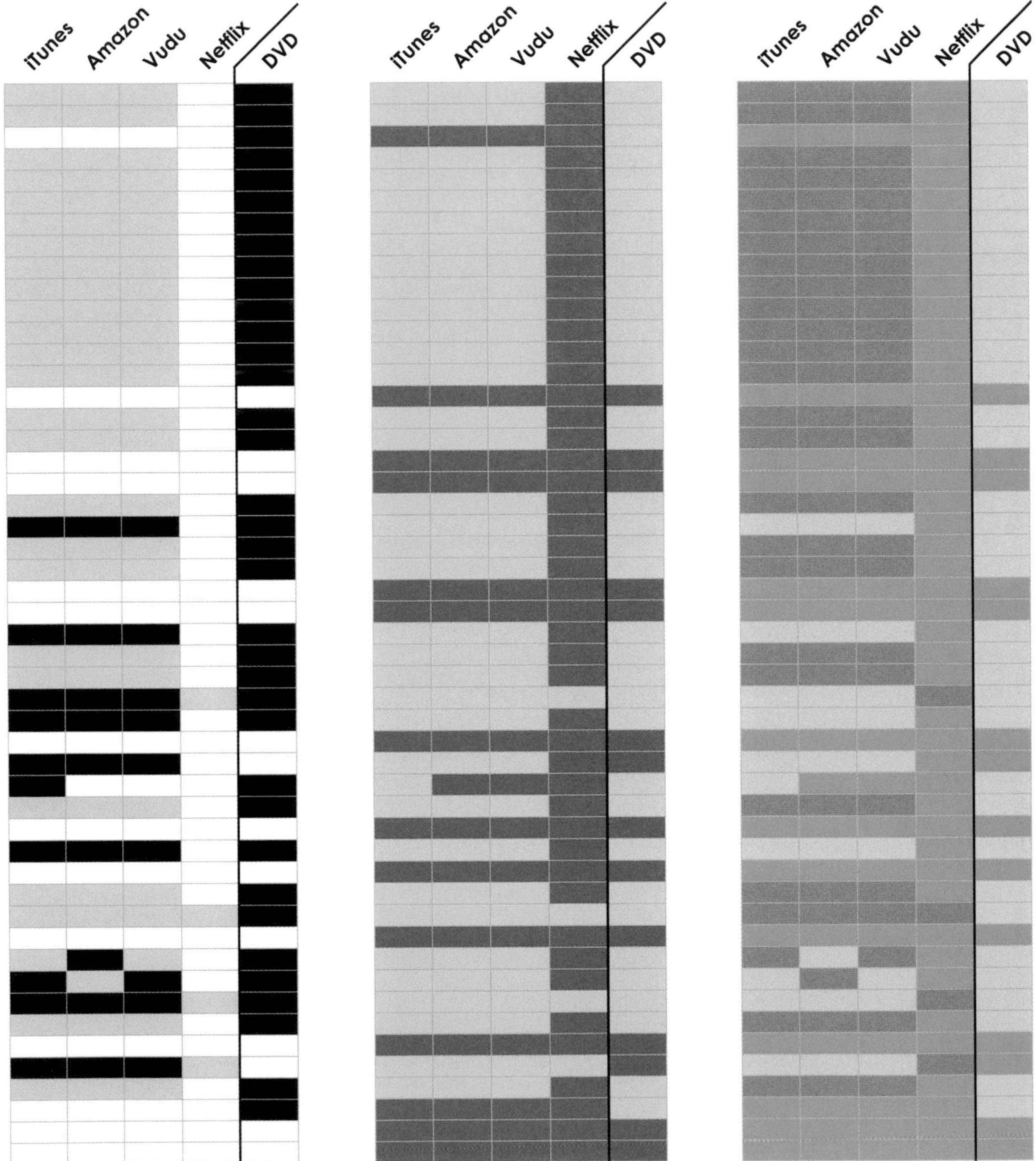

FIGURE 5-28 *Various color choices change the visual hierarchy.*

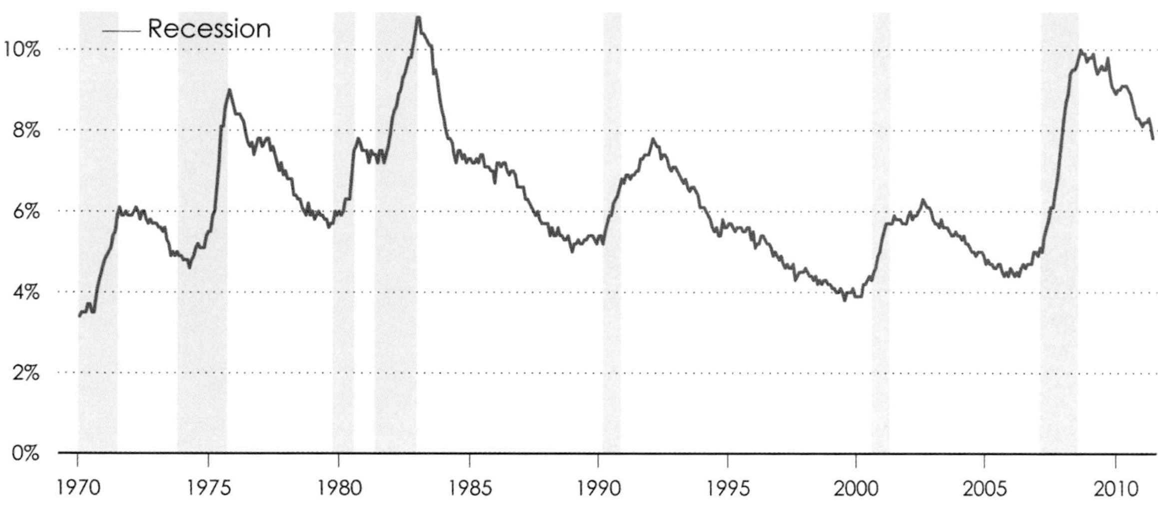

FIGURE 5-29 *Highlighting in the background*

However, wherever your highlighting fits in the hierarchy, be sure the new visual cues don't conflict with existing ones. If you have a bar chart that uses length as a visual cue, you obviously do not highlight with length, too. Have a scatter plot? Don't highlight with position. Heat map? Highlight with the color palette rather than introduce hues that change visual patterns.

Ask yourself how people decode information via shapes and colors in a visualization, and then don't get in the way. For example, in Figure 5-30 on the left is a bar graph with no highlighting, and the charts on the right show unsuccessful attempts to highlight. Why don't they work? A bar chart uses length as its visual cue, so when you extend a bar, the new length changes the value. Change width, and the new bar fills more area. (Bar charts actually use area to encode data, but because width stays constant, you can decode values via bar height.) Then on the far right: A shift up doesn't exactly change the value, but it makes the chart less readable.

Note: These conflicting visual cues are in the context of how you use bar charts, but you of course must consider conflicts within the context of your own visualization and how you encode your data.

In contrast, Figure 5-31 shows highlighting with visual cues not used by the bar graph. The color, border, and pointer send focus to the bar of interest but don't change the overall visual pattern.

Note: Highlight with unused visual cues. Otherwise, you change perceived patterns and make it more difficult to interpret the visualization.

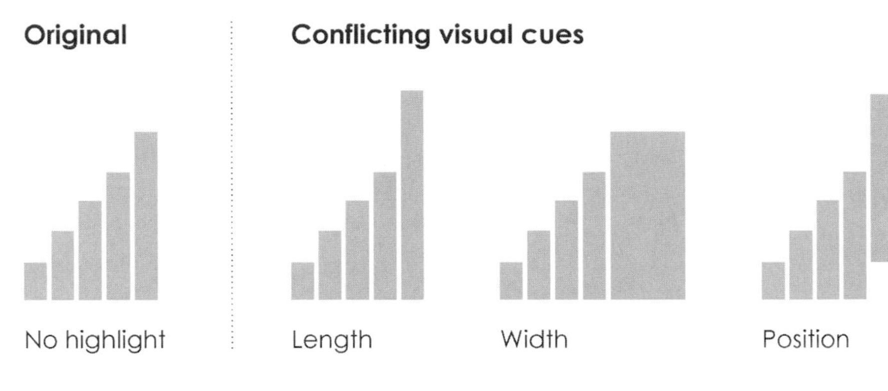

FIGURE 5-30 *Using conflicting visual cues to highlight*

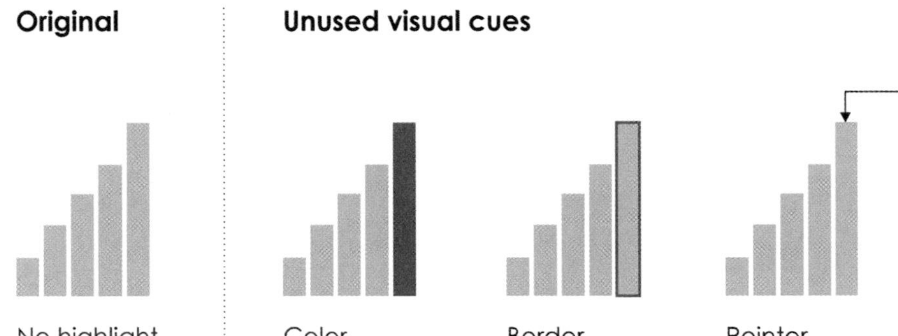

FIGURE 5-31 *Using unused visual cues to highlight*

ANNOTATION

When you highlight elements, it is not always obvious why, especially when readers aren't familiar with the data. (And they aren't most of the time.) Annotation within a visualization can help clearly explain what a visualization shows. What is that outlier? What does that trend mean? This might be left to text outside of a visualization, but when you put explanations within a graphic—as an additional layer of information—the visualization is self-encapsulated so that it's useful on its own.

EXPLAIN THE DATA

Like everything discussed so far, annotation follows a visual hierarchy. You have headers, subheaders, subsubheaders, and explanatory text. As shown in Figure 5-32, size, color, and placement dictate how much attention annotations receive.

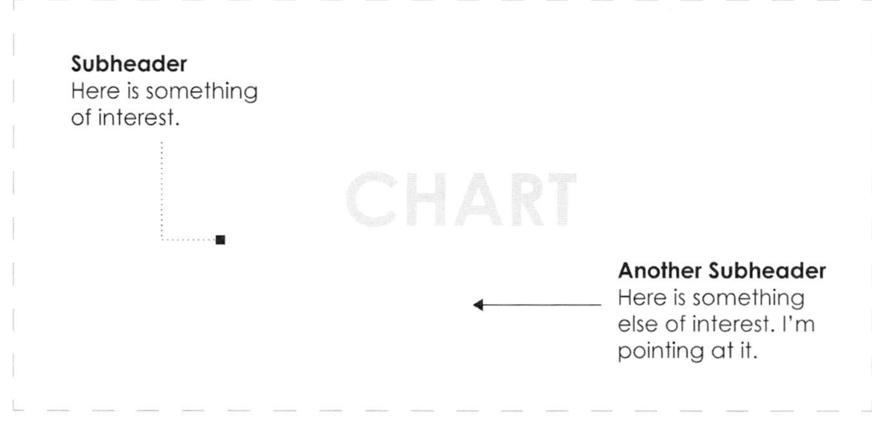
Header title that describes findings
Lead-in text is your chance to provide more details on what the data is about, where it's from, and what the audience should see or look at.
Subheader
Here is something of interest.
CHART
Another Subheader
Here is something else of interest. I'm pointing at it.
Source: This is where the data is from.

FIGURE 5-32 *Annotation for a chart*

The header is typically printed with larger and bolder fonts to set the stage or to describe what people should see or look for in the data. If the header is small and blends with everything else, people might skip it and look straight to the more visual elements. A descriptive title also helps. For example, "Rising Gas Prices" says more about a chart than just "Gas Prices." The former presents a conclusion immediately, and readers will look to the chart to verify and see details. The latter leaves data interpretation to readers and places them in the exploration phase. Then again, this might be your goal, so describe accordingly.

Lead-in text, like the header, is used to prepare readers for what a chart shows, but in further detail. The text is typically smaller than the header and expands on what the header declares, where the data is from, how it was derived, or what it means. Basically, it's information that might help others understand the data better but often doesn't directly point to specific elements.

To explain specific points or areas, you can use lines and arrows and use annotation as a layer on top of a chart. This places descriptions directly in the context of the data so that a reader doesn't have to look outside a graph for additional information to fully understand what you show.

For example, returning to the scatter plot in Figure 5-4, a layer of annotation is added, in addition to highlighting of specific points, as shown in Figure 5-33. Dark circles and pointers highlight specific players, and lines connect annotation to dots for the lowest scoring player with the lowest usage percentage, DeSagana Diop, and the highest scoring player with the highest usage percentage, Dwyane Wade. The point for Will Bynum, who somewhat strays from the trend, is also highlighted and annotated. There is also a pointer for the trend line and an explanation of usage percentage, which isn't common knowledge for most.

The key to useful annotation is to explain or highlight a chart as it relates to the data (and your audience). For example, the explanation for the trend line could be, "There is a positive correlation between points per game and usage percentage." This is true, but the generic statistical description doesn't relate to the context of the data. Similarly, you could describe Dwyane Wade as the player with the highest usage percentage and points per game, but

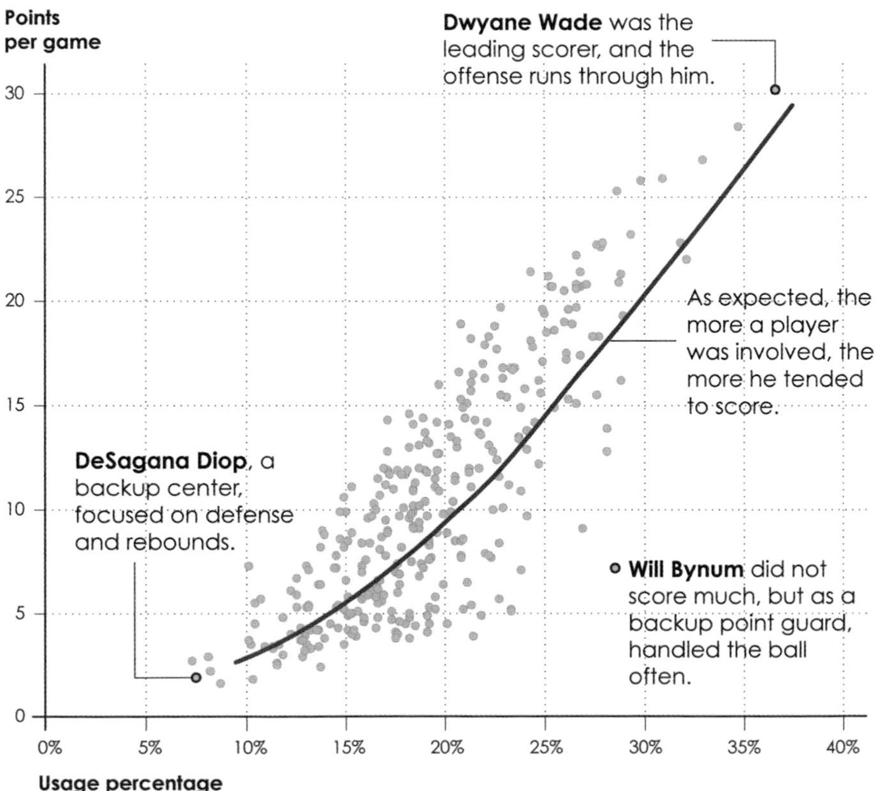

FIGURE 5-33 *Annotation added to scatter plot*

what does that say about him as a player? These are subtle changes that can greatly decrease or improve readability.

EXPLAIN STATISTICAL CONCEPTS

If a large proportion of your audience is unfamiliar with statistical concepts, you can annotate to explain or help them relate. The descriptions in the previous scatter plot of basketball players are an example. They don't just point out Dwyane Wade, DeSagana Diop, and Will Bynum. They also help explain what the corner positions, as well as a partial outlier, on an x-y plot mean so that readers can infer what positions in the middle represent. The pointer for the trend line is a description of correlation.

Figure 5-34 is another scatter plot, but it focuses on the gender pay gap in the United States, based on median salaries, according to the Bureau of Labor

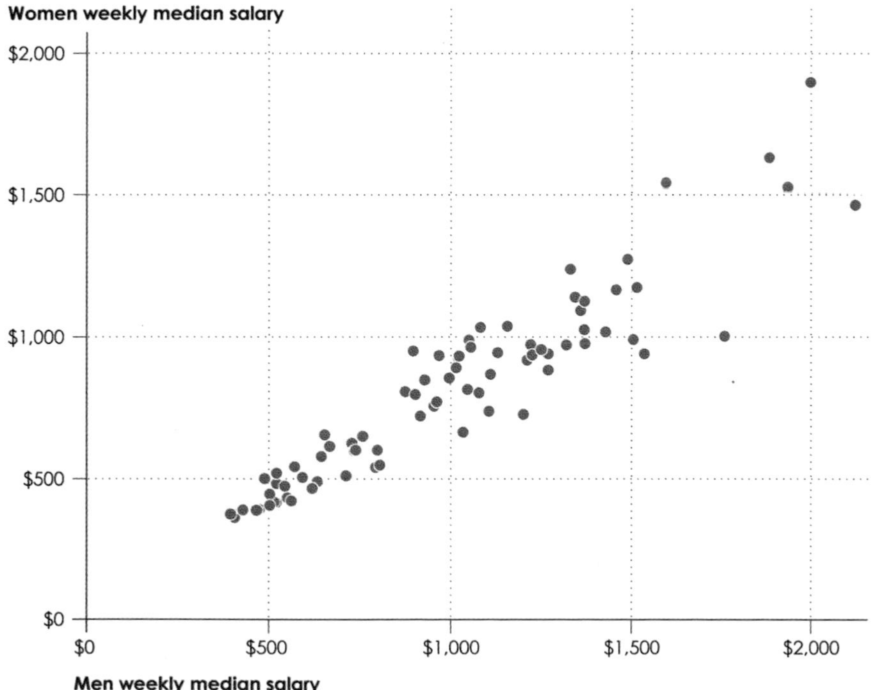

FIGURE 5-34 *Unannotated scatter plot*

Statistics. Each dot represents a profession, and men's median salary is plotted on the horizontal axis versus women's median salary on the vertical.

Without annotation, it's clear there is an expected upward trend between the two. With professions where men tend to make more, women tend to make more, too. If you look closely, you can also see that the dots tend toward the horizontal axis, which means men tend to make more with the same occupation.

The annotated chart in Figure 5-35 makes the pay difference clearer. A diagonal line through the middle represents equal pay, which is marked as such. Dots below the line are jobs where men make more than women, and dots above the line are where women tend to make more. These areas are also labeled.

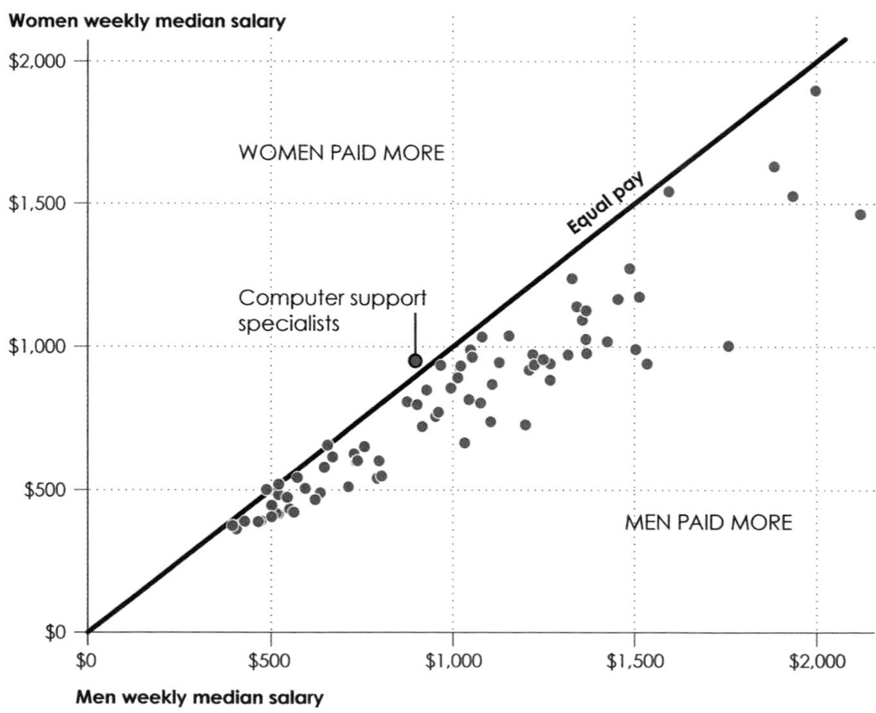

FIGURE 5-35 *Annotated scatter plot*

Computer support specialist is the only profession in this dataset where women tended to make more than men.

The annotation explains how to read the scatter plot and what the data means. Sure, many people know how to read a scatter plot and interpret relationships between two variables, but many don't, and it doesn't hurt to clarify.

Distributions are another challenging concept. People have to understand skew, mean, median, and variation, and that observations are aggregated across a continuous value scale when visualized.

For example, it is common for people to interpret the value axis of a histogram as time and the count or density on the vertical axis as a metric of interest. This leads to confusion, so it is useful to explain the various facets of a distribution.

In Chapter 4, "Exploring Data Visually," you saw distributions for flight arrival delays. Figure 5-36 shows the distribution of delays for Southwest Airlines. A negative delay means an early arrival, and a positive one means the plane arrived late to the destination airport. A delay of zero means an on-time arrival.

To clarify, simply add those descriptions as annotation on the histogram, as shown in Figure 5-37. Avoid jargon and explain in the context of the data.

In the end, you must consider what your audience will or might not understand graphically and statistically, and annotate based on that. Single variables, time series, and spatial data are easier to understand visually because they tend to be more intuitive than multiple variables or more complex relationships.

Note: Show people your visualizations to see how they interpret results. If they're confused, explain the data clearly.

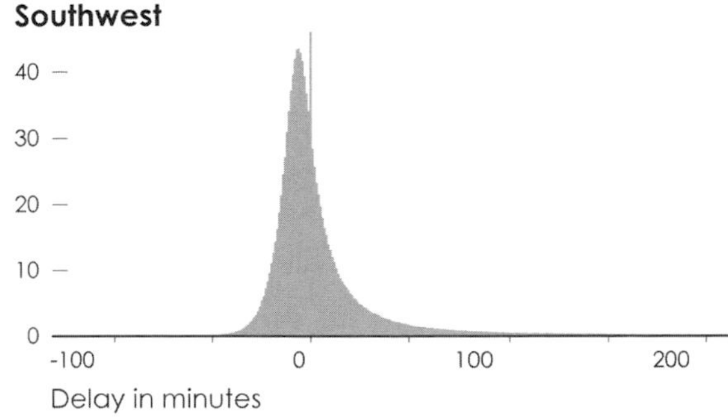

FIGURE 5-36 *Histogram showing distribution*

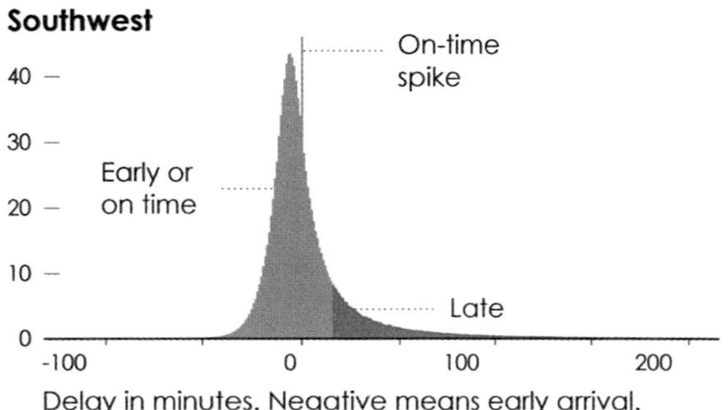

FIGURE 5-37 *Explanation of distribution*

EXPERIMENT WITH TYPOGRAPHY

There have been plenty of polls and questionnaires that ask what the best typeface is for visualization, but there's always a ton of variability, and there's never any consensus. This might be because taste in typefaces has a lot to do with personal preference. Nevertheless, it's worth exploring various fonts for labels and annotation, outside of software defaults, which are generic and less refined.

Note: A typeface is a design for text, such as Helvetica or Baskerville, and it pertains to the appearance of the characters. A font is a specification for a typeface, such as 10-point bold Baskerville.

The effect of typeface choice is most obvious at the extremes. Figure 5-38 shows the same graph with various typefaces, and you can see how readability and feel changes for each. For example, Helvetica, a sans-serif

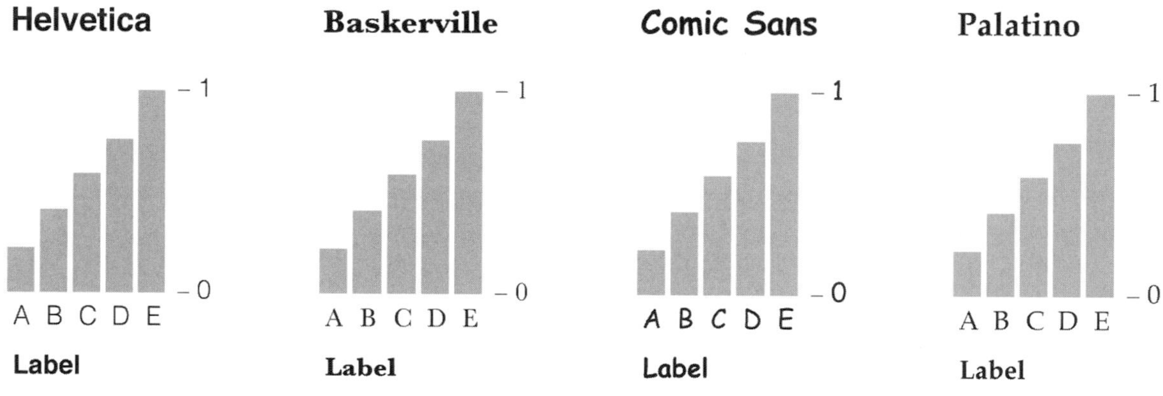

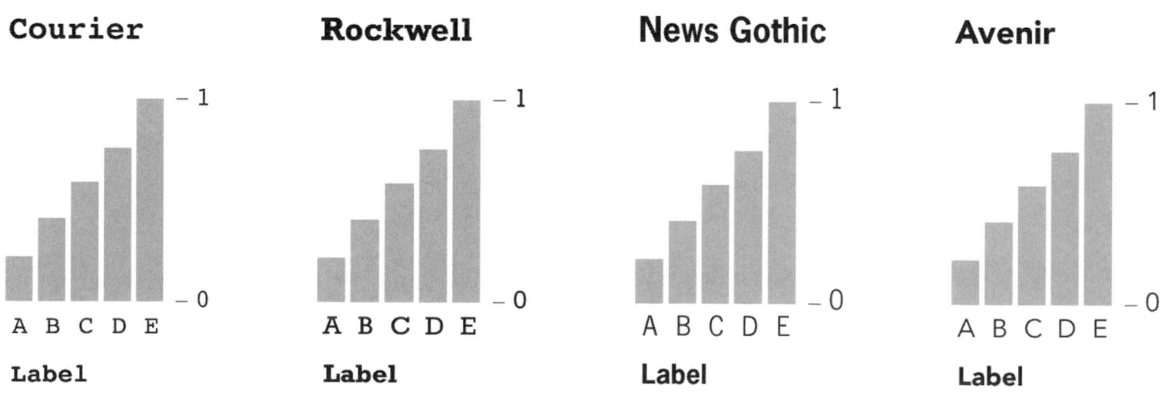

FIGURE 5-38 *Trying various fonts*

typeface, is known for its neutral look that lends to fact presentation, whereas Comic Sans has developed into a meme and a way to avoid being taken seriously. Serif typefaces such as Baskerville and Palatino are reminiscent of vintage graphics. You should probably avoid Wingdings for most practical uses.

Again though, a lot has to do with personal preference, so experiment and see what you like—especially because it's so easy to do with modern software. Remember the visual hierarchy, though. Headers typically stand out visually, so a larger, bold font often works, whereas tick labels are usually smaller and demand less attention, so the typeface should still be readable at a relatively small size. Sans-serif fonts often work well for the latter because serif fonts with a lot of flourish can be harder to read in confined spaces. Although, this is nowhere near a rule.

DO THE MATH

After you get data, the natural first step is to visualize it directly, but after that, it might be useful to do some math for a different point of view. This can shift focus toward something more interesting in the data and in some cases, avoid guesswork as readers try to interpret your graphics.

For example, summary statistics, such as mean or median, can serve as a quick point of reference or to provide a sense of scale, as shown in Figure 5-39. Violent crime rates for each state are shown, and bars are colored based on whether they are above the national average. The distributions of rates isn't especially complex in this example, but it helps you get a sense of where each state lies relative to the national average.

As an additional step, you can transform the data based on a reference point, rather than just show it in the context of the raw data. Figure 5-40 shows global gas prices, which you saw in the previous chapter, relative to average gas price in the United States. Purple indicates higher gas prices, and green indicates countries where gas prices were lower. The two maps show the same data but tell different stories via subtraction and division. The first map focused on worldwide comparisons, whereas this map provides a simple connection between the data and U.S. readers.

What about Figure 5-29 that shows the unemployment rate over time? Maybe you're more interested in annual changes than you are monthly unemployment

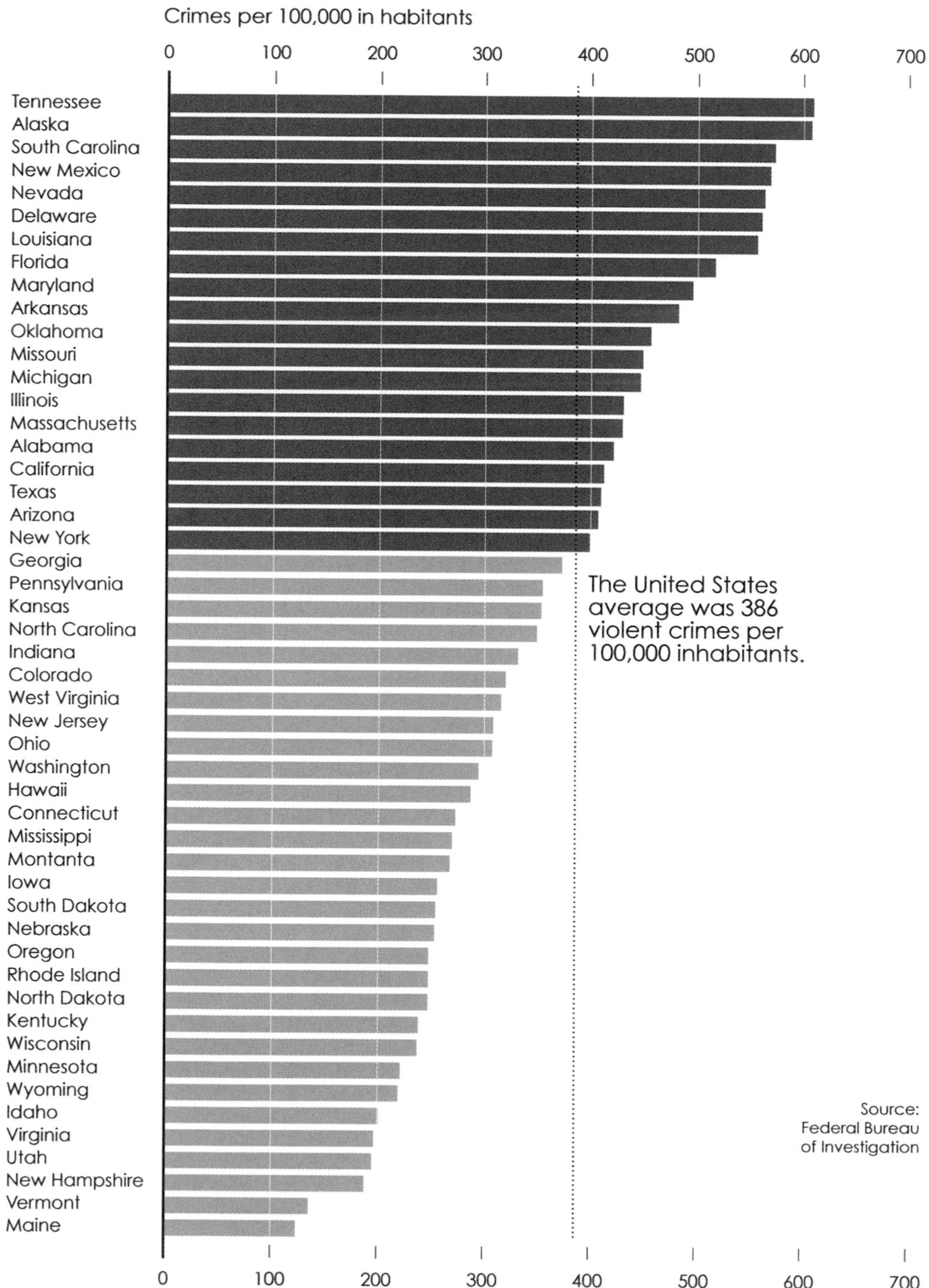

FIGURE 5-39 *Using mean as a point of reference*

rates. From each rate, subtract the rate that came the year before, as shown in Figure 5-41.

You can take it the other direction, and add values, as shown in Figure 5-42. A step chart shows the monthly cumulative cost of cable over a year versus the modest cost of Hulu Plus and Netflix. An aggregate at the end shows total annual savings if you were to switch to the latter.

Straightforward math operations can help you see your data from a different angle or bring focus to a graphic. Of course, the more statistics you know, the better you can process and analyze your data, which in turn can lead to more informative graphics. Account for how people might interpret a graphic, and if they have to do math in their head to make inferences, it might be worth the effort to do the math for them and translate the results visually.

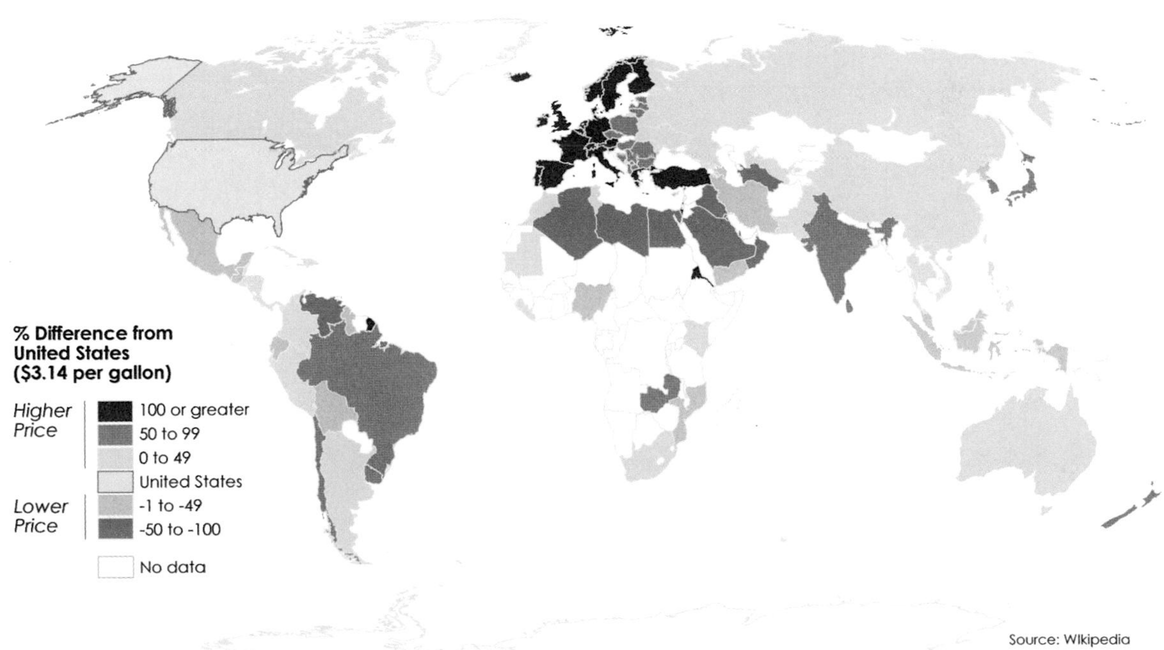

FIGURE 5-40 *Transforming data based on point of reference*

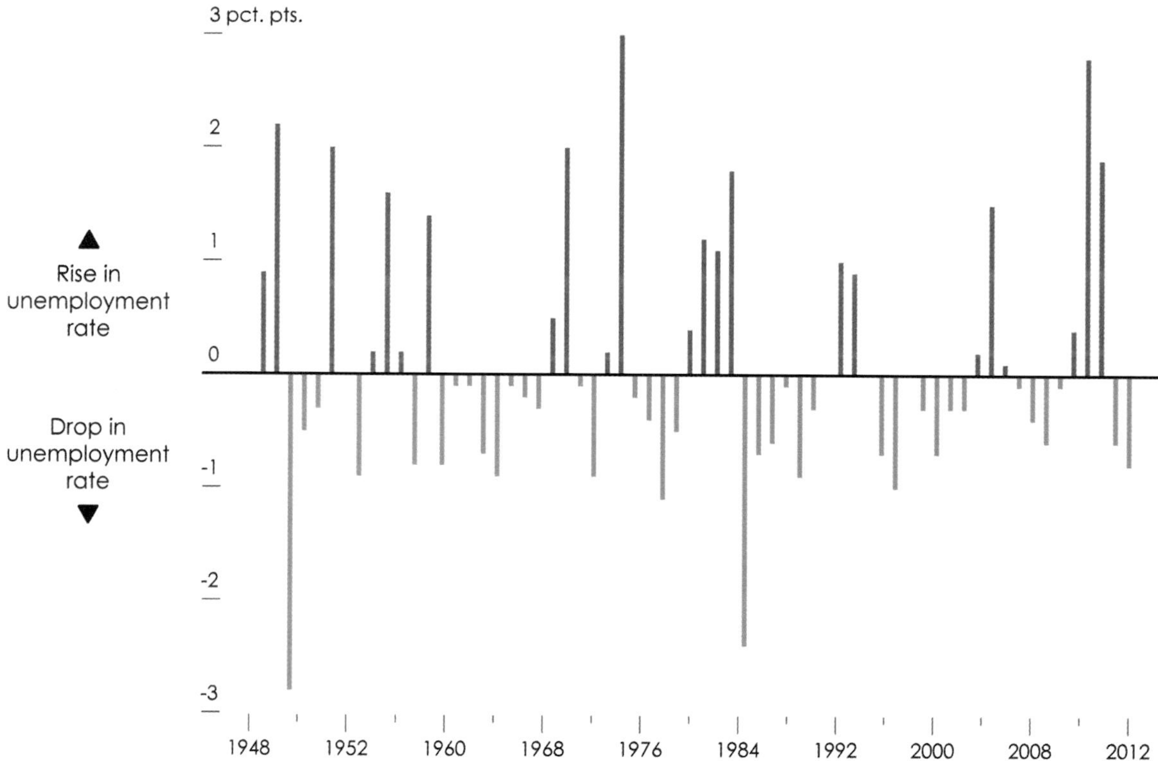

FIGURE 5-41 *Showing changes instead of raw values*

WRAPPING UP

You get leeway as you explore data on your own. However, if you want to make actionable insights based purely on your graphics, you must make sure you're seeing the right thing. You wouldn't want to make a poor decision because you visualized your data inaccurately.

Similarly, as you present your graphics to others or pass them onto the rest of the world, where others might make decisions based on what you show, it's your responsibility to display the data accurately. Differentiate elements, highlight important bits, and annotate to explain and describe the data.

Of course, as you saw in Chapter 2, visualization as a medium creates a wide spectrum of applications. How much you highlight, how much you explain to viewers and readers, and what you display depends on what you want to show and who you present to.

Warning: Make sure your data is comparable when you transform multiple datasets. Are they from the same source? Is the methodology the same? What level of uncertainty is attached to the estimates? If you're not sure, you should find out because incorrect math can lead to incorrect conclusions.

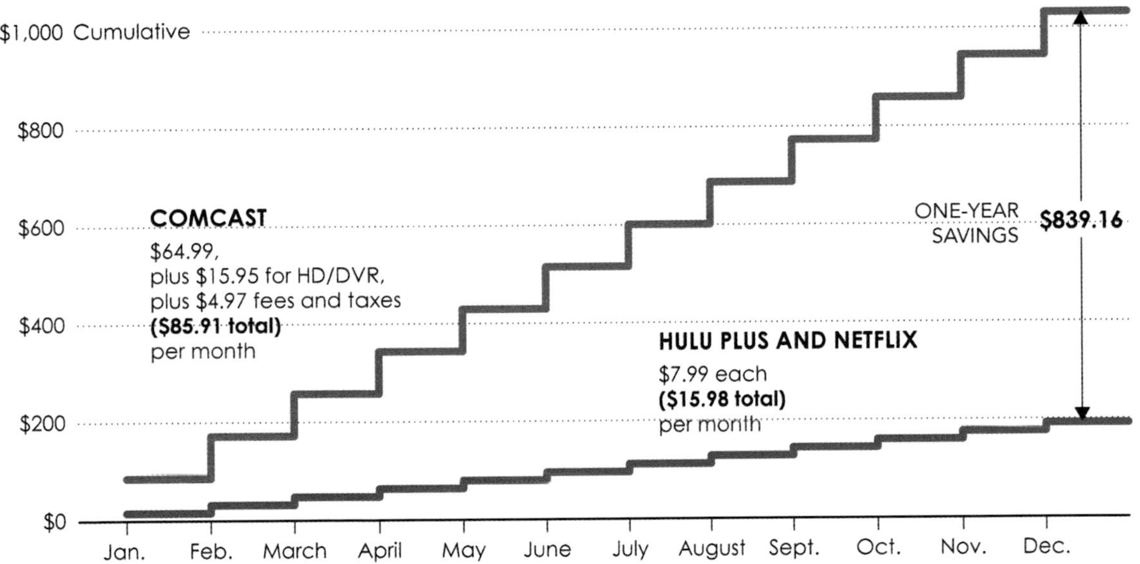

FIGURE 5-42 *Cumulative values and totals*

Pooh

Designing for an Audience

Many people have experience exploring data visually and making charts that are clear and concise, but when they present results, they don't take the extra step to communicate to a wide audience. Instead, they take screenshots or software output and put the raw graphs into a report or post them online.

This works under the assumption that your audience understands your data in the way that you do. Great if that's the case, but what if it isn't? People who don't know the background behind a dataset, or have the same technical expertise as you won't see the same thing as those who do.

When you design visualization for an audience, you must consider what your audience knows, what they don't know, and what you want them to know. How will they read your graphic? How will they interpret your data?

COMMON MISCONCEPTIONS

Before getting into specifics, it's best to clear up common misconceptions about designing data graphics for a wide audience. There are a lot of books and articles that provide suggestions as unyielding rules for various purposes, and these "rules" often conflict. This leads to a lot of confusion. So it's time to clear the air and start fresh.

NOVEL GRAPHICAL FORMS

There are visualization types that have been around for decades. Think bar charts, pie charts, dot plots, and the other usual suspects. People are accustomed to reading data through these traditional forms, but some see this as a negative. How can something traditional catch readers' eyes and keep them engaged? Some think you always have to use new and exciting graphical forms to make visualization interesting, but this idea misses the point of visualizing data (which is why I don't like visualization contests that weigh "innovation" as heavily as insight).

Note: Experimentation with new visualization methods is great, but you also want to make sure that others can decode your encodings. Often, traditional is the best route. Traditional methods have been around for a while because they work.

Visualization can be appreciated purely from an aesthetic point of view, but it's most interesting when it's about data that's worth looking at. That's why you start with data, explore it, and then show results rather than start with a visual and try to squeeze a dataset into it. It's like trying to use a hammer to bang in a bunch of screws.

For example, you've most likely seen the graphic in Figure 6-1. It's the periodic table of elements. Each square represents a chemical element, and they are arranged by their atomic number, or the number of protons in the nucleus of an atom, left to right and top to bottom. The row and column of each element depends on its electron configuration, which results in groups of elements with similar characteristics. The second column, for example, is alkaline earth metals, which are shiny and silver-white.

The data, which is the elements and their properties in this case, dictates the layout of the periodic table, which actually does represent periodicity and relationships of the elements. The table is specific to the data it shows.

It doesn't make sense to take the elements out, keep the structure, shove a different dataset into it and call it the periodic table of whatever, when there's no periodicity or natural grouping. People have made everything from the periodic table of typography, to poetry, to beverages. It's usually cleverer and useful to organize data based on the actual data instead of the atomic number of chemical elements.

Novel graphics are fine, but don't make them hard to read or nonsensical for the sake of uniqueness. Instead, make use of the uniqueness or relevance of your data.

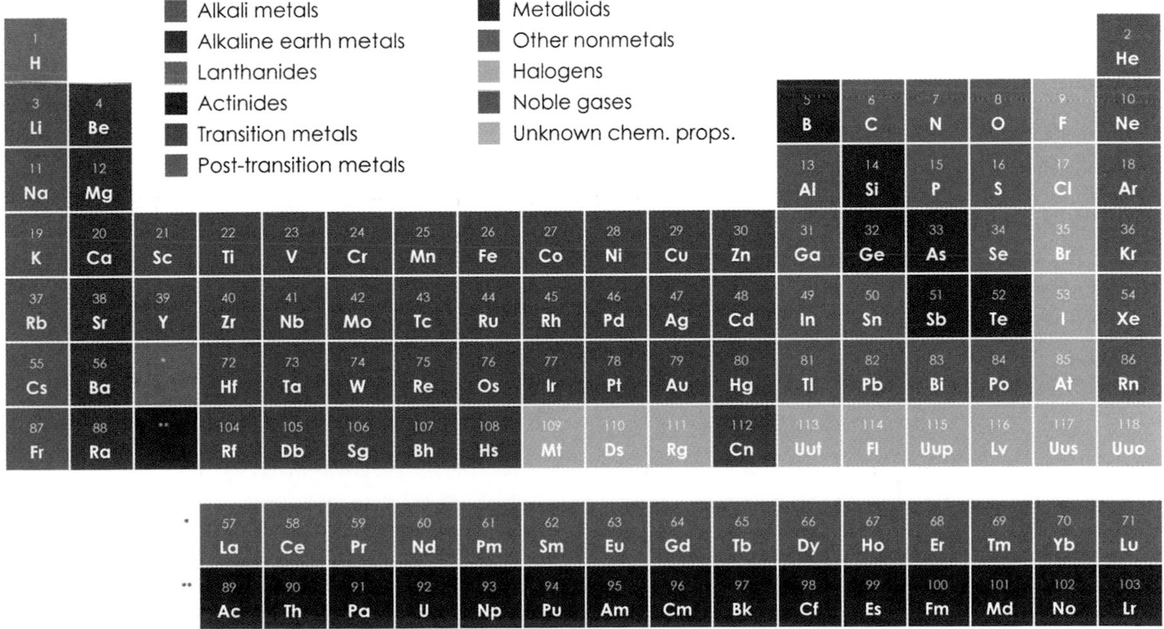

FIGURE 6-1 *Periodic table of elements*

For example, the *World Progress Report*, shown in Figure 6-2, represents the state of the world based on data from the United Nations Statistics Division. At the time, the United Nations had released a lot of data, but reports were mostly composed of reference tables. The goal of this project was to make the data accessible to a wider audience, and it got a lot of attention, but you can see that all of the charts are traditional ones. The content made the graphics interesting.

From the other side of the fence, you might often see comments such as, "This took me more than 5 minutes to understand. Fail."

Some assume that insight from every data graphic should be instant, and in many cases, this is a perfectly valid. A dashboard that updates in real time, used to quickly see the status of a system, requires instant readability. On the other hand, a tool designed to explore connections between millions of people can be complex and require time to understand. You might show simple views and aggregates, but you might also miss details that could lead to better understanding of the data.

Sometimes there's just a lot of data to show, and it takes a while to go through all of it. For example, the *Better Life Index* shown in Figure 6-3, created by the Organization for Economic Co-operation and Development (OECD), Moritz Stefaner, and creative consultancy Raureif, provides a way to explore the quality of life in OECD countries.

Each flower represents a country, and each has eleven petals, which represent topics that the OECD collects data on, such as housing conditions and spending and household income. An index is created based on these factors, and the higher a country is positioned on the vertical axis, the better the estimated well-being in that country.

Note: The OECD promotes policies to help improve economic and social well-being and compiles data for participating countries to estimate progress. There are currently 34 member countries.

The challenge is to decide what makes the quality of life in one country better than another. A better way of life is defined differently based on who you ask. Some might care little about how much money they can make but care a lot about overall health. Others might switch that. The interactive enables you to specify what's important to you, and the countries adjust height based on your picks. So it's your own better life index.

FIGURE 6-2 *(facing page)* World Progress Report *(2010), http://www.flowingprints.com/*

Although Figure 6-3 weights all topics equally, Figure 6-4 takes an extreme and places work-life balance as the most important topic and doesn't consider anything else. Some countries move up and others move way down.

WORLD PROGRESS REPORT

WHERE WE HAVE BEEN, WHERE WE ARE, WHERE WE ARE HEADED

BIRTH

SKILLED ATTENDANT AT DELIVERY

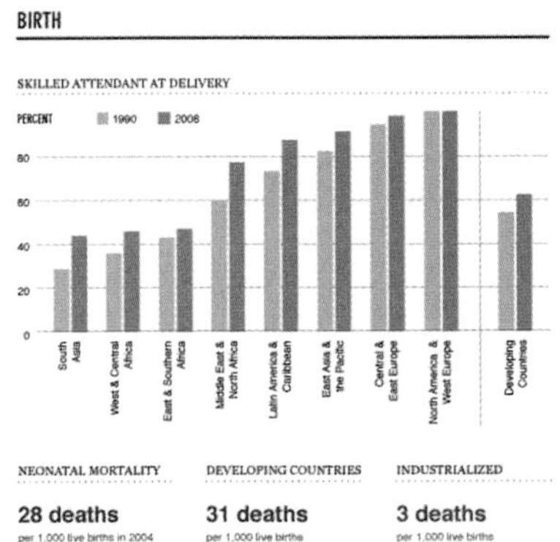

NEONATAL MORTALITY

28 deaths

per 1,000 live births in 2004

DEVELOPING COUNTRIES

31 deaths

per 1,000 live births

INDUSTRIALIZED

3 deaths

per 1,000 live births

POPULATION

PAST AND PROJECTED

CURRENT

6.7 billion

10, 8, 6, 4, 2, 0

1950, 2000

GENDER BREAKDOWN

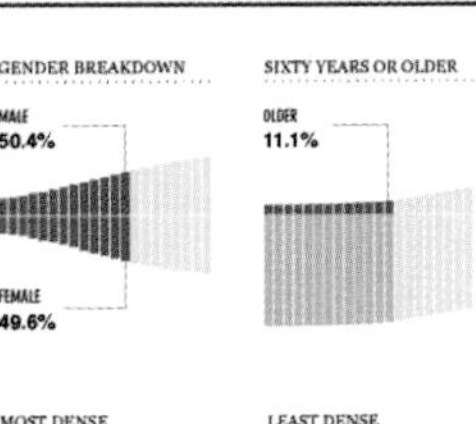

SIXTY YEARS OR OLDER

OLDER

11.1%

GROWTH RATE

CURRENT

1.1%

2.0, 1.5, 1.0, 0.5, 0.0

1950, 2000

MOST DENSE

MACAU

18,534 people per sq. km.

LEAST DENSE

GREENLAND

0.026 people per sq. km.

MOST POPULATED

CHINA

1.34 billion people

LEAST POPULATED

PITCAIRN ISLANDS

67 people

EDUCATION

LITERACY RATES

ENROLLMENT IN TERTIARY EDUCATION

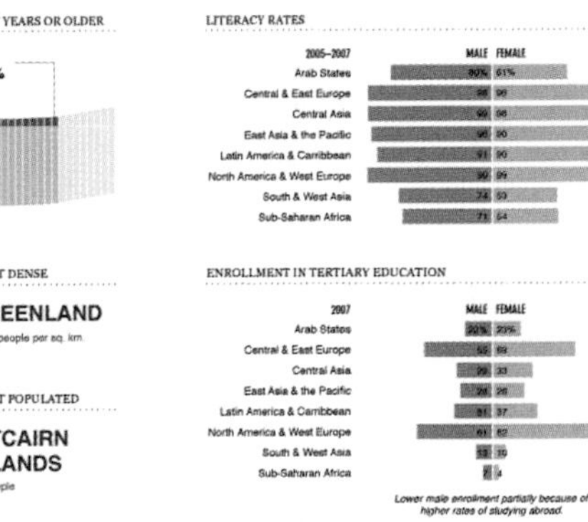

Lower male enrollment partially because of higher rates of studying abroad.

LIFE EXPECTANCY

WORLD AVERAGE

66 years old

UNDER AVERAGE

65 countries

AVG. FIFTY YEARS AGO

49 years old

LIFE EXPECTANCY AT BIRTH

65 or older

60-65

45 or younger

LONGEST LIFE

MACAU

84.36 years old

SHORTEST LIFE

SWAZILAND

31.99 years old

LABOR

MORE WORKING HOURS

LESS WORKING HOURS

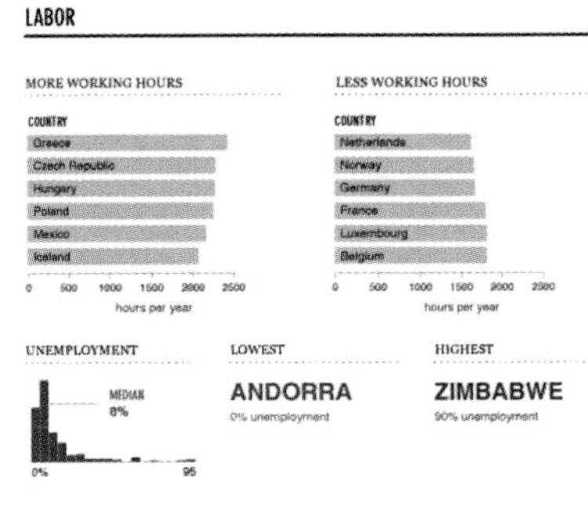

UNEMPLOYMENT

MEDIAN

8%

0% 95

LOWEST

ANDORRA

0% unemployment

HIGHEST

ZIMBABWE

90% unemployment

ENVIRONMENT

GREENHOUSE GAS EMISSIONS

HYDRO ELECTRICITY

WIND ELECTRICITY

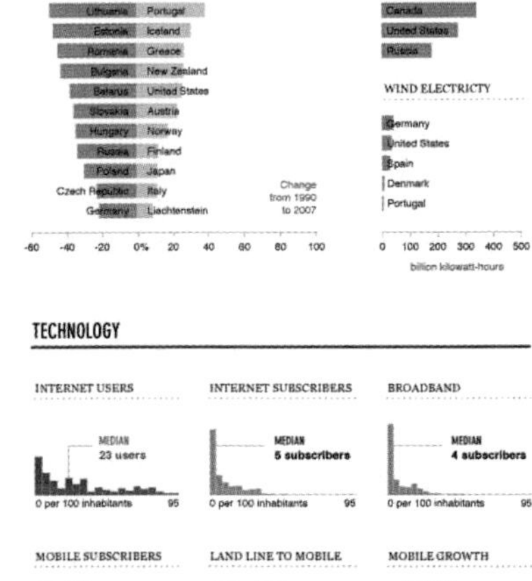

AGRICULTURE

FRUITS AND VEGGIES

1,463.6 million

tons produced in 2007

MEAT

248.0 million

CEREAL

2,351.4 million

SUGAR

1,838.3 million

COFFEE AND TEA

11.7 million

TOBACCO

6.2 million

DRY BEANS

1.1 million

NUTS

0.8 million

POVERTY

LIVING UNDER $1.25 PER DAY

Sub-Saharan Africa

South Asia

Oceania

Southeast Asia

East Asia

West Asia

Latin America and Caribbean

North Africa

YEARS

1997

2007

2008

0% 10 20 30 40 50 60 70

UNDERNOURISHED

Sub-Saharan Africa

South Asia

Oceania

Southeast Asia

East Asia

West Asia

Latin America and Caribbean

North Africa

YEARS

1990-1992

2004-2006

2008

0% 10 20 30 40

TECHNOLOGY

INTERNET USERS

MEDIAN

23 users

0 per 100 inhabitants 95

INTERNET SUBSCRIBERS

MEDIAN

5 subscribers

0 per 100 inhabitants 95

BROADBAND

MEDIAN

4 subscribers

0 per 100 inhabitants 95

MOBILE SUBSCRIBERS

59.74

per 100 inhabitants

LAND LINE TO MOBILE

1 to 3.2

MOBILE GROWTH

+285%

since 2003

TOURISM

PURPOSE OF VISIT

Leisure, Other, Not specified, Business

MEANS OF TRANSPORT

Air, Road, Rail, Water

MARKET SHARE

Europe, Asia & Pacific, Americas, Africa, Middle East

MOST ARRIVALS

FRANCE

79.3 million in 2008

TOURIST SPENDING

$944 billion

internationally in 2008

AVG. ANNUAL GROWTH

+2.8%

in arrivals from 2000 to 2008

Sources: United Nations Statistics Division, Food and Agriculture Organization, United Nations Framework Convention on Climate Change, International Labour Organization, Organization for Economic Co-operation and Development, United Nations Children's Fund, UNESCO Institute for Statistics, World Health Organization, United Nations Population Division, International Telecommunications Union, and World Tourism Organization i.e. UNdata, http://data.un.org

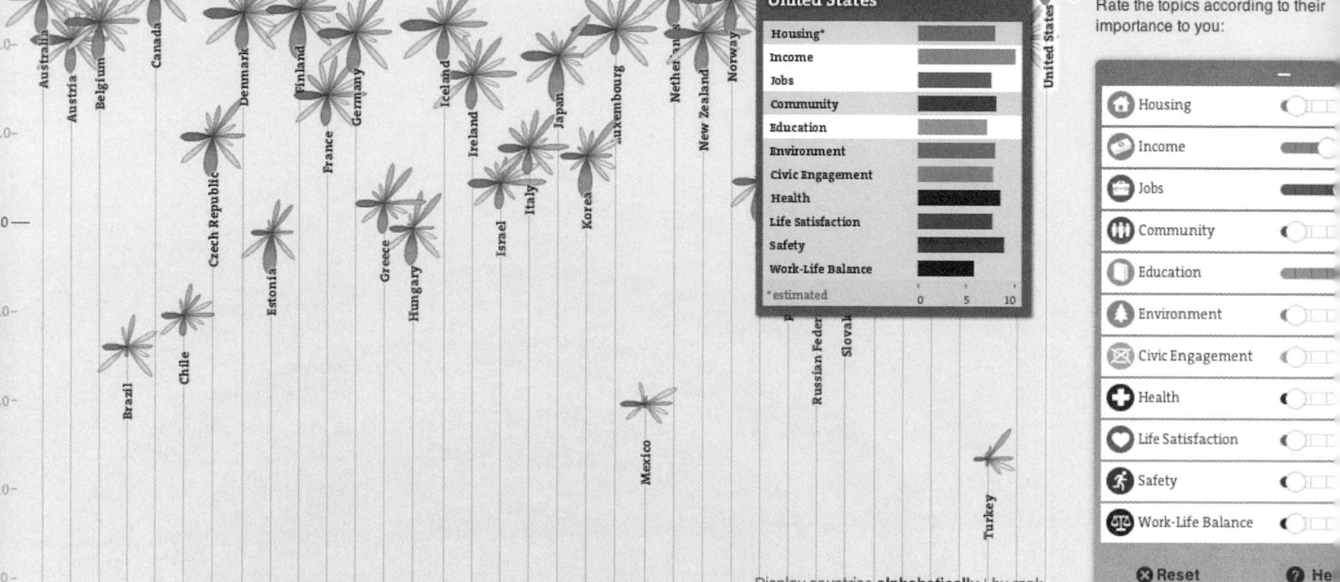

FIGURE 6-3 Better Life Index *(2011) by OECD, Moritz Stefaner, and Raureif, http://www.oecdbetterlifeindex.org/*

The *Better Life Index* is meant to be explored and fiddled with. It can be scanned quickly but there's a much better payout when you spend time with the interactive. The flower metaphor might not fit in with traditional statistical graphics, but it does make the data feel more tangible and relatable. You can also mouse over a country to see specific estimates for each topic, as shown in Figure 6-5.

So you need to strike a balance between function and uniqueness. Novel simply for the sake of novel can often make understanding of the data (which should always be your goal) difficult. However, unique forms that feed off the uniqueness of the data can help show what the numbers represent, other than their quantitative values.

VISUALIZING EVERYTHING

Sometimes a table is better. Sometimes it's better to show numbers instead of abstract them with shapes. Sometimes you have a lot of data, and it makes more sense to visualize a simple aggregate than it does to show every data point.

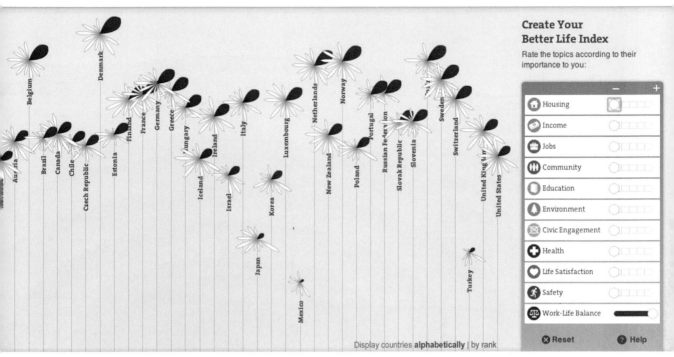

FIGURE 6-4 *A custom Better Life Index*

Imagine you run a fundraiser with a financial goal, and you receive hundreds of donations, so you have a data point for every donation: the dollar amount, who it's from, and where each generous contributor lives. Although it might be interesting to see the distribution of donations, the total amount raised could be all that people care about. You don't need to visualize data just because you have it.

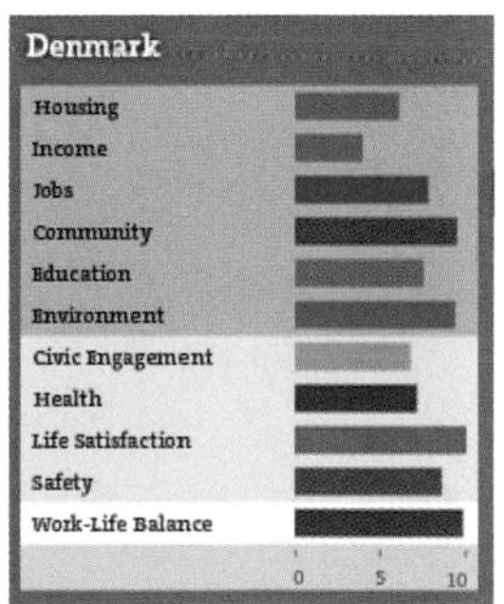

FIGURE 6-5 *Topic breakdowns for a country*

You might also not have a lot of data to show. It's possible (and likely) that the only numbers available are aggregates instead of raw data. You see this in magazines a lot, labeled as such and such "by the numbers" in a sidebar feature. It's a bunch of numbers that represent different but related things, with different units. It's okay to print the actual numbers.

For example, consider three estimates about the state of the world. Life expectancy at birth is 70 years, the literacy rate of youth females ages 15 to 24 is 87 percent, and the gross domestic product is approximately $70 trillion. You might be inclined to visualize this simply because there are numbers. Maybe you make something like Figure 6-6, but there isn't a good reason for it because there aren't any other values to compare to for each of the estimates.

Note: Some might argue that the figure with shapes and colors is more "visually compelling," but in this case, it's just fluff that fills space. Actually, you might argue whether three unrelated data points require a graphic at all.

Instead, you could do something like Figure 6-7. Simply display the estimates. The point of visualization is to understand relationships in data and patterns in data, so when you don't have the data to do that, you don't need to squeeze out something visual.

Random numbers about the world

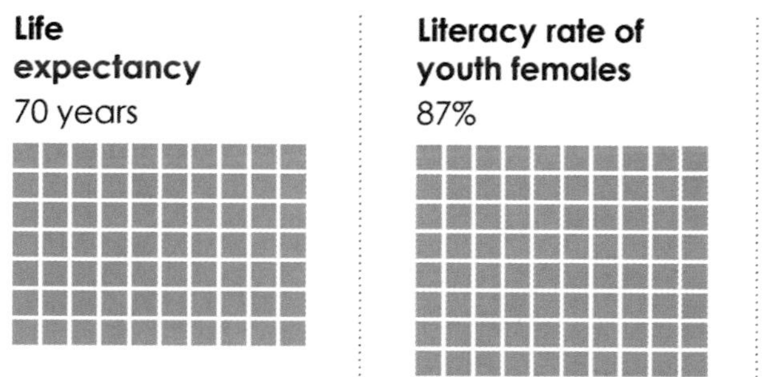

FIGURE 6-6 *Visualization without comparison*

Random numbers about the world

LIFE EXPECTANCY	LITERACY RATE OF YOUTH FEMALES	GROSS DOMESTIC PRODUCT
70 years	87%	$70 trillion

FIGURE 6-7 *Just showing the numbers*

MAKING THINGS PRETTY

It's easy to say that a graphic is nice to look at but doesn't provide insight. "It's just a pretty picture." The statement often assumes that there's little return value when you make something beautiful, and that mechanical function is the only thing that matters when you design data graphics. It assumes that the only goal of visualizing data is analysis, but as a medium, visualization can also evoke feelings about a subject and encourage readers to ponder or simply appreciate the context of the data. Maybe memorability or nostalgia is the goal.

Aesthetics isn't just a shiny veneer that you slap on at the last minute. It represents the thought you put into a visualization, which is tightly coupled with clarity and affects interpretation.

For example, as shown in Figure 6-8, Nicolas Garcia Belmonte visualized wind patterns in the United States based on data from the National Weather Service. The interactive animation shows the last 72 hours of wind motion. Lines represent wind direction, the radius of circles represents speed, and hue represents temperature. Each mark is a weather station, which you can mouse over for additional details.

In contrast, Martin Wattenberg and Fernanda Viégas also visualized wind patterns using the same data, but with a different look and feel, as you saw in Chapter 1, "Understanding Data," and shown in Figure 6-9. Higher line density and longer segments represent greater wind speeds.

The first map is modular showing a circle for each of 1,200 weather stations. It feels like an exploratory tool. The second map interpolates wind paths and feels more like an art piece that you sit back and digest. Both provide similar insights and help you infer current wind patterns, but because the former is more tool-like, you might approach the data with an analytical mindset, whereas you might approach the latter like you would a painting in an art gallery.

FIGURE 6-8 United States Wind Patterns *(2012) by Nicolas Garcia Belmonte, http://www.senchalabs.org/philogl/PhiloGL/examples/winds/*

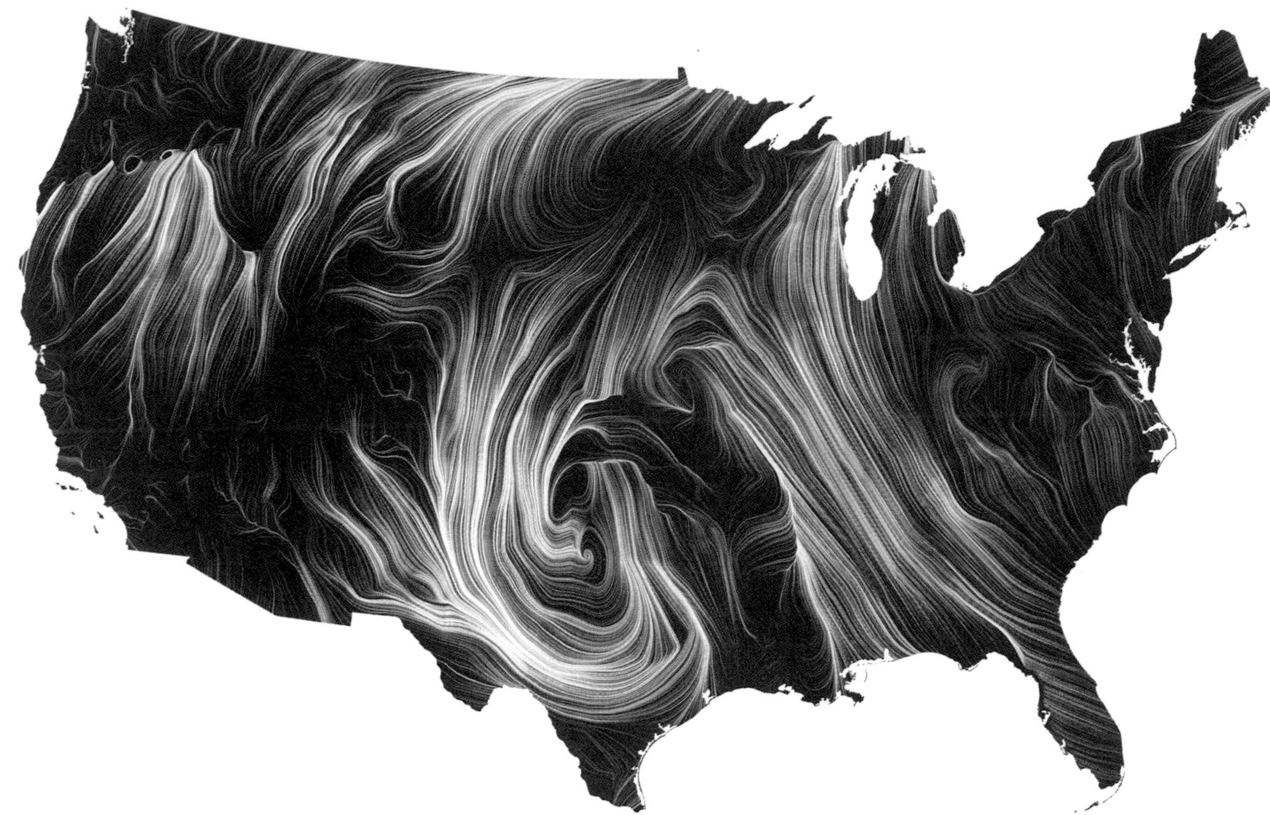

FIGURE 6-9 Wind Map *(2012) by Fernanda Viégas and Martin Wattenberg, http://hint.fm/wind/*

This applies to more basic charts, too. I used to make graphics for a data-ish comic of sorts called *Data Underload* on FlowingData, and it was mostly an excuse to play with colors and shapes. They were more conceptual than they were based on real data but were designed as if they were based on real data.

For example, Figure 6-10 is a chart that shows imaginary sleep schedules based on age. The information from WebMD was used for the average amount of sleep, but the start and end times were just guesses. It was mostly for my amusement, but some confused the chart for one that showed real data. A major news publication even wanted to publish a version of the graphic, until I explained that it was a comic.

Had the chart been drawn with pen on the back of a napkin, as shown in Figure 6-11, the reader interpretation would be different. It looks far less serious and would confuse fewer people.

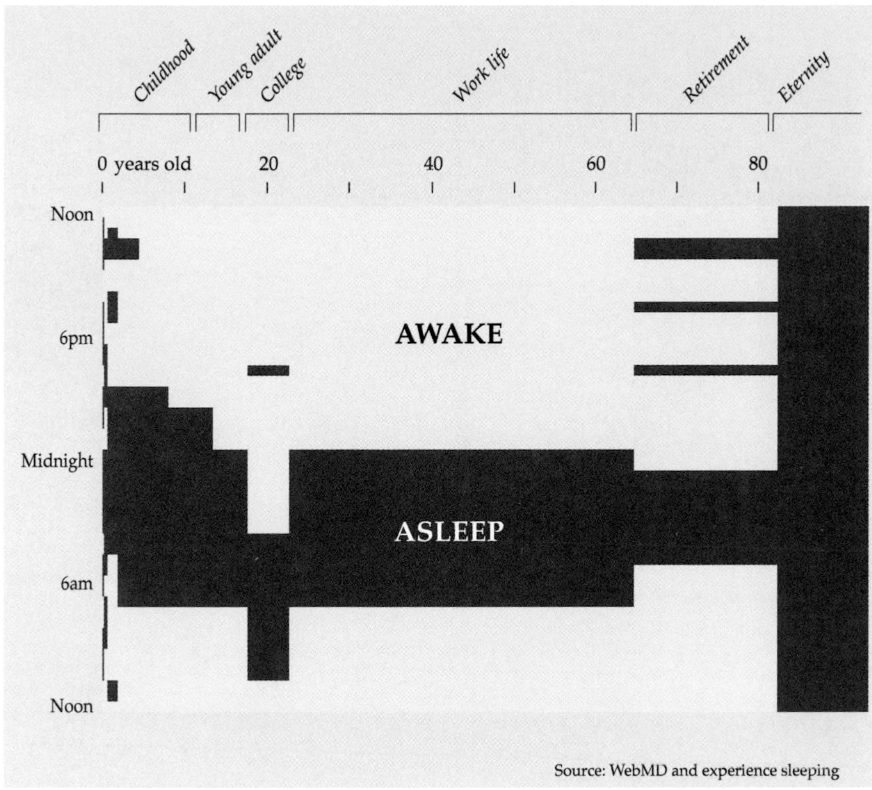

FIGURE 6-10 Sleep Schedule *(2010), http://datafl.ws/22*

At the end of the day, looks matter. After all, visualization is, well, visual, so people judge based on what you show and how you show it. Often, a poor chart doesn't mean poor analysis, but many people see it this way, regardless of whether it's right.

People buy things (or at least look at them more closely) in a store based on looks, and the same thing applies to visualization. For example, Figure 6-12 shows graphs with the same data, but they look different. If you had to read a report filled with graphs, which aesthetic would you choose?

That said, thoughtful aesthetics doesn't compensate for visualization with a poor foundation (the data). You need both sound analysis and design that considers your objectives and your audience. Without the former, you just have pretty pictures, and without the latter, you have software output.

Note: As you know, beauty is in the eye of the beholder, so a visualization with software defaults can look great. The beauty is in the data. A text file of numbers can be beautiful, but that inner beauty isn't obvious to everyone.

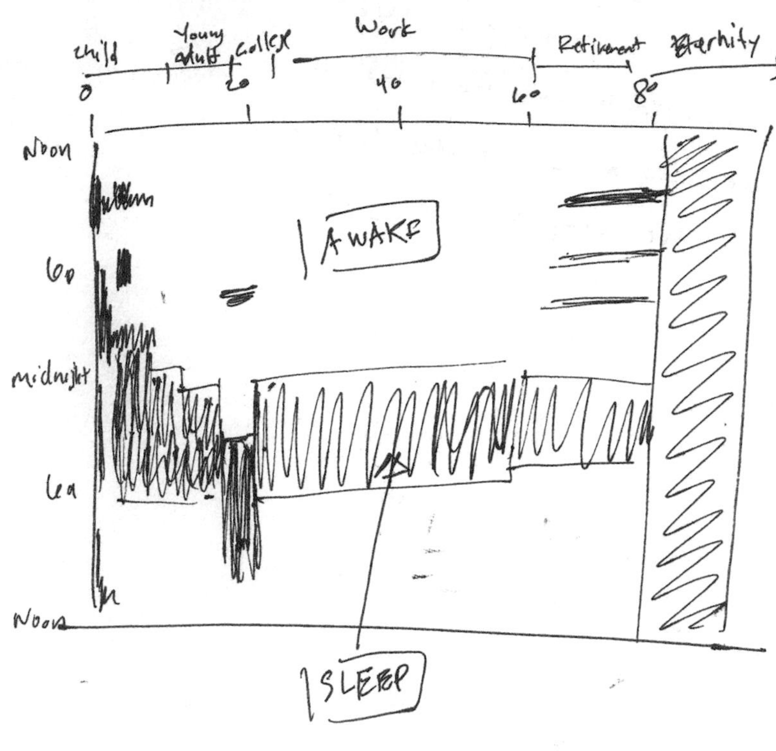

FIGURE 6-11 *A rough sketch rather than a polished graphic*

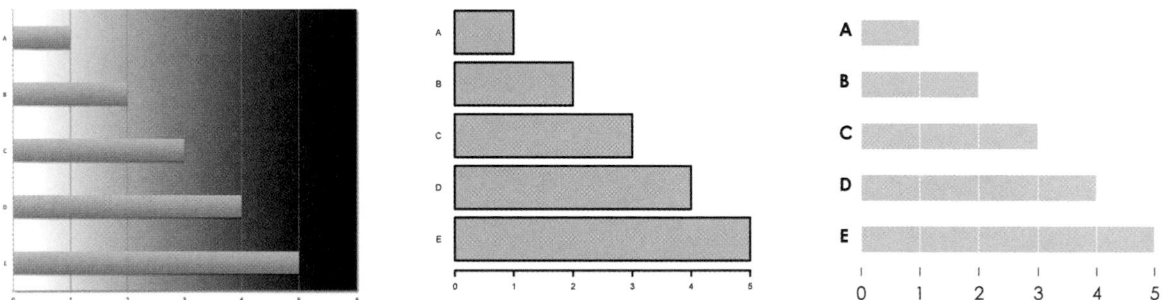

FIGURE 6-12 *Same data and graph with different aesthetics*

THE RULES

If you've looked up the how-to's of visualization, you've no doubt come across plenty of "rules." There are books that are nothing but do's and don'ts for your graphics, and a common mistake is to assume that these apply to all visualization when they are written for specific applications, such as analysis, reports, or presentations.

> *"There are no hard rules, but, as in any skill-based profession, there are some things that work better than others, and as for any cultural activity, you are not acting in the void, but in an established set of conventions, expectations and common knowledge."*
>
> —*Moritz Stefaner*

This isn't to say that these rules are wrong. You just have to know when they apply and keep in mind that most visualization rules are generalizations. Don't follow them blindly, or you might as well get a computer to do all the work for you.

One frequently voiced rule is to show only what is useful and to strip out the rest, which often leads people to believe that you should always make stripped down, almost barebones charts. However, what is considered useful? That depends on your goals. For example, Santiago Ortiz visualized six seasons of the show *Lost*, as shown in Figure 6-13.

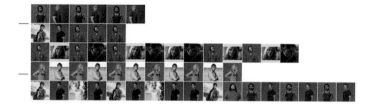

FIGURE 6-13 Lostalgic *(2012) by Santiago Ortiz, http://moebio.com/lostalgic*

Each row represents an act and each profile picture is shown whenever the corresponding character says something. It might have been more efficient to represent each act as a bar chart, but that was not the goal of *Lostalgic*. Although you can certainly draw insights about character relationships and interactions, the purpose of the interactive is to provide a new look into the show that fans can use to rewind and fast forward to the episodes they enjoyed most.

The details made the show fun to watch, so without them, the exploratory visualization would not be as interesting to the audience it was intended for.

Note: More often than not visualization rules are actually strong suggestions or make assumptions about what you want to visualize your data for. Keep these suggestions in mind as you design data graphics, but also use your own judgment to decide what works best for your goals.

So although visualization rules are fine for guidance and to ensure that others interpret your work correctly, you still must decide when the rules apply and to what degree you should follow them.

When you first learn visualization, it's good to learn the rules of perception, structure, and layout. When you learn to write, there are rules for grammar, sentence and paragraph structure, and punctuation. This ensures that people can understand your words. Whereas with visualization, you want to help people decode visual cues to understand data; with writing you want to help people translate your words to complete thoughts. However, as you write more or read your favorite authors—even though a lot of the grammar stays intact—the rules don't always need to be followed with an iron fist. Visualization, which is essentially a visual language of data, works the same way.

PRESENT DATA TO PEOPLE

With misconceptions out of the way, now consider the groups of people who look at your graphics. Visualizing data for various audiences means different goals that depend on what you want people to see and interpret. Accuracy and truth should always top your list of objectives, but this comes in many visual forms as you present to one person, a hundred people, a thousand, or a million.

JUST YOU

This is an exciting part of the visualization process. Not only is it when you get to explore your data to figure out what it represents and what it means, but it's also when you get to play with shapes, colors, and layouts to see what works and doesn't. It's always a little different for each dataset, so you learn something new every time.

You might tend to create charts in a rapid fire fashion during the exploration phase. You want to know what the data is about and if there are any angles you should explore in depth. Layout and aesthetics is of little concern at this point, whereas efficiency and speed is what you need to look for. If there's a lot of data or the dataset is complex, you shouldn't be surprised to spend most of your time in this phase.

When you know what you want to show, you can figure out how you want to show it. You can start with paper before you mess around on the computer, so you need to have a notebook within arm's reach. Sketch, scribble, and jot down notes for what you think might be useful, and then try to translate that through the computer.

Note: More formal analyses happen during this time, too. It's an iterative process in which visualization informs statistical methods and vice versa. The more you learn about your data during this phase, the more you have to say to an audience.

Sketching on paper can provide a separation between what you want to do and the limitations or technical challenges imposed by a computer. It's good to have limitations in mind, but it's better (and easier) to have a lot of ideas and then dial back to fit within digital limits and time constraints than it is to restrict work to what you know how to do on a computer.

Use the tools that work best for you. Then learn all you can about your data through exploration and analysis. Use your findings to guide design.

Note: You tell the computer what to do—not the other way around.

A SPECIFIC AUDIENCE

The main challenge of visualization for an audience is that you must make sure readers can make the leaps from encoding to decoding to understanding the data. If your audience is already familiar with the background behind your data or has perhaps even worked with it, the barriers to get to each step are lower.

You can take readers through your analysis or the graphics you created during the exploration phase, and it probably won't be a stretch that most can follow your logic.

Note: You have some leeway if your audience is familiar with your data and analyses, but it's better to explain too much than too little.

However, because you are presenting data to others, consider how they will examine your work. When you're the only audience, you design for one person, one distance, and one computer screen or piece of paper. When there are others in the mix, there is variability. People have different backgrounds, printers, and screen resolutions, and although it's often hard to accommodate for everyone, you can at least try to account for as many situations as you can—within reason, of course.

For example, when you give a slide presentation, there are people who sit in the front, in the middle, and in the back. Start with the suggestions from the previous chapter on clarity, but consider how people in the back will see your graphics versus those in the front. As shown in Figure 6-14, a graphic might look fine up close but unreadable from far away.

Design your graphics for just those in the front, and those in the back might not see, but design for those in the back, and everyone can see. This seems obvious, but how many times have you seen a presentation where the speaker says, "You probably can't see this, but...." and then goes on to explain as if you did?

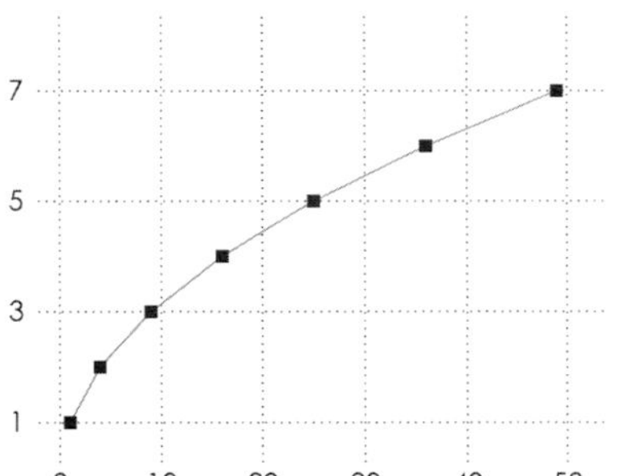

This too, if you squint.

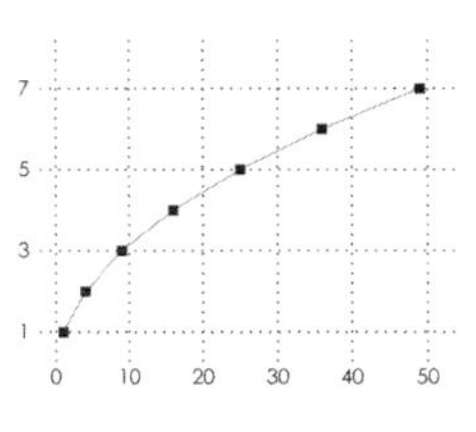

Um, what?

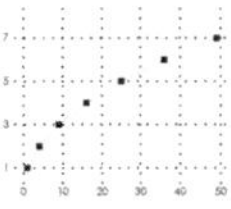

FIGURE 6-14 *How a chart might look from variable distance*

A WIDER AUDIENCE

Designing data graphics gets tricky as your audience grows. You have the same variables to consider as you do when you present to colleagues, but the range of each variable also increases. Screen resolution can go from mobile phone to giant monitor, and people can easily take your graphics outside the context of a report or a slide deck and post it on their website.

Most important, the range of data literacy and familiarity with your data's context is much wider when you design graphics for a general audience.

This doesn't mean you have to dumb down your visualization or limit what you show, but you do have to make sure you explain complex concepts within a human context. Avoid jargon and present the data in a way that is relatable and doesn't require a doctorate in statistics to understand.

For example, information visualization firm Periscopic visualized polar bear population, habitat, and threat information in the interactive *State of the Polar Bear*. It shows the data from various angles and lets you explore the data by space and time, or rather provides an overview story that you can interact with for more details. Figure 6-16 is the initial view, which shows subpopulations recognized throughout the circumpolar Arctic by the Polar Bear Specialist Group.

You can see detailed information when you click on a region, as shown in Figure 6-16. There are other areas of the interactive to explore (which you should do) but the takeaway is that the source material, which is the results of a collaborative effort across several countries, could potentially confuse readers. However, layers of interaction take you through the data, and ample context from researchers help explain what you see.

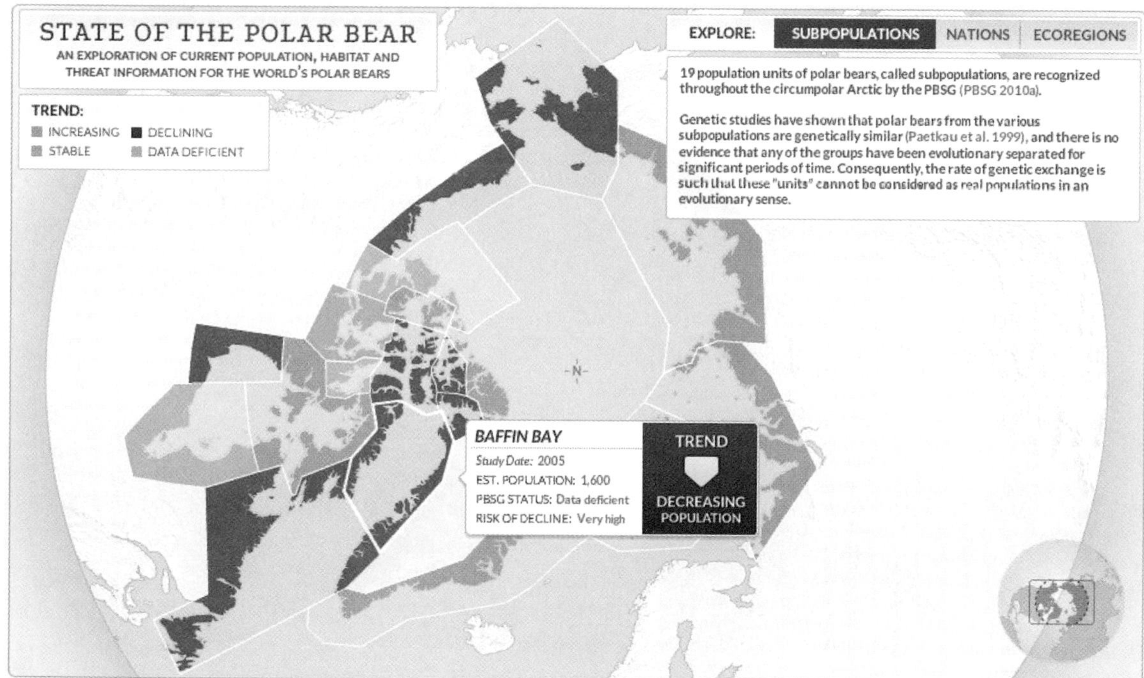

FIGURE 6-15 State of the Polar Bear *(2012) by Periscopic, http://pbsg.npolar.no/en/dynamic/app/*

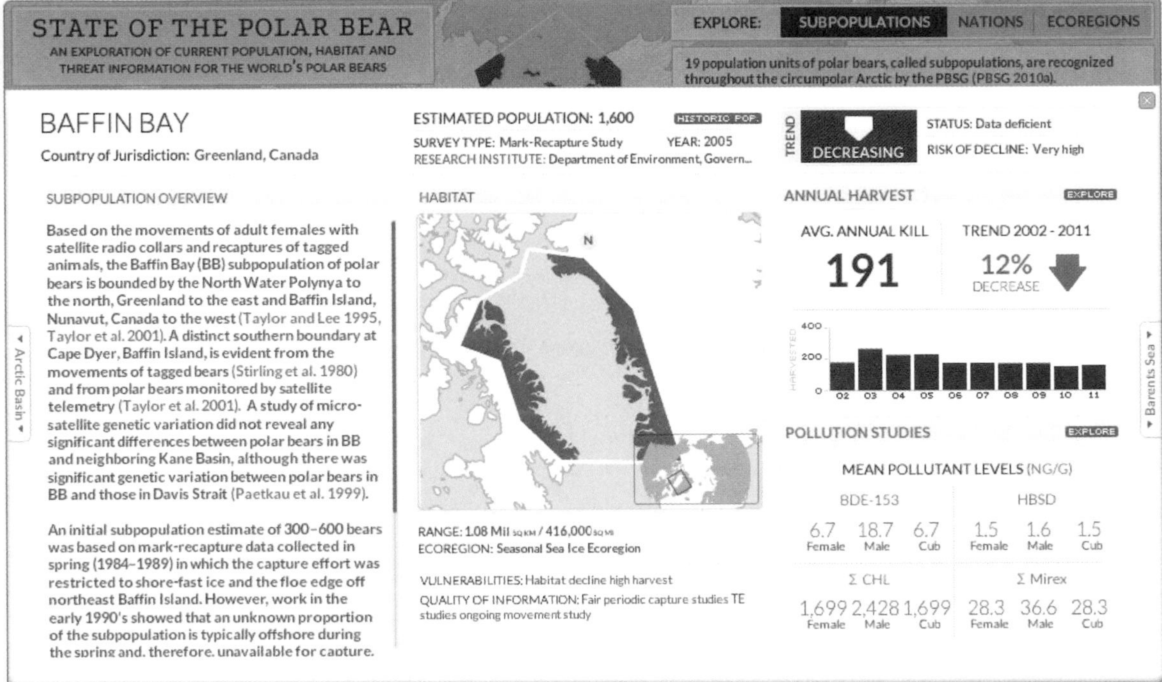

FIGURE 6-16 *More details on a region*

Although *Polar Bear* is interactive, the same can be done with static visualization. Create multiple graphics and present them sequentially. Think of it as a tour through data, and you're the guide who points to and explains interesting facets.

THINGS TO CONSIDER

Put yourself in the tourists' position. You're on a tour of a city where historic things have occurred over the centuries. What do you want out of your tour guide? You want him to tell you about when and where events occurred, maybe what the people were like at the time, or why that building is colored and labeled in that weird way. You don't mind that the guide injects his personality into the tour, but you want him to stay on course and stick to the subject that you paid to hear about. Above all else, you hope that the guide tells you the truth and doesn't make up content on the spot. If he doesn't know an answer to a question, he should say so.

As the tour guide of data, you assume similar responsibilities. It's your job to point out the direction of interest, provide background, make sure you don't confuse, and keep focus. The extent that you provide these to readers varies by who your readers are, but above all else, you must show the truth.

DATA BACKGROUND

Remember my wedding photos in Chapter 1? At first it was foreign—a picture plucked out of context—but as I told you more, it felt real, and you had a better sense of what happened that day. Apply this more generally to visualization, and it's easier for people to understand the data. Show people a graph out of context, and it's more difficult to interpret.

For example, look at Figure 6-17. This is a map from the live flight tracking site FlightAware. From the flight information page, I can tell you this was flight N48DL on April 19, 2012, from Slidell, Louisiana to Sarasota, Florida. The duration of the flight was 4 hours and 23 minutes.

Other than what looks like a broken flight tracker, there doesn't seem to be anything notable about this map. However, the full story is that this is the flight path of a small plane that circled over the Gulf of Mexico for more than 2 hours and eventually sank in the ocean. The pilot was unresponsive. Suddenly, the map means something else.

Sometimes, when you work with a dataset for a while, it's easy to forget that others aren't as familiar with it as you are. When you know all the intimate details, it's hard to step back and remember what it was like when you first opened up a file or database—just a bunch of numbers. However, that's where a lot of people are when they look at a visualization, so bring them up to speed.

Note: The best way to learn where people are is to show your work to those who don't know your data. You get an immediate sense of understanding just from first impressions.

GUIDANCE FOR CONCEPTS

This is mentioned in the previous chapter, but it's worth another look, more broadly speaking. Statistical graphics used for data exploration and designed by statisticians can be great tools and are the foundation of plenty of fine graphics. The challenge with a general audience though is that you have to put yourself in the place of those who don't work with data for a living. It's not enough to simply expect people to know what you do.

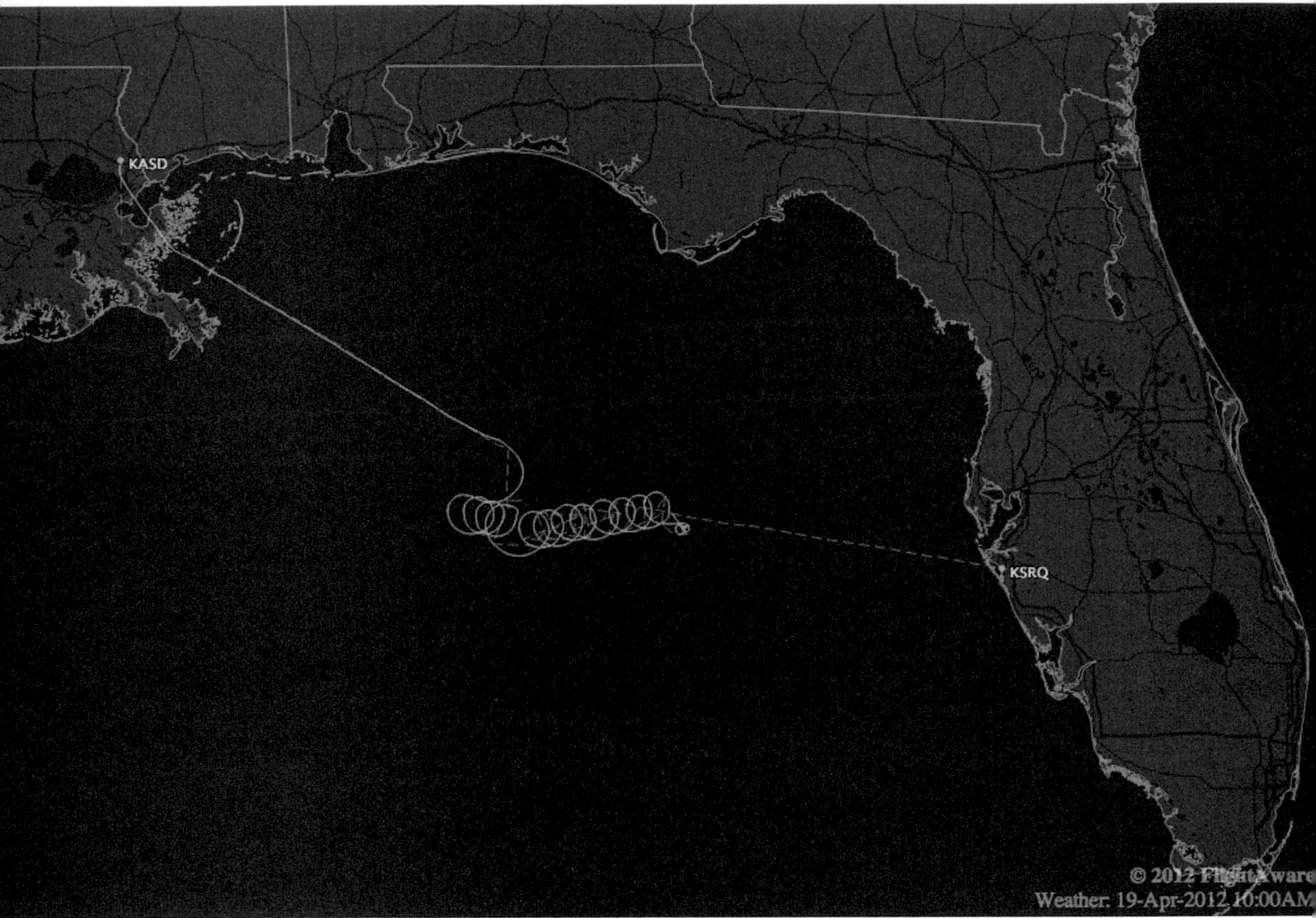

FIGURE 6-17 *A flight from Slidell, Louisiana to Sarasota, Florida, according to FlightAware, http://flightaware.com/live/flight/N48DL*

Even those who do work with data might not be familiar with a chart type or look for the same visual features that you might. During my second year of graduate school, my professor showed us a line chart with an upward trend, and he asked if there was anything peculiar about it. No one had an answer because all we saw was a line headed to the upper right. But then he pointed out a small blip in the line, and that turned out to be the focal point of the chart. By then I had seen my fair share of charts, so I was surprised that the blip didn't occur to me, but after the professor pointed it out, the spot on the chart seemed obvious.

So consider what your audience knows and doesn't, and what they need to know to understand your visualization. Basically, make sure they can decode the geometry and colors. This is vital because if they can't decipher what a visualization means, purely from a bits to numbers standpoint, they can't even get to the part where they relate to the data and form insights (see Figure 6-18).

Statistics education research shows that when grade school students first read graphs, they tend to focus on individual values. That is, they might comprehend that the height of a bar in a bar chart corresponds to a value, but they don't make an immediate connection between all the bars. Comparison comes later and then aggregation. Distributions and multivariate relationships are advanced concepts.

Note: Check out the article "Making Sense of Graphs" in the *Journal for Research in Mathematics Education* by Susan Friel, Frances Curcio, and George Bright for a good summary of how students comprehend graphs.

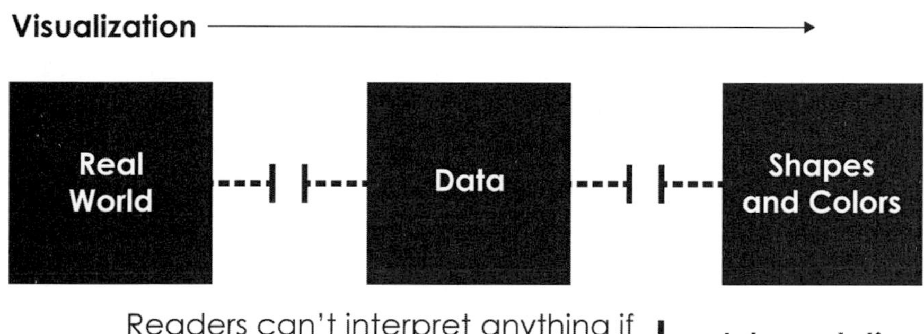

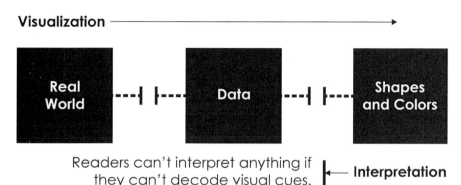

FIGURE 6-18 *Broken connections between visual cues, data, and what it represents*

You can see how this corresponds to data literacy. It's relatively easy to understand minimums, maximums, and anecdotes, but it is trickier as you try to make comparisons within a population and form clusters. Finally, you must understand distributions before you can properly interpret variability.

Note: See the Annotation section in Chapter 5, "Designing for an Audience," for examples of describing statistical concepts.

Although findings are in the context of education, visualization for an audience is essentially teaching. This doesn't mean you should avoid more advanced visualization methods completely, but when you do use them, explain how to read them and what they show. There will be fewer misunderstandings.

DATA NARRATIVE

Visualization is often framed as a medium for storytelling. The numbers are the source material, and the graphs are how you describe the source. When referring to stories or data narrative, I don't mean novels (but great if that's what you're after). Rather, I mean statistical stories, as shown in Figure 6-19.

It often helps to ask a question about the data and then try to answer that through the visualization. It gives you a place to start with the data and provides focus in a graphic. Also, a simple question often leads to other questions and insights that you might not have thought of.

FIGURE 6-19 *(following page) Statistical questions and possible outcomes*

Possible questions Fill in the blanks	**Statistical concepts**	**Possible visuals**
What ________ is the best and worst?	Maximums and minimums	
How has ________ changed over time?	Temporal patterns	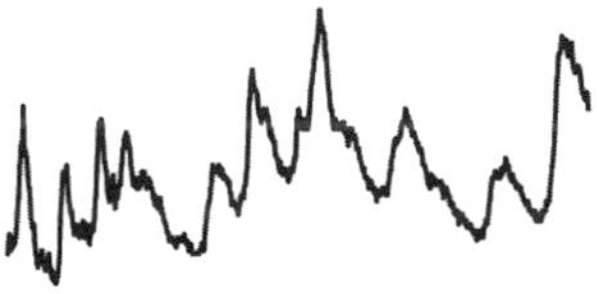
What ________ stands out from the rest?	Outliers	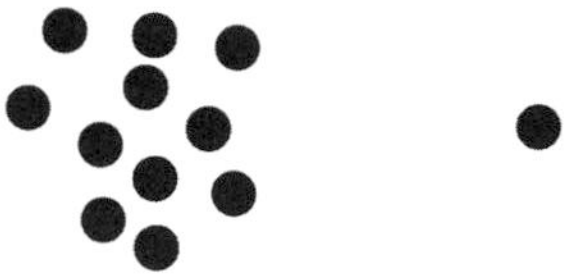
What makes ________ different from ________?	Clustering	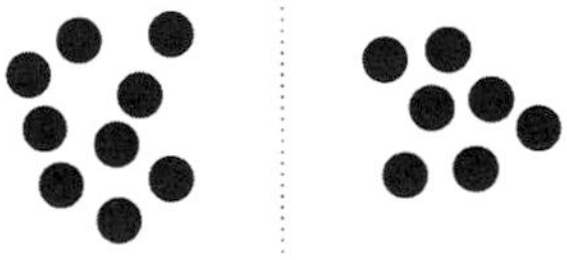
How are ________ and ________ related to each other?	Correlation	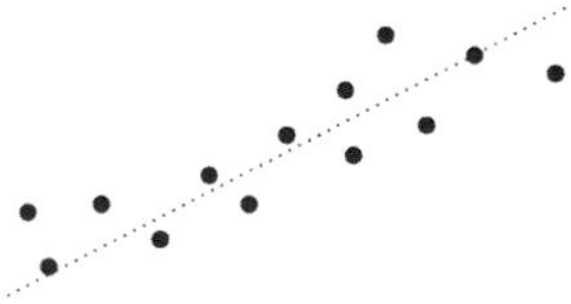
What's the breakdown for ________?	Distributions	

When you have your story, instead of just arranging graphics in a way that looks good, talk about the data in a way that fits with the context of the report. For example, you can take readers through your analysis, from big picture down to details and notable data points, or you can go the other way, from case study to overall summary.

If the work is more like a reference than a story, you can split up the data into groups or categories. Maybe you have country-specific data. Categorize by region or development level, or you might just want to show the profile of each country arranged in alphabetical order.

Note: In Chapter 1, you first went from single observation to an overview with the wedding photos, and later in the chapter, you went from the overview to more specific with crashes in the United States.

Again, it is all about how you think your audience will read your graphics, coupled with how you want your audience to read them.

The same goes for charts embedded within text, such as with reports or articles. Create continuity between your visualization and words. Often people make a hodge-podge of graphs, with little thought to how they relate, and crudely stitch them together. You end up with separate modules that you look at individually, but you usually want to read the series of graphs like you would an article.

Sometimes continuity is as simple as paying attention to where a graphic is placed within a body of text. For example, Figure 6-20 shows a generic layout for a report where graphics are placed in areas that are convenient space-wise, but references to the graphics within the text are on different pages. This requires that a reader flip forward in the report to see a figure after reading about it on the previous page or to go backward to reread the figure's explanation.

In this case, it's useful to design the figure so that it's self-encapsulated. Readers can then glean information directly from a chart rather than referring to a previous page for what it means.

This makes sense if you think about how people might read your report. Do they read the report from front to back, or do they scan the report first, looking at headers and figures, and then look closer if something seems interesting? Self-encapsulated charts make it easier for the scanning reader.

On the other hand, your paper might use a one-column layout, as shown in Figure 6-21. This layout is common online. It's easier to place figures within the flow of the text, rather than off to the side or on a separate page, so it might make more sense to explain in the body text rather than within the figure. This might provide a better flow to the report and make it more readable as a whole, whereas heavy annotation in the figure might actually break continuity.

Note: Flipping a page to read about a figure might not seem like much, but it can grow to be a bigger burden with longer reports and more figures.

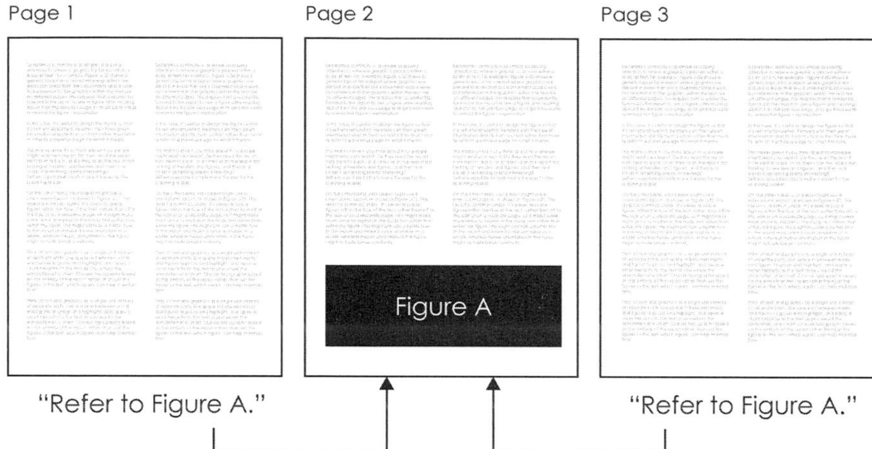

FIGURE 6-20 *Generic report layout with separated figures and text*

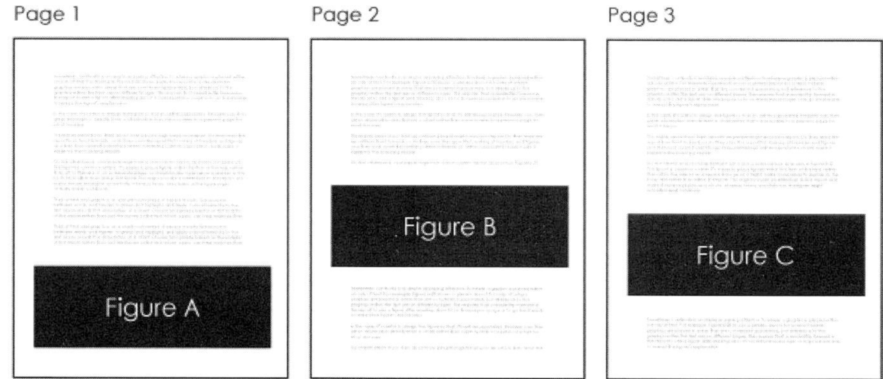

FIGURE 6-21 *One-column layout for a report with embedded charts*

Think of text and graphics as a single unit instead of separate parts. Use space in between words and figures to group and highlight, and apply a visual hierarchy to the text as you would the annotation of a chart. Choose typography based on the entirety of the report rather than just the figures or the text, which again, can help maintain flow.

RELATABILITY

A great way for readers to connect with a visualization is to let readers see the data as it pertains to them. It's not surprising that people spend more time interacting with data that they somehow see themselves in or can relate to their surroundings.

For example, the line graph in Figure 6-22 shows the probability of death within a year in the United States given your age. One line is for women and the other for men. There's an upward trend overall that shows an increased probability of death the older that you get, and for those who reach more than 120 years, the probability is near certainty.

There's an immediate connection to the data because most people are aware of age and mortality. When you note the overall trend, you most likely looked at your own age and gender to find the corresponding probability. Maybe you noted the age of family members or friends. It's easy to relate to the data.

Of course, you could take it a step further, and present the data completely in the context of the individual who looks at the graphic. Figure 6-23 is a different angle that asks for your age before you see the overall picture and then shows only your estimated probability of death. The graph is shown below the individual percentage, so the same data is used and visualized the same way (eventually), but it focuses on the individual, which makes the data personal.

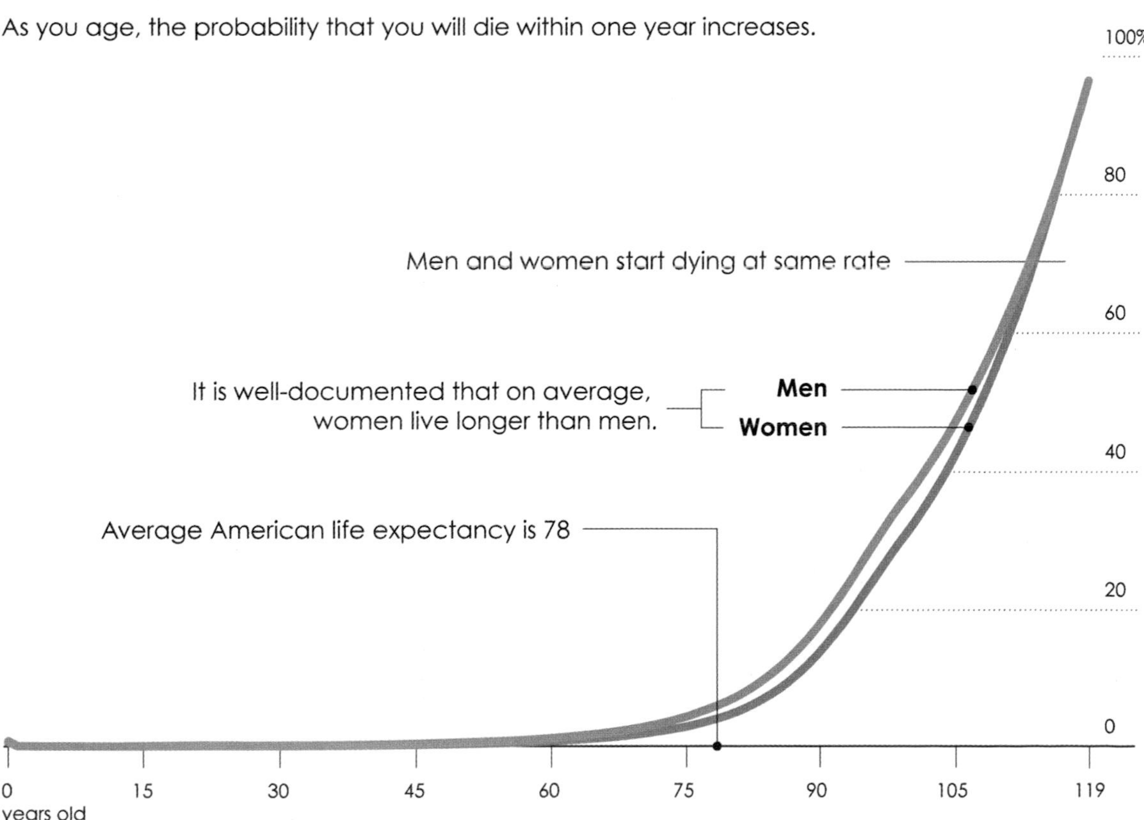
Probability of death
As you age, the probability that you will die within one year increases.
100%
80
60
40
20
0
Men and women start dying at same rate
It is well-documented that on average, women live longer than men.
Men
Women
Average American life expectancy is 78
0
15
30
45
60
75
90
105
119
years old
Source: Social Security Administration

FIGURE 6-22 *Probability of death in a year*

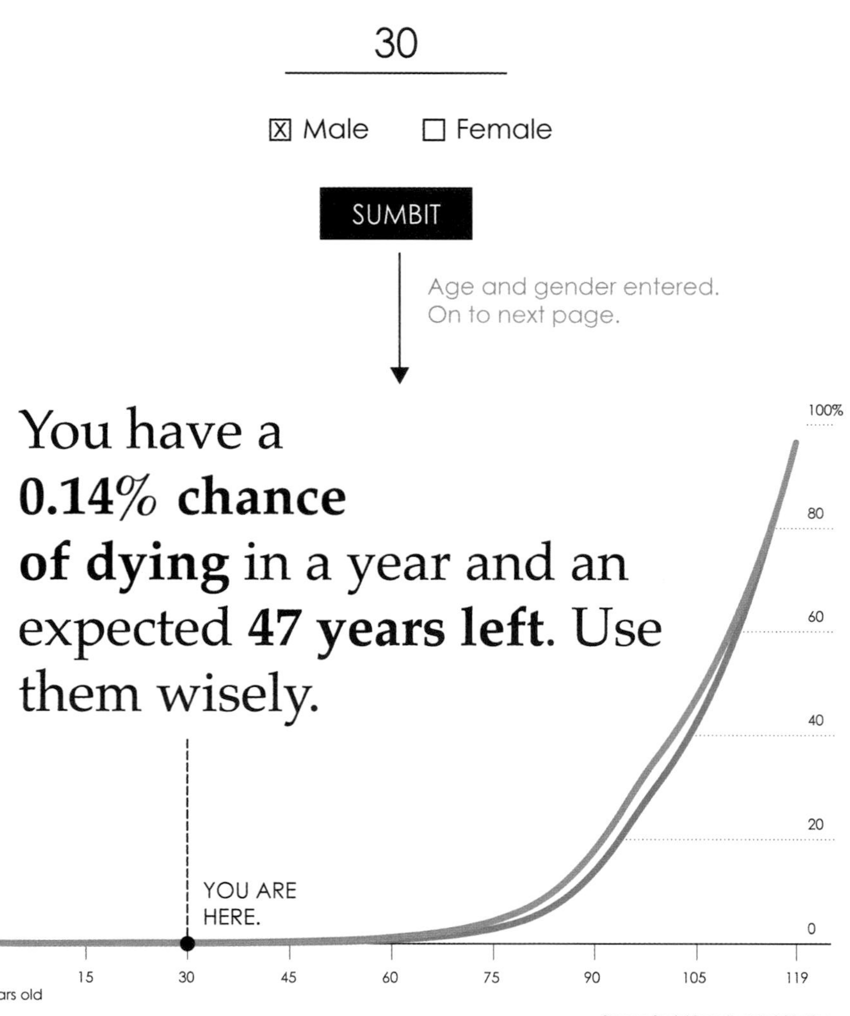
How old are you?
30
Male
Female
SUMBIT
Age and gender entered.
On to next page.
You have a 0.14% chance of dying in a year and an expected 47 years left. Use them wisely.
100%
80
60
40
20
0
YOU ARE HERE.
0
15
30
45
60
75
90
105
119
years old
Source: Social Security Administration

FIGURE 6-23 *Individual focus for greater relatability*

Spatial data is also often easy to relate to. For example, Figure 6-24 shows screenshots from an interactive map that animates the growth of Wal-Mart stores in the United States. As with most online maps, you can pan and zoom, so as the animation plays, many people can zoom into where they live and wait for the nearest store to appear. It is like a form of verification, where people used themselves as a point of reference. Then they replay the animation to see the pattern of store openings.

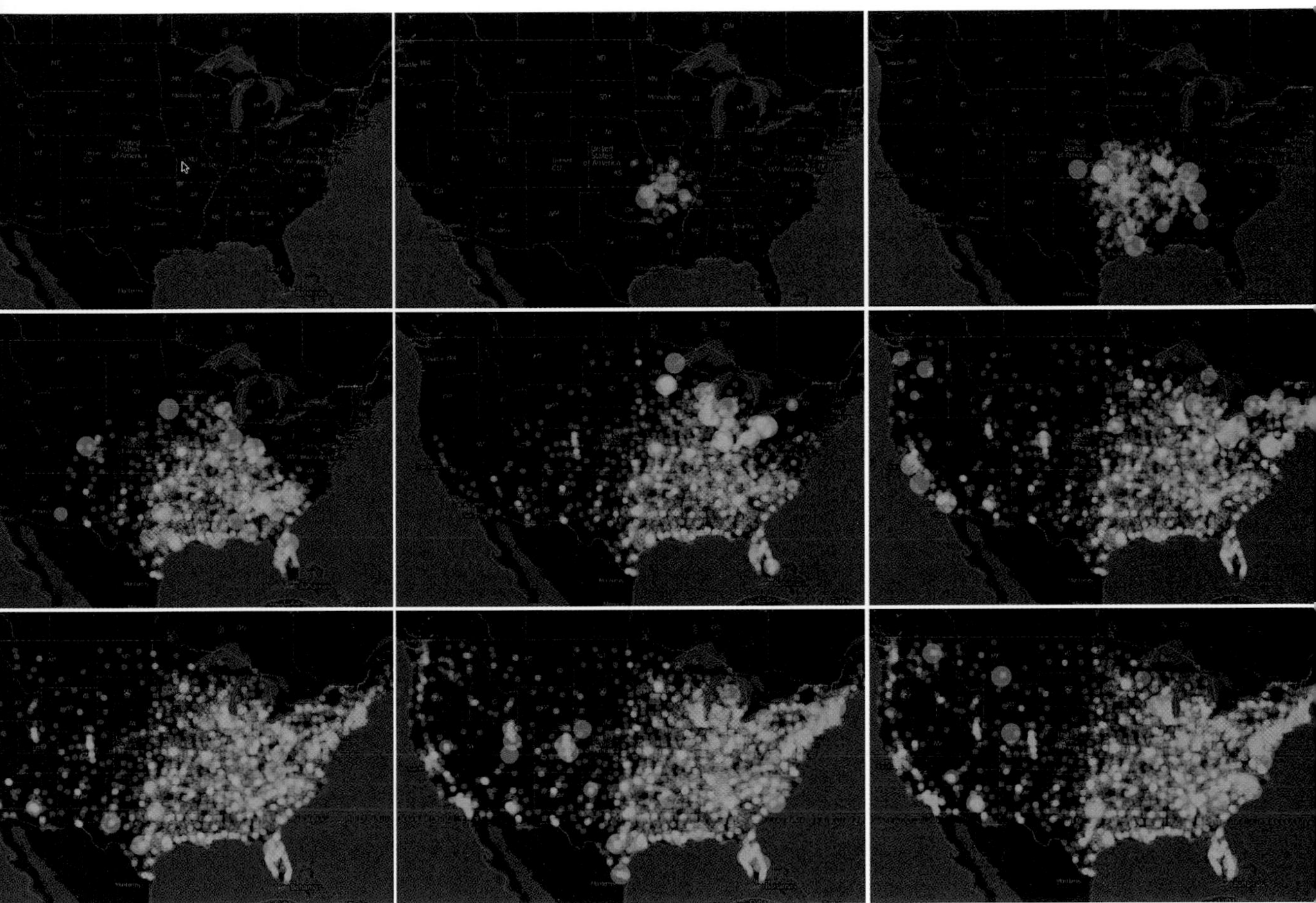

FIGURE 6-24 *Relatability through physical location*

During the exploration phase, it's often beneficial to see an overview of the data first and then decide where to go from there. But you can also benefit moving the opposite direction. You actually see this often in news articles, where the writer starts with an anecdote to draw readers in and then zooms out for a broader point of view.

Of course, not all data offers an immediate connection. Maybe you have data about server status or voltage in a battery. Sometimes metaphors can be useful, but a point of reference is the main thing you're after—something familiar to compare to.

PUTTING IT TOGETHER

Put everything together—from understanding data, to exploration, clarity, and adapting to an audience—and you get a general process for how to make data graphics. The amount of time and effort you spend in each phase varies by the data and what you want to do with it. For example, if there's not much data to look at, you spend less time exploring, but if you have a lot of data, you most likely spend more time in exploration and iterate between the phases.

Now try this with an actual dataset. The Centers for Disease Control and Prevention (CDC) keeps track of obesity rates at the county level in the United States. You can download data from the site, which includes obesity rates from 2004 through 2009 for each county. There are approximately 3,100 counties.

How have rates changed over the years? You probably know rates have increased, but by how much and with what variation in between years? Start with box plots to see the changing distributions, as shown in Figure 6-25. There's clearly an annual increase (and a good bit of variation).

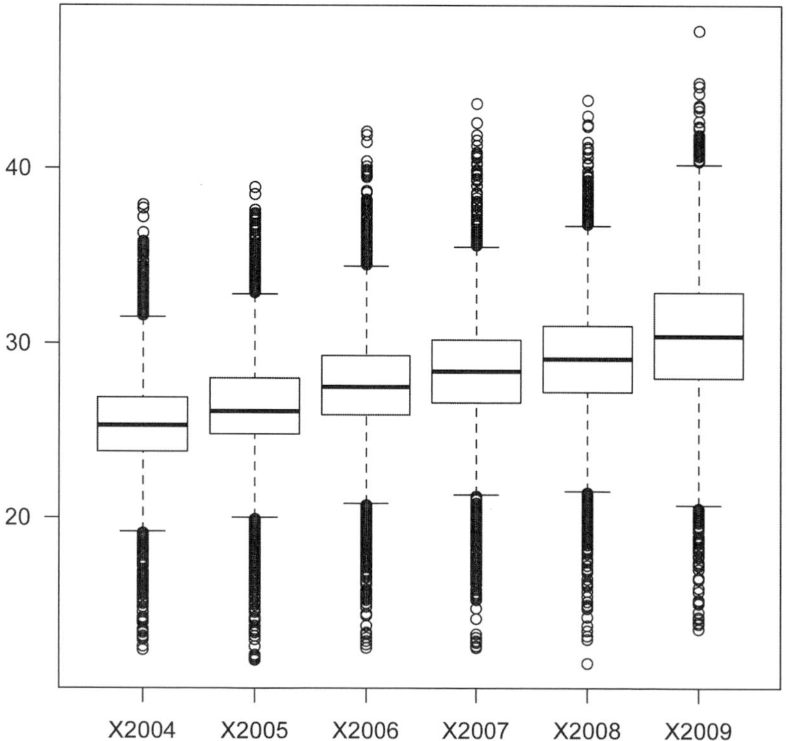

FIGURE 6-25 *Comparing box plots to see changes over time*

You might want to see the specifics of each distribution though, so instead use histograms stacked on top of each other to compare, as shown in Figure 6-26 Again, you can see that as a whole, obesity rates increased. Each distribution looks bell curve-ish, and the peak shifts right as you scan from top to bottom.

Figure 6-27 is specific and shows a time series line for each county. Some transparency is used so that areas in which there are many lines appear darker. So you see the distribution around the 20 to 30 range and an increasing trend. The black line shows the annual county median.

Because you do have geographic data, it might be useful to see the data in map form, as shown in Figure 6-28. Darker shades of red indicate higher obesity rates. Are there any geographic patterns or clustering? It looks like the southeast has higher obesity rates than the rest of the country, and that pattern is more pronounced in later years.

However, the nationwide increase isn't especially obvious year-to-year. For example, the map for 2004 and 2005 look similar, except of course the 2005 map is a little darker. There are two things that get in the way of making the differences more clear: county borders and a color scheme that doesn't provide enough contrast. The former makes the map look lighter, especially on the east coast where counties are smaller, and the latter makes it harder to see differences between counties and by year. Figure 6-29 provides more contrast and removes county borders for a clearer look.

Are there counties that improved, as in decreased obesity rate, during the years? There must be. Figure 6-30 shows such counties highlighted in blue, but there aren't any regional patterns. The blue looks randomly scattered across the country.

The key with weight of course is about the long term. There were only 51 counties out of 3,138 that had lower obesity rates in 2009 than in 2004. Figure 6-31 highlights the improved counties in blue, and the results aren't impressive.

It looks like it's best to focus on the increase of obesity rate across the country. There wasn't much decrease. However, even with the color scheme that has more contrast and removed borders, the maps shown in Figure 6-30 don't make the annual changes that obvious. There's change each year, but not that much. However, compare the map for 2004 against the one for 2009. The differences are much easier to see. What if the final graphic uses the map from 2004 and the one from 2009? That seems to make for a better comparison, so go with that.

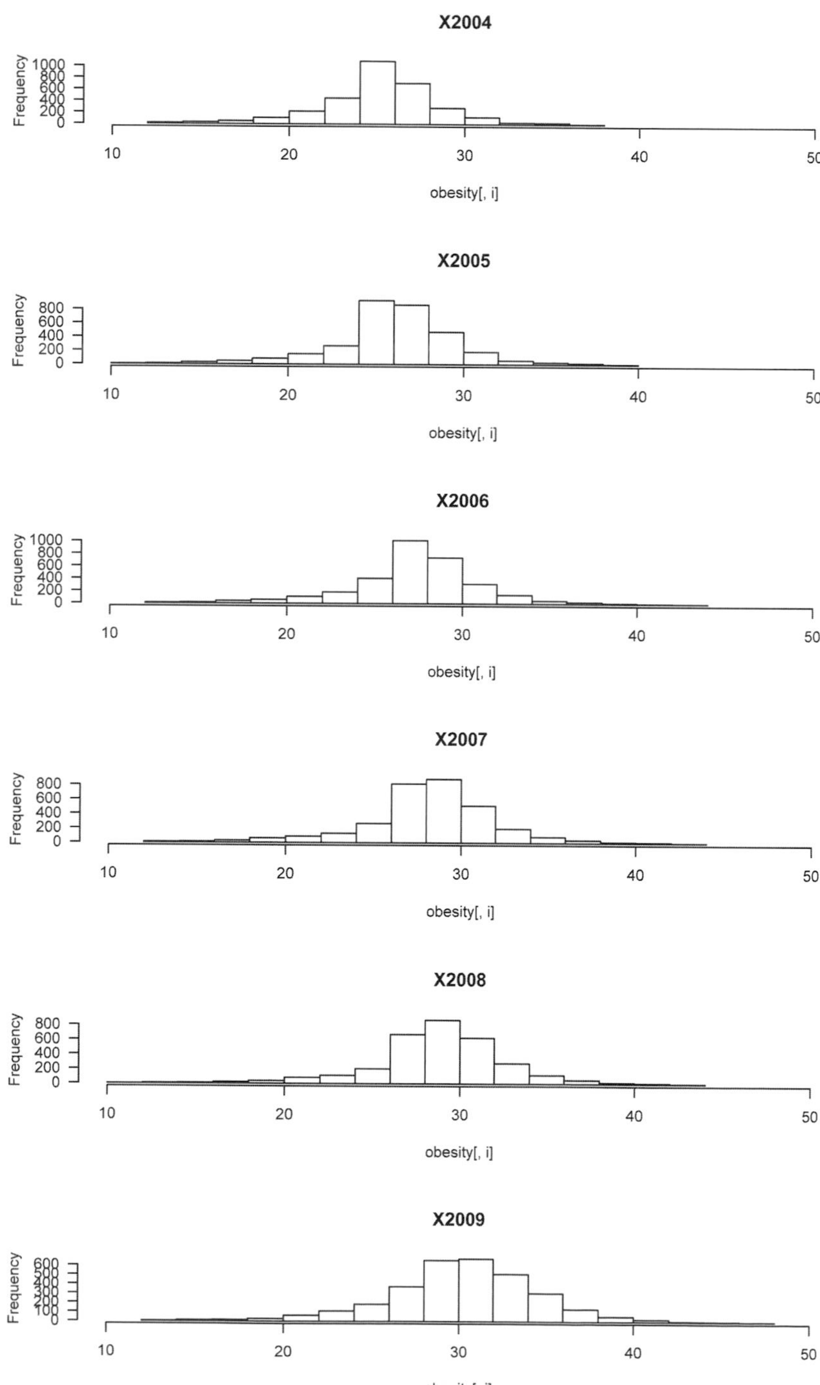

FIGURE 6-26 *Exploring distributions to see changes over time*

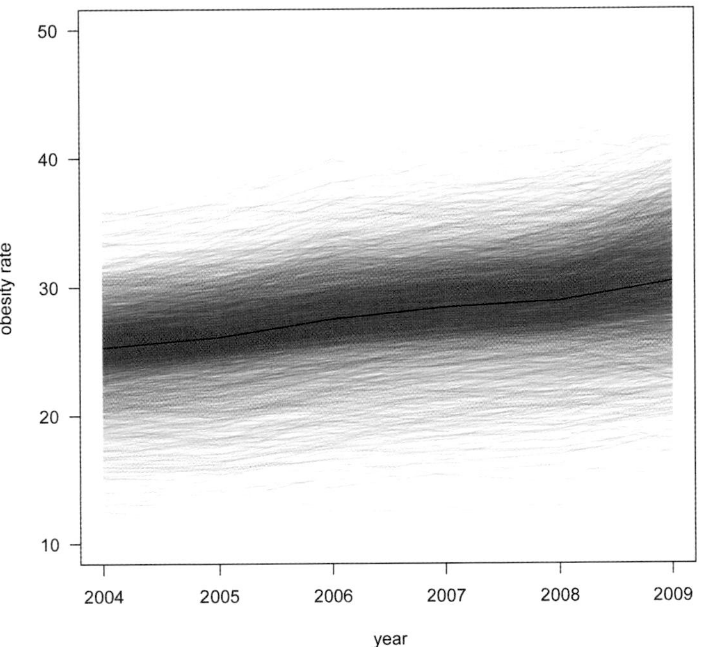

FIGURE 6-27 *Time series plot with transparency and exploration of visual form*

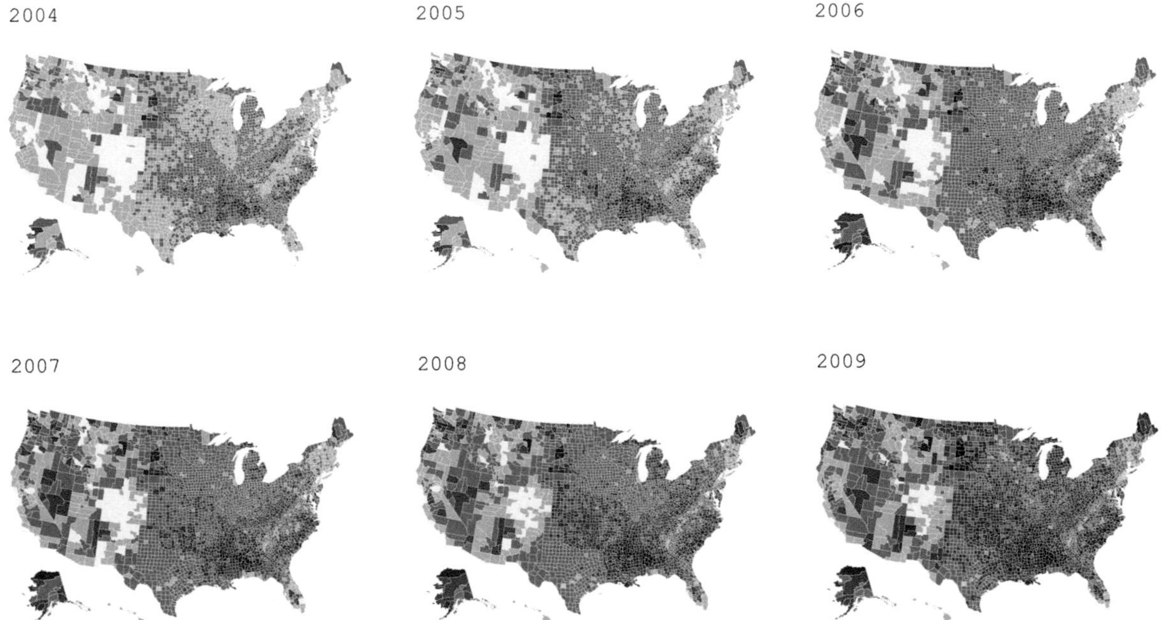

FIGURE 6-28 *Annual maps for spatio-temporal comparisons*

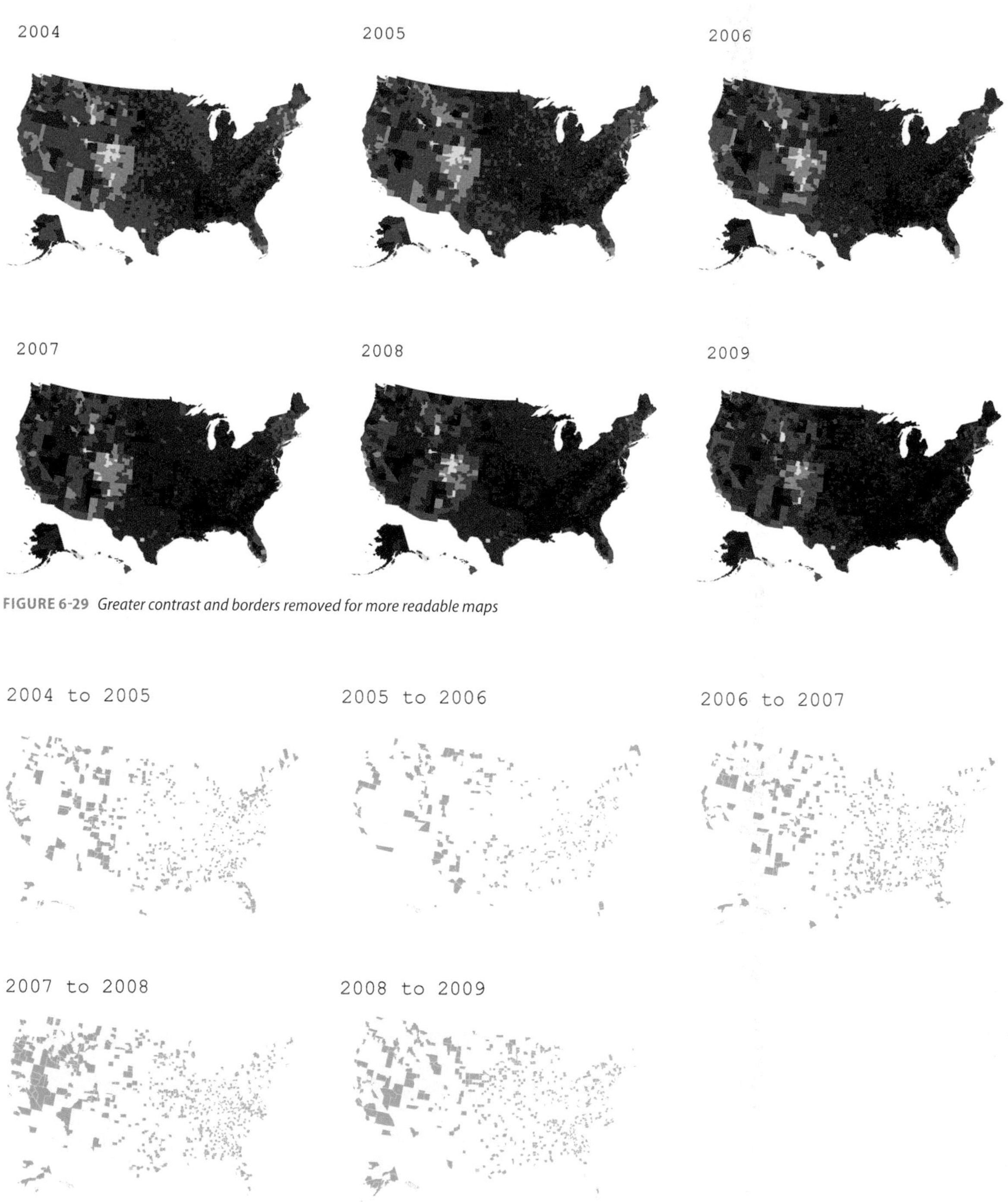

FIGURE 6-29 *Greater contrast and borders removed for more readable maps*

FIGURE 6-30 *Decreased obesity rate over the years, exploring binary color scheme*

Figure 6-32 shows some quick sketches for layout. I thought about using the time series plots, but this data seemed better fit for maps over time. People most likely want to examine their own regions, and the time series obscures that.

Finally, Figure 6-33 is the final graphic. There's a map for 2004 and another for 2009. The histogram underneath each map doubles as a legend for what range each shade of red represents.

Notice the process that resulted in the final graphic. Not every chart was used, and most of the time was spent figuring out what the data had to show. On FlowingData, readers often see a visualization they like and then ask me what software was used or how they can (quickly) make the same thing with their data. It's just not that straightforward a lot of the time. You poke around, understand the data the best you can, and then produce.

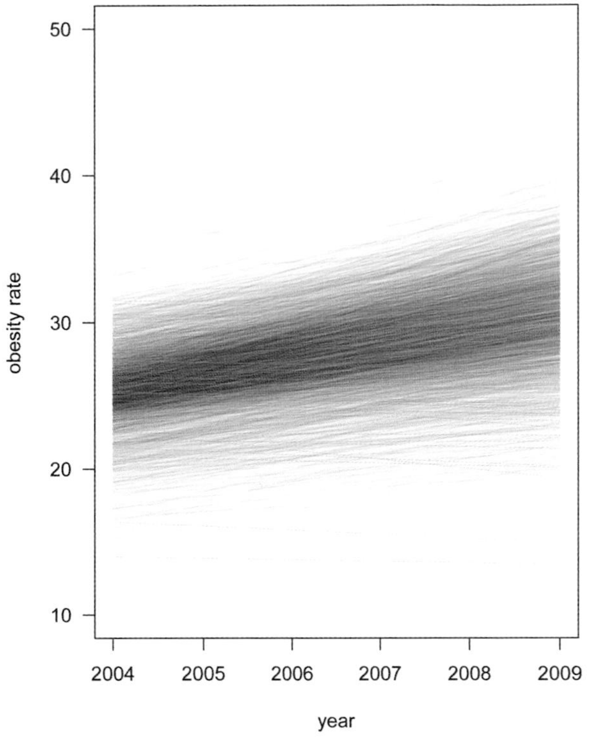

FIGURE 6-31 *Time series version of maps in Figure 6-30 to look for patterns*

WRAPPING UP

Regardless of the audience, you must step outside the dataset and consider what it represents in the real world. Don't fall into the trap of thinking that data is just digital output that sits alone on a hard drive and isolated in a spreadsheet.

There's a saying popularized by Mark Twain that perhaps gives statistics a bad name: "There are three kinds of lies: lies, damned lies, and statistics." People often misinterpret this as that there's dishonesty in numbers, but the lies don't come from the numbers themselves. They stem from those who use data incorrectly or irresponsibly. As you present data to others, it's your responsibility to show the truth.

Learn all you can about your data, and design your graphics based on those findings. After that, the smaller things tend to fall in place.

FIGURE 6-32 *(following page) Sketches for layout before moving on to final graphic*

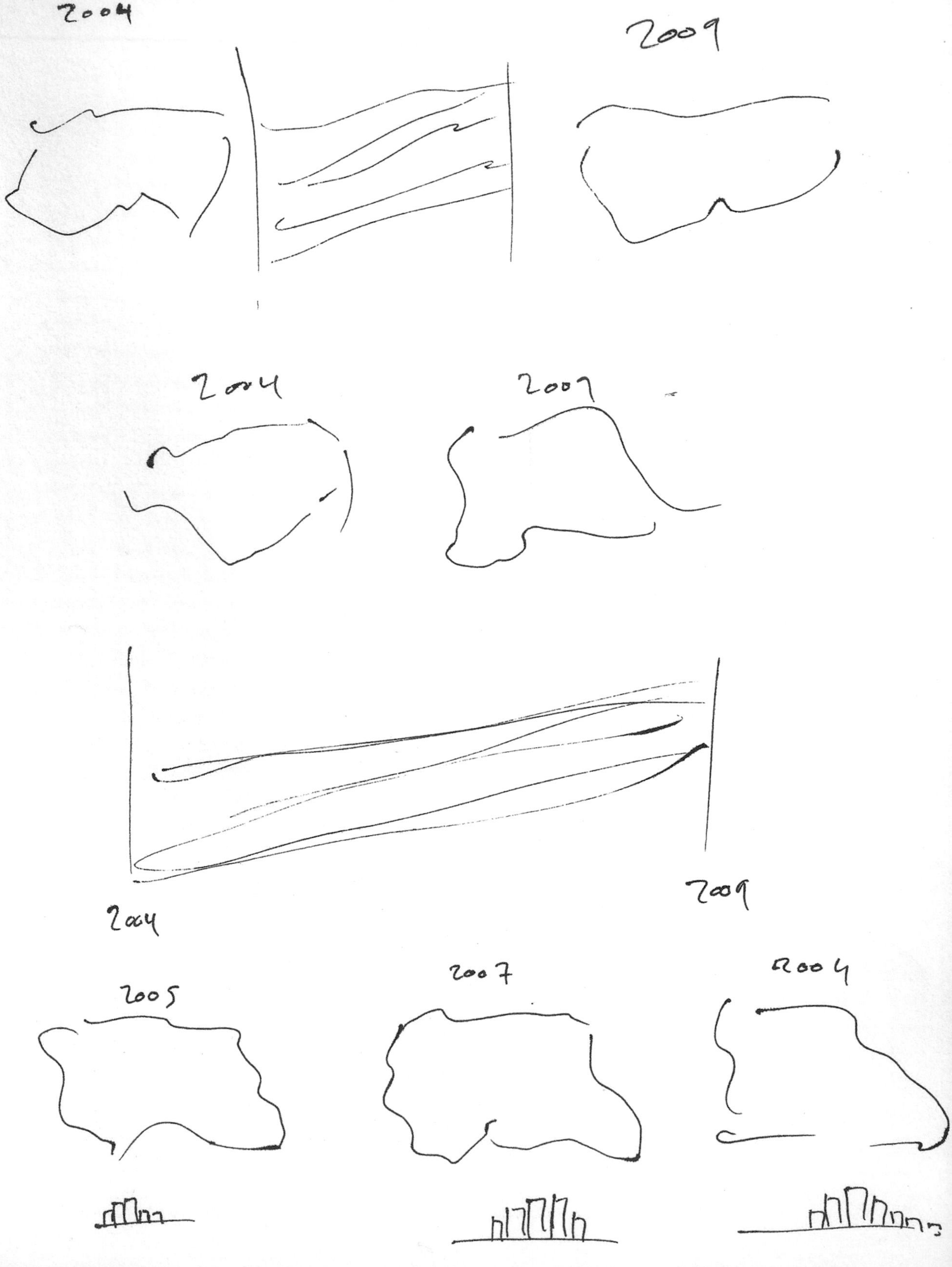
2004
2009
2004
2007
2004
2009
2005
2007
2004

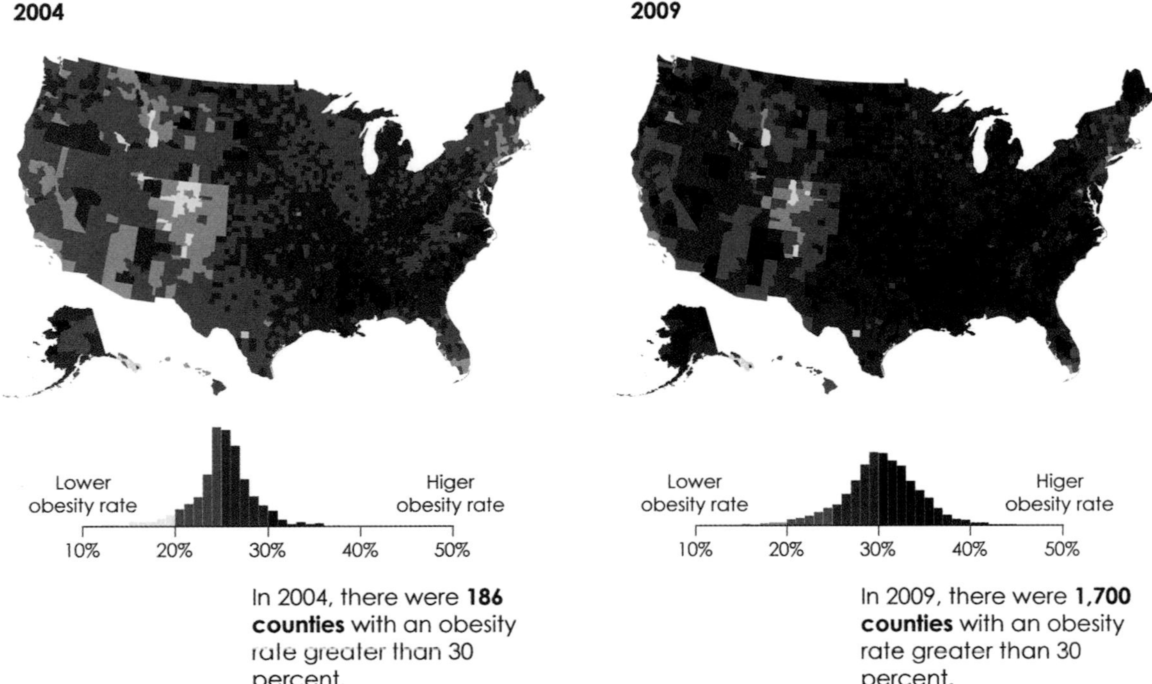
Rising obesity rates in the United States, by county
Based on estimates from the Centers for Disease Control and Prevention, 36 percent of American adults and 17 percent of youth were obese in 2009 and 2010. That's over 90 million people.
2004
2009
Lower
obesity rate
Higer
obesity rate
10%
20%
30%
40%
50%
Lower
obesity rate
Higer
obesity rate
10%
20%
30%
40%
50%
In 2004, there were 186 counties with an obesity rate greater than 30 percent.
In 2009, there were 1,700 counties with an obesity rate greater than 30 percent.
*Maps use age-adjusted estimates

FIGURE 6-33 *Annotated and finalized graphic using findings from exploration*

Where to Go from Here

Now that you've learned about data, how to represent it, explore it visually, enhance clarity, and design for an audience, the obvious next step is to put it into practice. Grab your data and visualize.

Visualize with what though? There are lots of tools at your disposal. The one that's best for you depends on your data and what you want to do with it, but most likely it'll be some combination of the ones in this chapter. Some are good for quick looks at your data, whereas others are better for wider audiences.

VISUALIZATION TOOLS

There are two main groups of visualization solutions: nonprogrammatic and programmatic. The offerings for the former used to be restricted to a handful of programs, but as data resources grow, more click-and-drag tools spring up to help you understand your data.

MICROSOFT EXCEL

The familiar spreadsheet software is universal and has been around for decades. Heck, the first time I used Excel was a couple of decades ago, and I still use it sometimes, if only because a lot of data is made available as an Excel spreadsheet. It's easy to highlight columns and make a few charts, so you can get a quick idea of what your data looks like.

Note: If you use Excel for the entire visualization process, work with chart settings to improve clarity. Default settings rarely fit the bill.

That said, I wouldn't use Excel for thorough analysis or graphics made for publication. It's limited by the amount of data it can handle at once, and unless you know Visual Basic for Applications (VBA), the programming language built in to Excel, it can be a chore to reproduce charts for different datasets.

GOOGLE SPREADSHEETS

This is essentially Google's version of Microsoft Excel, but it's simpler and online. Figure 7-1 shows some of the charting options, but the online feature is the main plus because you can quickly access your data across different machines and devices, and you can collaborate via built-in chat and real-time editing.

You can also import HTML and XML files from the web using the `importHTML` and `importXML` functions, respectively, which is the main thing I use the application for. For example, you might find an HTML table on Wikipedia, but you want the data as a CSV file, so you use `importHTML` and then export the data from Google Spreadsheets. See more at http://drive.google.com/.

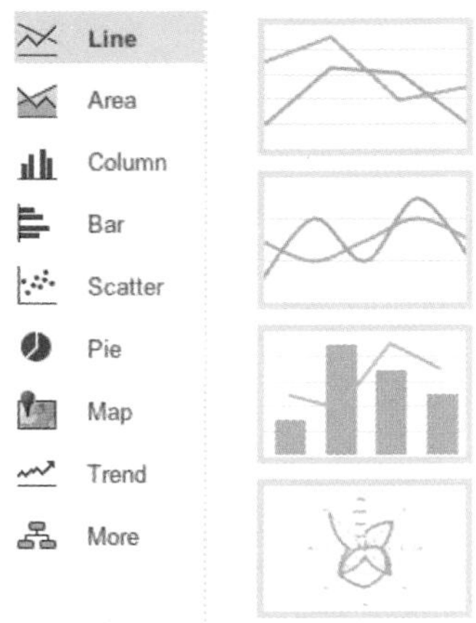

FIGURE 7-1 *Chart options with Google Spreadsheets*

TABLEAU SOFTWARE

At the time of this writing, Tableau Software is the up-and-coming analysis software. If you want to dig deeper into your data than you can in Excel, without programming, this is a good place to look. As shown in Figure 7-2, the program is visually-based, and you can easily interact with your data as you find interesting spots to look at. The downside is that the software is pricey (with special pricing for students and nonprofits). It also runs only on Windows at this time, but there is an OS X version in the works.

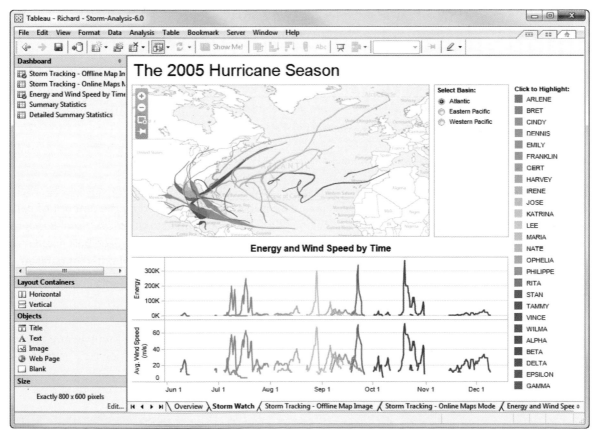

FIGURE 7-2 *Tableau Software*

Tableau Public is free and enables you to put together dashboards with a variety of charts and publish online. Although, as the name suggests, you must make your data public and upload it to Tableau servers. See more at http://tableausoftware.com.

MANY EYES

Many Eyes is a project by IBM Research that enables you to upload datasets and explore your data via a wide variety of visualization tools, some traditional and others experimental. The site started in 2007 and updates stopped in 2010 but remain online for public use. Like Tableau Public, the downside is that your data becomes public after you upload it to the servers. See more at http://many-eyes.com.

DATA-SPECIFIC TOOLS

The following software options cast a wide net in that they try to handle multiple data types and offer a wide variety of visualizations. This is great for analysis and exploration because they enable you to quickly see your data from different angles. However, sometimes it's better to specialize and do one thing well.

Gephi

If you've ever seen a network graph or a visualization with a bunch of edges and nodes (that might or might not have looked like a hairball), most likely it was made with Gephi (shown in Chapter 2, "Visualization: The Medium"). It's open-source graphing software that enables you to interactively explore networks and hierarchy. See more at http://gephi.org.

ImagePlot

As you saw in Chapter 2, ImagePlot Software Studies Lab at California Institute for Telecommunication and Information enables you to explore large sets of images as if they were data points. Plot by color, time, or volume to, for example, explore the trends and changes of an artist or photo collection. See more at http://lab.softwarestudies.com/p/imageplot.html.

Treemap

There are a number of ways to make treemaps, but the interactive software by the University of Maryland Human-Computer Interaction Lab is the original and is free to use. As you saw in Chapter 3, "Representing Data," treemaps (developed by Ben Shneiderman in 1991) are useful for exploring hierarchical data in a small space. The Hive Group also develops and maintains a commercial version for businesses. See more at: http://www.cs.umd.edu/hcil/treemap/.

TileMill

Not that long ago, custom maps used to be hard to make and highly technical, but various programs have made it relatively straightforward to design maps to your liking, for your needs, and with your data. TileMill, by mapping platform MapBox, is open source desktop software available for Windows, OS X, and Ubuntu. Just download and install the program, and then load a shapefile, as shown in Figure 7-3.

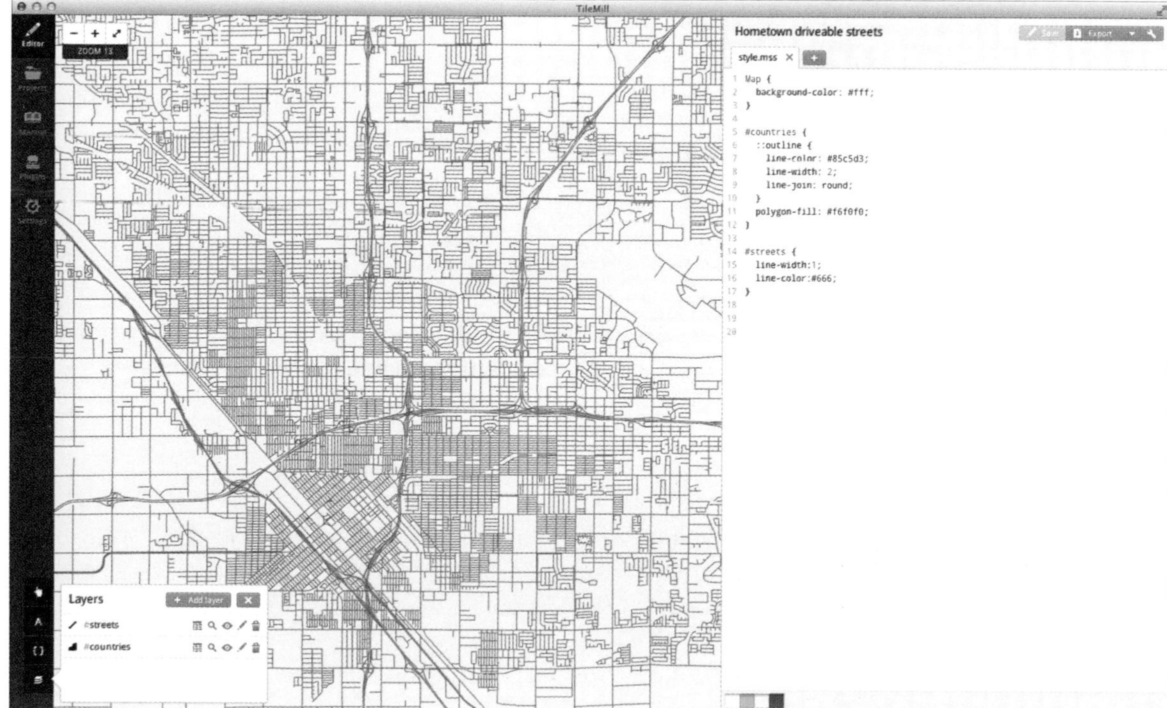

FIGURE 7-3 *TileMill by MapBox*

In case you're unfamiliar with shapefiles, it's a file format that describes geospatial data, such as polygons, lines, and points, and they're easy to find online. For example, the United States Census Bureau provides shapefiles for roads, bodies of water, and blocks. See more at http://mapbox.com/tilemill/.

indiemapper

Figure 7-4 shows indiemapper, which is a free service provided by cartography group Axis Maps. Like TileMill, it enables you to create custom maps and map your own data, but it runs in the browser rather than as a desktop client. It's straightforward to use, and there are plenty of examples to help you begin.

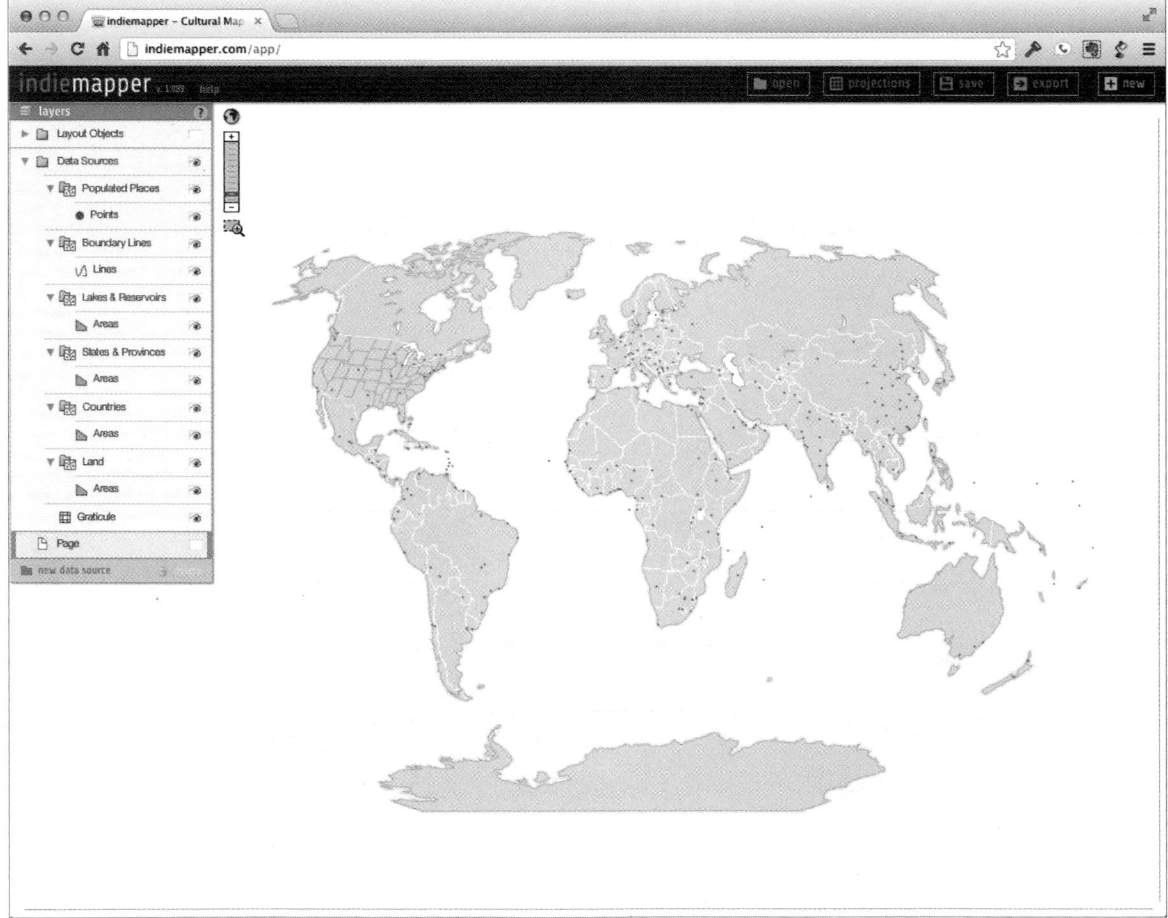

FIGURE 7-4 *indiemapper by Axis Maps*

A favorite part of the application is that you can easily change map projects, which provide some guidance on what projection works better for your situation. See more at: http://indiemapper.com/.

GeoCommons

GeoCommons is similar to indiemapper, but it's more focused on exploration and analysis. You can upload your own data or draw from the GeoCommons database and then interact with points and areas. You can also export data in a number of common data formats to import into other software. See more at: http://geocommons.com/.

ArcGIS

Before the previously mentioned mapping tools were available, ArcGIS was the primary mapping software for most people (and still is for many). It's a feature-rich platform that enables you to do just about anything with maps. For most though, the basic subset of features is enough, so to avoid the hefty cost of the software, it's probably best to try the free options first, and if those aren't enough, try ArcGIS. See more at: http://arcgis.com/.

PROGRAMMING

Out-of-the-box software gets you up and running in a short amount of time, but the trade-off is that you're using software that's generalized in some way so that more people can use it with their own data. Also, if you want a new feature or method, you need to wait for someone else to implement it for you. On the other hand, you can visualize data to your specific needs and gain flexibility when you can code. You're the one who tells the computer what to do.

Obviously, the trade-off when you code is that it takes time to learn a new language, but after you get over the learning hump, the visualization process tends to move faster. It's also grows easier to reproduce your work and apply it to other datasets as you build up your library and learn new things.

Note: Learning to program can seem intimidating at first because the code seems odd and foreign, but think of it as learning a new language. It's confusing at first, but when you're fluent you can communicate your thoughts clearly.

R

R is a language and environment for statistical computing and graphics. It was originally used mostly by statisticians but it has expanded its audience in recent years. There are plotting functions that enable you to make graphics with just a few lines of code, and often, one line can do the trick. Figure 7-5 shows examples of what you can do.

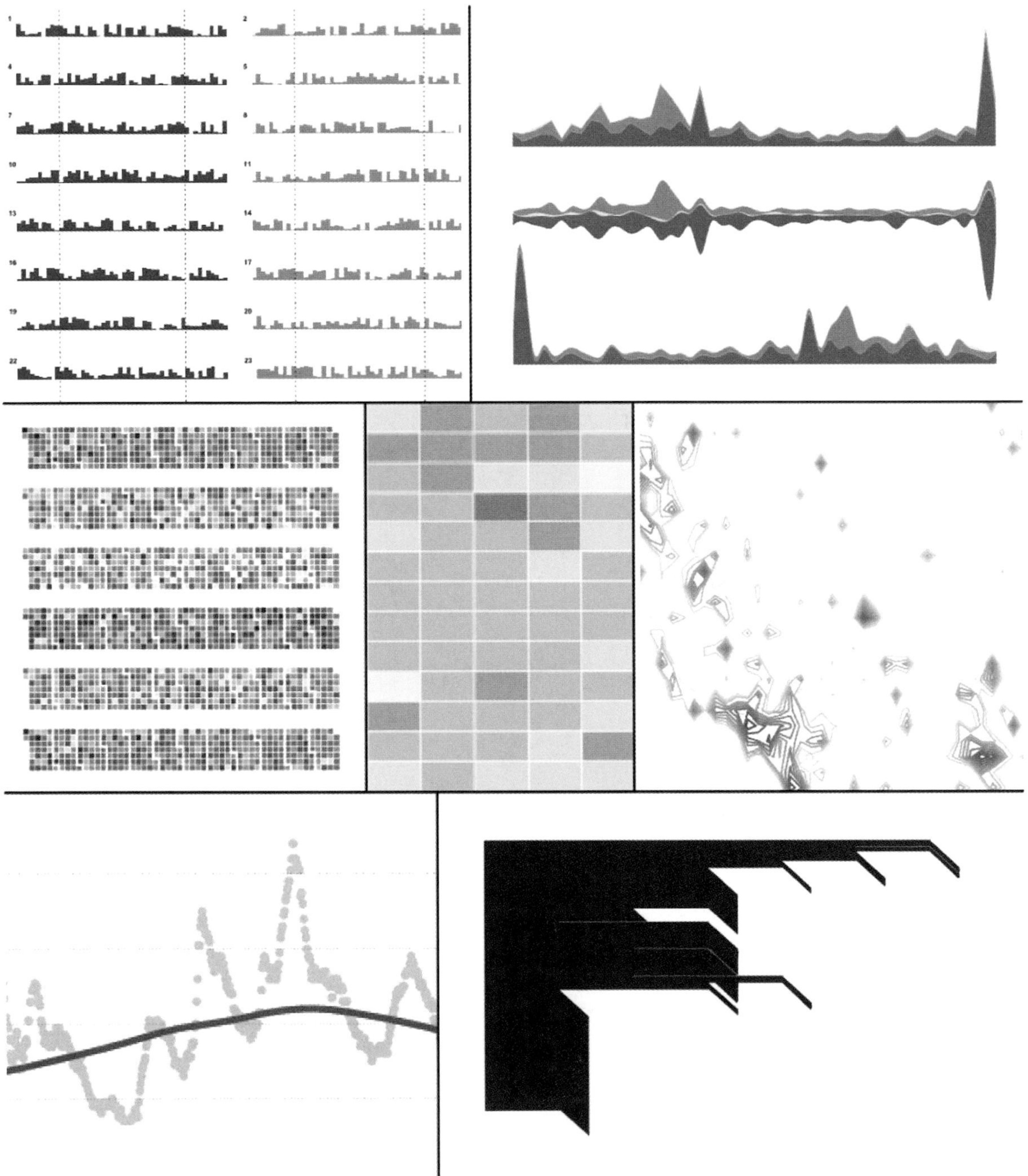

FIGURE 7-5 *Graphics created in R*

R's main strength is that it's open source and many packages expand on the base distribution, which makes statistical graphics (and analysis) more straightforward, such as:

Note: Although you can write code and create graphics with the base distribution of R, many find the RStudio IDE to be helpful to keep code organized (http://www.rstudio.com/).

- **ggplot2:** A plotting system based on the Leland Wilkinson's grammar of graphics, which is a framework for statistical visualization.
- **network:** Create network graphs with nodes and edges.
- **ggmaps:** Visualization of spatial data on top of maps from Google Maps, OpenStreetMap, and others. It uses ggplot2.
- **animation :** Build a gallery of images and string them together for an animation.
- **portfolio:** Visualize hierarchical data with a treemap.

These are just a small sample. You can view and install packages easily via the package manager.

Most of the graphics for this book were produced in R and then refined in illustration software, discussed later in this chapter. In any case, if you're new to code, and you want to make static graphics programmatically, R is a great place to start. See more at: http://r-project.org/.

JAVASCRIPT, HTML, SVG, AND CSS

Not long ago, you couldn't do much visualization-wise that was native in the browser. You had to use Flash and ActionScript. But when Apple mobile devices didn't have Flash on them, there was a quick rush forward toward JavaScript and HTML. The former is used to manipulate the latter, in addition to Scalable Vector Graphics (SVG). Cascading Style Sheets (CSS) are used to specify color, size, and other aesthetic features.

Figure 7-6 shows a few examples of visualization in JavaScript, but there's a ton of flexibility to make what you want. You're limited by your own imagination more than you are by the technology.

Whereas support in various browsers was inconsistent before, functionality is available now in modern browsers, such as Firefox, Safari, and Google Chrome (Internet Explorer is getting there) to make interactive visualization online.

Note: Things get even more interesting when you introduce interaction and animation to the mix. R was intended for static graphics, but it's a different story in the browser with JavaScript.

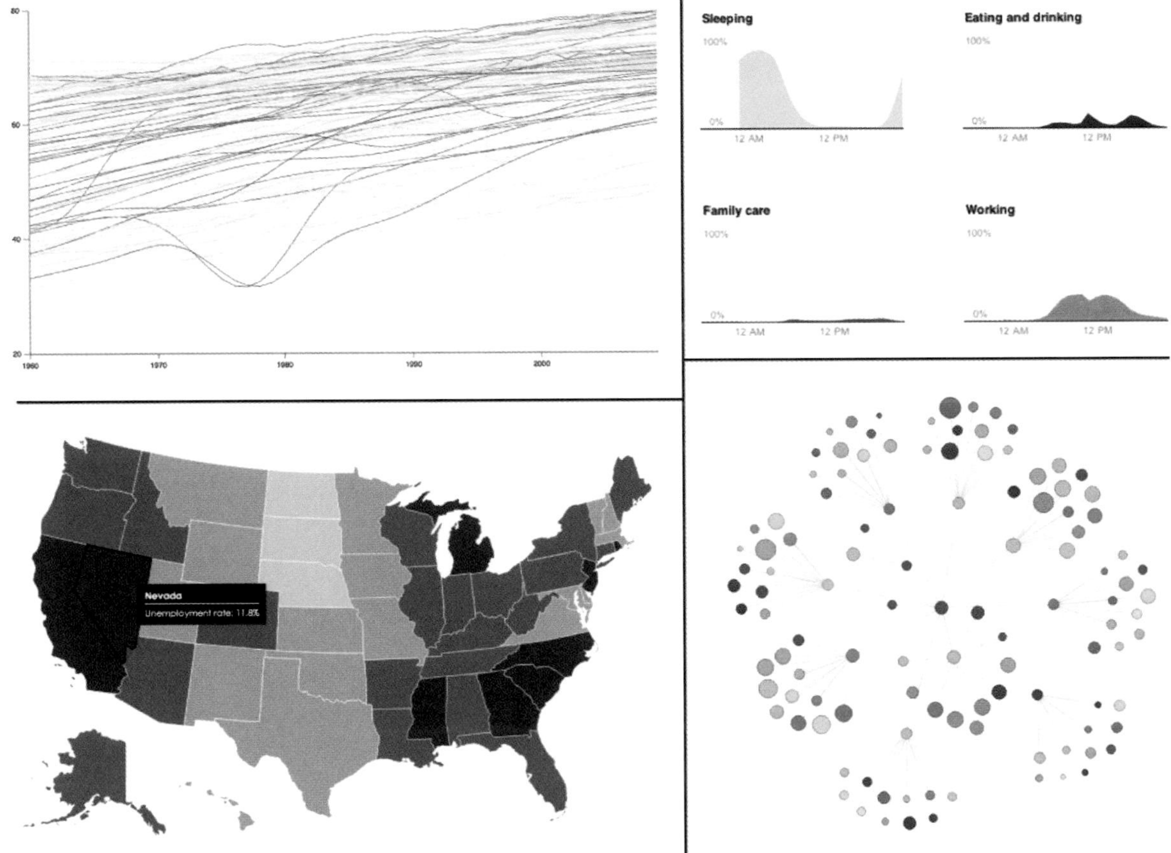

FIGURE 7-6 *Charts made with JavaScript*

If it's online and interactive, most likely the creator used JavaScript. You can start from scratch, but there are several visualization libraries that can make your life a lot easier:

- **Data-Driven Documents (D3)**, http://d3js.org/: This is one of the most, if not the most, popular JavaScript library for visualization. It was created by Michael Bostock and is actively developed. There are lots of examples to work off of and a growing community to find help.

- **Raphaël**, http://raphaeljs.com/: It's not as data-centric as D3, but it's lightweight and makes drawing vector graphics in the browser straightforward.
- **JavaScript InfoVis Toolkit**, http://philogb.github.com/jit/: The documentation and examples aren't as plentiful as the preceding two, but there's enough available for you to start.

These libraries are the main ones, but there are many specialized libraries that can help with certain data types. Just do a search for the visualization that you're interested in followed by "JavaScript" and you should find it quickly.

PROCESSING

Originally designed for artists, Processing is an open source programming language that uses a sketchbook metaphor to write code. If you're new to programming, this is also a good place to start because a few lines of code can get you far, with lots of examples, libraries, books, and a large and helpful community that make Processing inviting.

Note: Processing compiles to Java applets, but there is also a JavaScript version of Processing at http://processingjs.org/. See more at http://processing.org/.

FLASH AND ACTIONSCRIPT

This isn't a popular solution anymore, but most computers have Flash installed on them, so at the time of this writing, it's not completely bizarre to visualize data with Flash and ActionScript. However, technology seems to be shifting away from Flash for online applications, so if you're new to visualization and programming, you might as well go with JavaScript. That said, if you do want to try your hand with ActionScript, check out Flare at http://flare.prefuse.org/.

PYTHON

Python is a more general programming language that isn't geared toward graphics, but it's commonly used for data processing and web applications.

So it makes sense to explore data visually with Python if you're already familiar with the language. Although the visualization offerings aren't as extensive, the matplotlib library is a good place to start: http://matplotlib.org/. See more at http://python.org.

PHP

Like Python, PHP is a more general scripting language than R and Processing. It's mostly for web programming. However, most web servers already have PHP installed, which takes care of the installation step for you. Plus, PHP has a graphics library, which means you can use it for visualization. Basically, if you can load data and draw shapes based on that data, you can create a visualization. See more at http://us.php.net/gd.

ILLUSTRATION

Static graphics that look polished, especially ones that you see in newspapers and magazines, most likely went through illustration software at some point. Adobe Illustrator is the most popular one, but it can be expensive for people who don't use it regularly or just want to touch up their charts. Inkscape is the open source alternative, and although it's not quite as usable as Illustrator, it's enough to get the job done.

I use Illustrator regularly, so although the software is mainly for designers and artists, it's worth it for me. My typical workflow is to use R to create the foundation of a graphic, save charts as PDF files, and then bring them into Illustrator to change colors, add annotation, and rework to maximize clarity. You can, of course, customize in R, but I like to shift elements with point, click, and drag to see immediate changes.

See more at the following sites:

- Adobe Illustrator, http://www.adobe.com/products/illustrator.html
- Inkscape, http://inkscape.org

STATISTICS

Finally, remember that regardless of application, the goal is to understand data, and when it comes to designing for an audience, it's to help others understand. You can get a lot out of visualization, and often, it's all that you need to figure out what your data is about.

However, as your data grows more complex in size, dimensions, and granularity, visualization only gets you part of the way there. After all, there are only so many pixels on a screen, and eventually you run out of space. Statistics is helpful in dealing with sparse and missing data, too.

> *"For big data: Visualisation is fundamentally limited by the number of pixels you can pump to a screen. If you have big data, you have way more data than pixels, so you have to summarise your data. Statistics gives you lots of really good tools for this."*
>
> *—Hadley Wickham*

When I tell people that I'm a statistician, their gut reaction is to tell me how much they hated their introduction to statistics course in college. But before your eyes roll out of your head as hypothesis testing and bell curves run through your imagination, trust me that statistics is a lot more than that. At the very least, it provides you with a wider point of view for what your data is about and how to sift through the text files and databases of numbers. Plus, it's another tool to have in your back pocket.

Most schools offer statistics courses, but some open education sources also have a few offerings:

- Coursera, https://www.coursera.org/
- Udacity, http://www.udacity.com/

WRAPPING UP

I learned to cook the summer before my second year of college. I knew how to make only rice, which is a standard for most Asian children, and how to microwave premade frozen goods. During my first year of college, I survived on dorm food and an unlimited supply of tater tots and soft-serve ice cream, so my mom was worried that I would have nothing to eat after I moved into an apartment the next year. My daily schedule also revolves around feeding time, so needless to say, I was eager to learn how to cook.

Each night that summer I stood in the kitchen with my mom, and she would demonstrate a technique, and then I would mimic her. I learned how to use a wok, manage heat, prepare my ingredients ahead of time, and magically ended up with a meal that didn't taste horrible at the end of each session.

During the first weeks of learning, I carefully watched my mom's spatula movements, took note of ingredient proportions, and where to put dishware for plating. I tried to do everything exactly the same, and this worked well for the specific dishes that my mom had planned. Everything was laid out for me so that she could clearly explain each step.

Then it was time for me to cook on my own. I looked up recipes, sometimes ones that used only the ingredients I had in the refrigerator or ones that required that I just grab a few from the grocery store. With measuring cup, spoon, and timer in hand, I followed each recipe to the teaspoon and the second.

As I followed more recipes, I learned what ingredients tasted good (and gross) together, when a "teaspoon" meant "around one or three teaspoons, depending on what's in the pot," and most important, I learned why so many recipes end with "season to taste." Why can't someone just tell me the exact amount of seasoning to put? Because it varies by dish, even when you cook the same thing twice.

That little change, based on taste, can make the food taste amazing, subpar, or even inedible. These little changes happen throughout the cooking process, and the sum of these changes is why you like that dish so much at that one restaurant across town.

Learning how to visualize data works in the same way. There are general rules and suggestions that you learn at first. You might follow them to a cue in the beginning, but as you work with more data and make more things, you shift and adjust based on what you have and what you see. Those shifts and adjustments are what make great visualization stand out from the rest.

The goal is to get to the point where you can take any ingredients—your data—and understand what they represent. The better you understand your data, the better you can help others understand. That's how you get visualization that means something.

Index

C

D

E

F

G

H

I

O

P

Q

R

S

T

W–Z